DIFFERENTIAL CALCULUS

**Problems and Solutions
from Fundamentals to Nuances**

DIFFERENTIAL CALCULUS

Problems and Solutions
from Fundamentals to Nuances

Veselin Jungić
Petra Menz
Randall Pyke

Simon Fraser University, BC, Canada

World Scientific

NEW JERSEY · LONDON · SINGAPORE · BEIJING · SHANGHAI · HONG KONG · TAIPEI · CHENNAI · TOKYO

Published by

World Scientific Publishing Co. Pte. Ltd.

5 Toh Tuck Link, Singapore 596224

USA office: 27 Warren Street, Suite 401-402, Hackensack, NJ 07601

UK office: 57 Shelton Street, Covent Garden, London WC2H 9HE

Library of Congress Cataloging-in-Publication Data

Names: Jungić, Veselin, author. | Menz, Petra, author. | Pyke, Randall, author.
Title: Differential calculus : problems and solutions from fundamentals to nuances /
    Veselin Jungić, Petra Menz, Randall Pyke, Simon Fraser University, BC, Canada.
Description: New Jersey : World Scientific, [2024] | Includes bibliographical references and index.
Identifiers: LCCN 2023009931 | ISBN 9789811272981 (hardcover) |
    ISBN 9789811273896 (paperback) | ISBN 9789811272998 (ebook for institutions) |
    ISBN 9789811273001 (ebook for individuals)
Subjects: LCSH: Differential calculus. | Differential calculus--Problems, exercises, etc.
Classification: LCC QA305 .J86 2024 | DDC 515/.33--dc23/eng20230711
LC record available at https://lccn.loc.gov/2023009931

British Library Cataloguing-in-Publication Data
A catalogue record for this book is available from the British Library.

Cover photo by Olivia Gutenberg.

For any available supplementary material, please visit
https://www.worldscientific.com/worldscibooks/10.1142/13324#t=suppl

Desk Editors: Soundararajan Raghuraman/Rosie Williamson

Typeset by Stallion Press
Email: enquiries@stallionpress.com

Printed in Singapore

To my sons, my best teachers.

—Veselin Jungić

For Papa, who set me on the road of mathematics, for John, who travels the road with me, and for Eli, who is carving out his own road less traveled.

—Petra Menz

For my colleagues. From whom I've learned so much.

—Randall Pyke

Preface

The primary purpose of this collection of problems is to serve as a supplementary learning resource for students who are taking a university level differential calculus course. A secondary aim is to provide another teaching resource for our colleagues who teach differential calculus. This book can also be used as a refresher in differential calculus or as a quick reference for any reader (e.g., high school teachers, employees in industry) who has taken calculus in the past but may have forgotten parts of the subject or needs to further their understanding in certain areas of calculus.

This collection contains close to 900 problems covering topics in differential calculus: limits, continuity, derivatives, and their applications. The applications are comprised of a variety of approximations, growth and decay, optimization, curve sketching techniques, and analytical tools to investigate properties of parametrically defined planar curves. The problems are sorted by topic into relevant chapters. Each chapter starts with a summary of the appropriate mathematical notions and their properties. All problems are accompanied by an answer and most frequently with either a sketch of the solution process to guide the reader or a detailed solution.

The content of the questions is broadly designed for students enrolled in either a science program (mathematics, physics, statistics, chemistry, earth sciences, kinesiology, life sciences) or applied science program (engineering, computing science). The expectation is that the student is familiar with the mathematical notion of a function and the related notation, terminology, and basic properties.

It goes without saying that the reader is already proficient with various algebraic techniques that are part of the high school mathematics curriculum.

This is a reader-friendly book with problems ranging from routine to not-so-routine. Most of the problems have been used, some of them multiple times in randomized versions, as final examination questions in a standard North American university differential calculus course. With about 260 true-false and multiple-choice questions, the book provides to its users an accessible way to test and practice their understanding of calculus related facts and their nuances. As an aid to the reader, more than 220 figures are woven throughout the collection to visualize properties of functions, illustrate word problems, depict solutions, and provide an extensive bank of polar curves.

We want to emphasize that this book is a mathematical text containing precise and consistent notation and vocabulary throughout. One of the main messages is that a crucial step in mathematical thinking is to first verify the underlying assumptions before applying certain techniques. Through a careful selection of appropriate problems in each chapter, the book clearly communicates big ideas and applications of calculus: the notions of a function, an infinitesimal, differentiability, and approximation, among others. Most notably, problems and their solutions are purposefully created to address nuances that may evade students' or instructors' attention in a general calculus course.

This book is written for students who are at the beginning of their academic careers. As part of their transition to the post-secondary educational environment, students may face non-academic challenges ranging from living on their own for the first time to being overwhelmed by fast-paced content-packed university courses. This is why the last chapter in this collection contains a detailed list of recommendations to guide students in their well-being and approach to learning to thrive academically.

In short, this book is a cross between a handbook, a learning guide, and a rich learning and teaching resource, something that has been missing in the otherwise crowded field of calculus titles.

The basis for this collection are problems from exams that have been put together by our colleagues at Simon Fraser University over the last two decades. We would like to express our gratitude for their contribution to our work.

No project such as this can be free from errors and incompleteness. The authors would be grateful to anyone who points out any typos, errors, or sends any other suggestion on how to improve this manuscript.

About the Authors

 **Veselin Jungić** is a Teaching Professor at the Department of Mathematics, Simon Fraser University. Dr. Jungić is a 3M National Teaching Fellow and a Fellow of the Canadian Mathematical Society. He is a recipient of several teaching awards. Veselin is one of only a few Canadian mathematicians who was awarded both the Canadian Mathematical Society Pouliot Award (2020) and the Canadian Mathematical Society Teaching Award (2012). Dr. Jungić is the author of the book *Basics of Ramsey Theory*.

 **Petra Menz** is a Senior Lecturer in the Department of Mathematics at Simon Fraser University. She enjoys teaching mathematics at the undergraduate level and supporting students in achieving their academic goals. To this end, Dr. Menz has been involved in creating several curricular documents, most prominently a published calculus textbook and an open educational resource geared for social science calculus for which she won the BC Campus Award for Excellence in Open Education. Her research interests are multifarious: embodiment, language, technology, well-being, and cultures in mathematics.

Randall Pyke is a Senior Lecturer in the Department of Mathematics at Simon Fraser University. Randall received his PhD in mathematical physics and spent the early years of his career in the research stream at various universities before moving into the teaching focused stream at Simon Fraser University. In addition to teaching courses in calculus, differential equations, mathematical modelling, and linear algebra Randall has been involved in the Operations Research program at the University and is also regularly lecturing and writing about fractals and discrete dynamical systems.

Contents

Preliminaries

1 Symbols

$\emptyset$	empty set
$\{a, b, c\}$	set containing a, b, and c
$\{x : F(x)\}$	set of all x such that $F(x)$
$x \in A$	x belongs to A
$x \notin A$	x does not belongs to A
$\mathbb{N}$, $\mathbb{Z}$, $\mathbb{Q}$, $\mathbb{R}$	sets of natural numbers, integers, rationals, and reals
$A \cup B$ and $A \cap B$	union and intersection of sets A and B
$A \backslash B$	$\{x : x \in A \text{ and } x \notin B\}$
$A \subset B$	A is a subset of B
$[a, b]$	$\{x \in \mathbb{R} : a \le x \le b\}$
$[a, b)$	$\{x \in \mathbb{R} : a \le x < b\}$
(a, b)	$\{x \in \mathbb{R} : a < x < b\}$
$f : A \to B$	function from A to B (often we will only write *a function f*)
$x \mapsto f(x)$	function that maps x to $f(x)$
$y = f(x)$	function that maps x to y
y' or $f'(x)$ or $\frac{dy}{dx}$	first derivative of the function $y = f(x)$
$\int f(x)dx$	antiderivative of the function $y = f(x)$
$a \approx b$	a and b are close to each other, by some given criterion
π	number *pi*, the ratio of a circle's circumference to its diameter

e	number e, also known as Euler's number and Napier's constant
$a \pm b$	expression that includes the choice $a + b$ and the choice $a - b$
$\sum_{i=1}^{n} a_i$	sum $a_1 + a_2 + \cdots + a_n$
$\triangle ABC$	triangle with vertices A, B, and C
$\triangle ABC \sim \triangle DEF$	similar triangles
1 rad	radian, a unit of measure for angles, 1 rad $= \frac{180°}{\pi} \approx 57.2958°$

2 Commonly Used Mathematical Facts

Whenever there is a labelled figure, the symbols are used in the accompanying definitions and descriptions.

2.1 *Identities*

If not stated otherwise, x and y are any real numbers.

$$x^2 - y^2 = (x - y)(x + y) \qquad x^3 \pm y^3 = (x \pm y)(x^2 \mp xy + y^2)$$

$$x^k - y^k = (x - y) \cdot \sum_{i=0}^{k-1} x^i y^{k-1-i} \qquad 1 + x + \cdots + x^k = \frac{x^{k+1} - 1}{x - 1}, x \neq 1$$

$$\sqrt{x^2} = |x| \qquad a^x \cdot a^y = a^{x+y}, a > 0$$

$$\frac{a^x}{a^y} = a^{x-y}, a > 0 \qquad (a^x)^y = a^{xy}, a > 0$$

$$\log_a(xy) = \log_a x + \log_a y, 0 < a \neq 1, x, y > 0$$

$$\log_a \frac{x}{y} = \log_a x - \log_a y, 0 < a \neq 1, x, y > 0$$

$$\log_a(a^x) = x, 0 < a \neq 1 \qquad a^{\log_a x} = x, 0 < a \neq 1, x > 0$$

$$\log_a(x^y) = y \log_a x, 0 < a \neq 1, \ x > 0$$

$$\sin^2 x + \cos^2 x = 1$$

$$\sin 2x = 2 \sin x \cos x \qquad \cos 2x = \cos^2 x - \sin^2 x$$

$$\cos^2 x = \frac{1 + \cos 2x}{2} \qquad \sin^2 x = \frac{1 - \cos 2x}{2}$$

$$\arcsin(\sin x) = x, \ x \in \left[-\frac{\pi}{2}, \frac{\pi}{2} \right] \qquad \sin(\arcsin x) = x, \ x \in [-1, 1]$$

$$\cosh^2 x - \sinh^2 x = 1$$

2.2 Plane figures

Triangle. A *triangle* is a three-sided polygon.[a]

The perimeter of $\triangle ABC$ is $P = a + b + c$.

The area of $\triangle ABC$ is $\mathcal{A} = \frac{ch}{2}$.

The sum of angles in $\triangle ABC$ is $\alpha + \beta + \gamma = \pi$ radians.

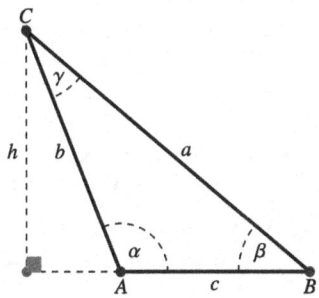

The *Law of Sines* is $\frac{\sin \alpha}{a} = \frac{\sin \beta}{b} = \frac{\sin \gamma}{c}$. The *Law of Cosines* is $a^2 = b^2 + c^2 - 2bc \cos \alpha$.

If $\alpha = \frac{\pi}{2}$, then $\triangle ABC$ is a *right triangle* for which $\beta + \gamma = \frac{\pi}{2}$.

In a right triangle, the Law of Cosines becomes the *Pythagorean Theorem* $a^2 = b^2 + c^2$.

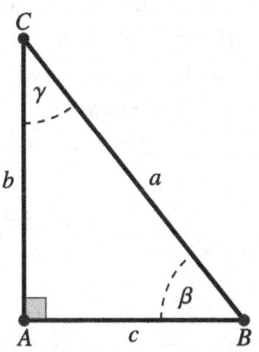

In a right triangle, the *primary* trigonometric ratios are $\sin \beta = \frac{b}{a}$, $\cos \beta = \frac{c}{a}$, and $\tan \beta = \frac{b}{c}$. The *secondary* trigonometric ratios are $\csc \beta = \frac{1}{\sin \beta} = \frac{a}{b}$, $\sec \beta = \frac{1}{\cos \beta} = \frac{a}{c}$, and $\cot \beta = \frac{1}{\tan \beta} = \frac{c}{b}$.

---

[a]E. J. Borowski and J. M. Borwein (1989) define a polygon as "a closed plane figure bounded by three or more straight sides that meet in pairs in the same number of vertices, and do not intersect other than at these vertices."

If $a = b = c$, then $\triangle ABC$ is an *equilateral triangle* for which $\alpha = \beta = \gamma = \frac{\pi}{3}$, $h = \frac{a\sqrt{3}}{2}$, and $\mathcal{A} = \frac{a^2\sqrt{3}}{4}$.

We say that triangles $\triangle ABC$ and $\triangle DEF$ are *similar* if their corresponding angles are congruent.

In similar triangles, corresponding sides are in the same ratio: $\frac{a}{d} = \frac{b}{e} = \frac{c}{f}$.

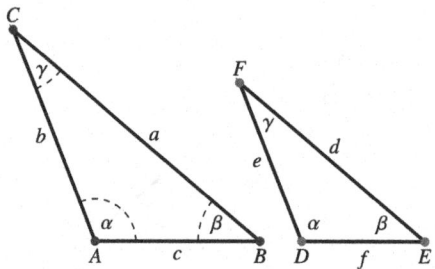

Quadrilateral. A *quadrilateral* is a four-sided polygon. We will consider only *convex* quadrilaterals, i.e., quadrilaterals with the property that if X and Y are points in the interior of the quadrilateral, then the line segment $\overline{XY}$ is in the interior too.

A quadrilateral with two sides parallel is called a *trapezoid*. A quadrilateral with opposite pairs of sides parallel is called a *parallelogram*. A parallelogram with all sides equal is called a *rhombus*. A parallelogram with four right angles is called a *rectangle*. A rectangle with all sides equal is called a *square*. The area of a parallelogram is $\mathcal{A} = ah$. The area of a rectangle is $\mathcal{A} = ab$. The area of a square is $\mathcal{A} = a^2$.

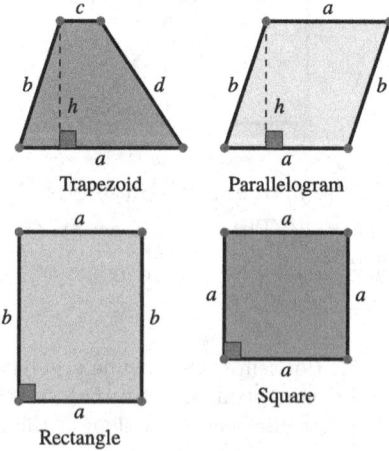

The area of a trapezoid is $\mathcal{A} = \frac{(a+c)h}{2}$.

Circle. A *circle* C is the set of all points in a plane at a given distance $r > 0$ from a given point O in the plane. We say that r is the radius and that O is the centre of the circle C. In the xy-plane, the circle C with the radius r and the centre (a, b) is determined by the equation $(x - a)^2 + (y - b)^2 = r^2$. It is common to use the term *circle* to describe the closed plane region that is bounded by a circle.

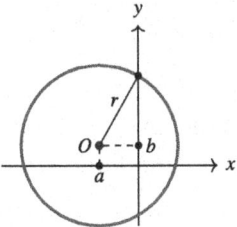

Circle $(x - a)^2 + (y - b)^2 = r^2$

The area of the circle C is $\mathcal{A} = r^2\pi$ and its circumference is $c = 2r\pi$.

Ellipse. An *ellipse* is the set of all points in a plane with the property that the sum of the distances to two fixed points is a given constant. Each fixed point is called a *focus*.

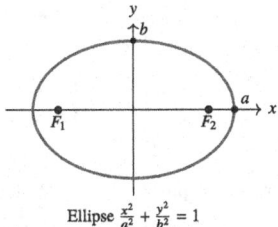

Ellipse $\frac{x^2}{a^2} + \frac{y^2}{b^2} = 1$

In the xy-plane, the ellipse C with centre at the origin, *width a*, and *height b* is determined by the equation $\frac{x^2}{a^2} + \frac{y^2}{b^2} = 1$. Assuming $a > b$, the foci are $(\pm c, 0)$ for $c = \sqrt{a^2 - b^2}$. The area of the ellipse C is $\mathcal{A} = ab\pi$.

2.3 Solids

Prism. A *prism* is a polyhedron[b] with an n-sided polygonal base, the face which is a translated copy of the base, and n other faces,

---

[b]E. J. Borowski and J. M. Borwein (1989) define a polyhedron as "a solid figure consisting of four or more plane faces (all polygons), pairs of which meet along an edge, three or more edges meeting at a vertex."

necessarily all parallelograms, that connect the base and its copy to form a *mantle*.

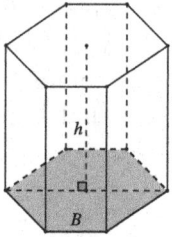

The volume of a prism is $V = hB$, where h is the height and B is the area of the base of the prism. The surface area of a prism is the sum of the areas of its faces.

A *rectangular parallelepiped* or *orthogonal parallelepiped* is a prism with six rectangular faces such that opposite faces are congruent. This type of polyhedron is also known as a *cuboid* and a *rectangular prism*. A rectangular parallelepiped with all edges equal, i.e., square faces, is called a *cube*.

The volume of the rectangular parallelepiped is $V = abc$ and its surface area is $S = 2(ab + ac + bc)$. The volume of the cube is $V = a^3$ and its surface area is $S = 6a^2$.

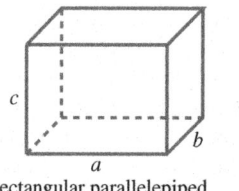
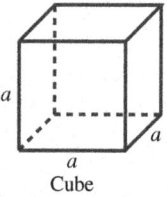

Rectangular parallelepiped   Cube

Pyramid. A *pyramid* is a polyhedron with an n-sided polygonal base and n triangular faces meeting at a point, called the apex. The volume of a pyramid is $V = \frac{hB}{3}$, where h is the altitude of the apex and B is the area of the base of the pyramid.

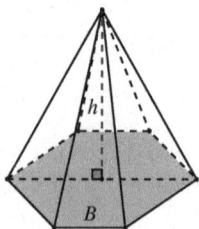

The surface area of a pyramid is the sum of areas of its faces.

Right Circular Cylinder. A *right circular cylinder* is a solid with a circular base, the circle that is a vertically translated copy of the base, and the surface formed by line segments joining every point on the boundary of the base to its translation on the boundary of the other circular face.

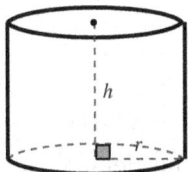

The term "cylinder" is commonly used not only to refer to the cylindrical solid but also to the cylindrical surface. The volume of the right circular cylinder is $V = hr^2\pi$. The surface area of the cylinder is $S = 2\pi r^2 + 2h\pi r$.

Right Circular Cone. A *right circular cone* is a solid with a circular base and the surface formed by congruent line segments joining every point on the boundary of the base to a common point, called the apex. The term "cone" is commonly used not only to refer to the conical solid but also to the conical surface.

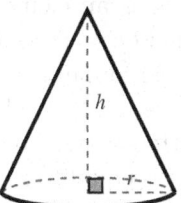

The volume of a right circular cone is $V = \frac{hr^2\pi}{3}$. The surface area of a cone is $S = \pi r^2 + \pi r\sqrt{r^2 + h^2}$.

Sphere. A *sphere* S is the set of all points in the space at a given distance $r > 0$ from a given point O. We say that r is the radius and that O is the centre of the sphere S.

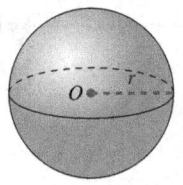

The volume of a sphere is $V = \frac{4r^3\pi}{3}$. The surface area of a sphere is $S = 4r^2\pi$. The circumference of a sphere, $C = 2\pi r$, is the circumference of a great circle.

3 Functions

3.1 *Definitions*

Function. A *function* $f : A \to B$ is a rule that assigns to each element in a set A exactly one element in a set B. The set A is called the *domain* of the function f. The set B is called the *codomain* of the function f.

When defining and manipulating functions, it is common to use the so-called *functional notation*, $y = f(x)$, which is read "y equals f of x," and means that the output y in the codomain correspondents to the input x in the domain by the rule f. In this setting, x is the *independent variable*, a variable that represents elements of the domain, and y is the *dependent variable*, a variable that represents elements of the codomain of the function.

If $f : A \to B$, then the *range* of the function f is the set of all elements y in B so that there is an element x in the domain A such that $y = f(x)$. Hence, range$(f) = \{y \in B : y = f(x), x \in A\}$, i.e., the range is a subset of the codomain.

For example, consider sets $A = \{-2, -1, 0, 1, 2\}$ and $B = \{0, 1, 2, 3, 4\}$ and the rule f that assigns to each element x in the set A the element in the set B that is equal to x^2. Hence, $f(-2) = f(2) = 4$, $f(-1) = f(1) = 1$, and $f(0) = 0$. The set A is the domain of the function f and the set B is its codomain. The range of the function f is the set $\{0, 1, 4\} \subset B$.

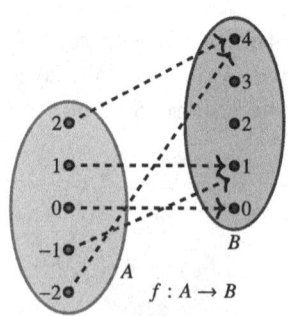

Occasionally, we will use the notation $x \mapsto f(x)$, which is read "x maps to $f(x)$," and means that the output $f(x)$ in the range correspondents to the input x in the domain by the rule f.

Piecewise Defined Function. If $f : A \to B$ and $g : A' \to B$ with $A \cap A' = \emptyset$ are two functions, then the function $F : A \cup A' \to B$ defined by

$$F(x) = \begin{cases} f(x) & \text{if } x \in A \\ g(x) & \text{if } x \in A' \end{cases}$$

is said to be a *piecewise defined function*.

For example, the *absolute value function* $y = |x| = \begin{cases} -x & \text{if } x \in (-\infty, 0) \\ x & \text{if } x \in [0, \infty) \end{cases}$ is a piecewise defined function.

Open and Closed Intervals and Interior Points. An *open interval* is a set of real numbers that can be written as $(-\infty, \infty) = \mathbb{R}$ or $(-\infty, a) = \{x : x < a\}$ or $(a, \infty) = \{x : x > a\}$ or $(a, b) = \{x : a < x < b\}$ for some real numbers a and b. A *closed interval* is a set of real numbers that can be written as $(-\infty, \infty) = \mathbb{R}$ or $(-\infty, a] = \{x : x \leq a\}$ or $[a, \infty) = \{x : x \geq a\}$ or $[a, b] = \{x : a \leq x \leq b\}$ for some real numbers a and b.[c] Intervals $(a, b] = \{x : a < x \leq b\}$ and $[a, b) = \{x : a \leq x < b\}$ are neither closed nor open.

The number c belongs to the *interior of a set* $D \subseteq \mathbb{R}$ if there are real numbers a and b such that $a < c < b$ and $(a, b) \subseteq D$. If c belongs to the interior of the set D, it is common to say that c is an *interior point* of D.

For example, if $D = [0, 2) \cup \{3\}$ then any $c \in (0, 2)$ is an interior point of D. Numbers 0 and 3 belong to D but are not interior points. The number 2 does not belong to D.

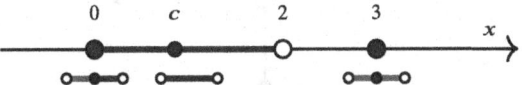

One-To-One Function. We say that a function f is one-to-one if different inputs have different outputs. In other words, f is *one-to-one* if, for any x_1, x_2 in the domain of f, $x_1 \neq x_2$ implies $f(x_1) \neq f(x_2)$ or, equivalently, if $f(x_1) = f(x_2)$ implies $x_1 = x_2$.

---

[c]The notion of open and closed sets and the reason why the set $\mathbb{R}$ is consider both open and closed are commonly studied in an Introduction to Topology course.

For example, the function $f : [0, \infty) \to [0, \infty)$ defined by $f(x) = x^2$ is one-to-one and the function $g : \mathbb{R} \to [0, \infty)$ defined by $g(x) = x^2$ is not.

Inverse Function. Let $f : A \to B$ be a one-to-one function and let the set B be the range of the function f. The *inverse function* of f, denoted f^{-1}, is the function defined on the domain B such that $f^{-1}(y) = x$ for any $x \in A$ and $y \in B$ if and only if $f(x) = y$.

For example, the *square root function* $x \mapsto \sqrt{x}$ is the inverse function of the function $f : [0, \infty) \to [0, \infty)$ defined by $f(x) = x^2$.

Composition of Functions. Let $f : A \to B$ and $g : B \to C$ be two functions. We define a new function $F : A \to C$ by $F(x) = g(f(x))$, $x \in A$. The function F is called the *composition of f and g* and is denoted by $F = g \circ f$.

For example, if $x \mapsto f(x) = 2x + 1$ and $x \mapsto g(x) = x^2$, then $x \mapsto F(x) = (g \circ f)(x) = g(f(x)) = g(2x + 1) = (2x + 1)^2$.

Properties of Functions. Let a be a positive real number and let $f : (-a, a) \to \mathbb{R}$. We say that f is an *odd function* if $f(-x) = -f(x)$ for all $x \in (-a, a)$. We say that f is an *even function* if $f(-x) = f(x)$ for all $x \in (-a, a)$.

For example, the *sign function*

$$x \mapsto \operatorname{sign}(x) = \begin{cases} -1 & \text{if } x < 0 \\ 0 & \text{if } x = 0 \\ 1 & \text{if } x > 0 \end{cases}$$

is an odd function.

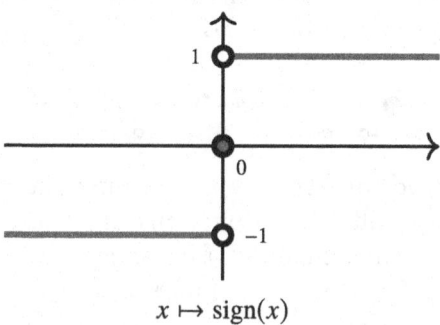

$$x \mapsto \operatorname{sign}(x)$$

Any *constant function* $x \mapsto f(x) = c$, $c \in \mathbb{R}$ is an even function. The function $x \mapsto 1 + \operatorname{sign}(x)$ is neither odd nor even.

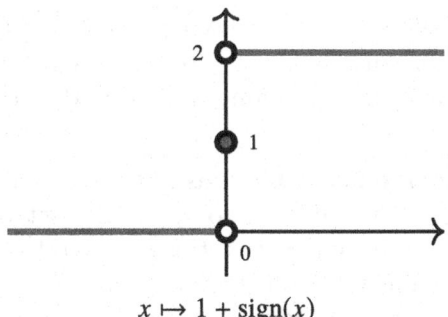

$$x \mapsto 1 + \text{sign}(x)$$

A function $f : \mathbb{R} \to \mathbb{R}$ is said to be *periodic* if there is a real number $T \neq 0$ such that $f(x+T) = f(x)$ for any real number x. The number T is called a *period* of the function f.

For example, the *fractional part function* $x \mapsto \{x\}$, a function that is defined as the difference between a real number and its greatest integer value, is periodic with a period $T = 1$.

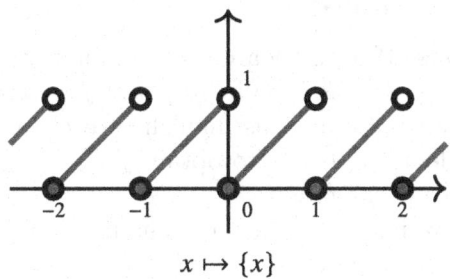

$$x \mapsto \{x\}$$

A function $f : D \to \mathbb{R}$, $D \subseteq \mathbb{R}$ is said to be *increasing on D* if for any $x_1, x_2 \in D$, $x_1 < x_2$ implies $f(x_1) < f(x_2)$. A function $f : D \to \mathbb{R}$ is said to be *decreasing on D* if for any $x_1, x_2 \in D$, $x_1 < x_2$ implies $f(x_1) > f(x_2)$.

For example, the *square function* $x \mapsto x^2$ is increasing on the interval $[0, \infty)$ and decreasing on the interval $(-\infty, 0]$.

Implicitly Defined Functions. We say that a function $x \mapsto f(x)$, $x \in D$ is *implicitly defined* by an equation $F(x, y) = 0$ if $F(x, f(x)) = 0$ for all $x \in D$.

For example, the function $x \mapsto \sqrt{1 - x^2}$, $x \in [-1, 1]$ is implicitly defined by the equation $x^2 + y^2 - 1 = 0$. Observe that $x \mapsto -\sqrt{1 - x^2}$, $x \in [-1, 1]$ is another function implicitly defined by the same equation.

Graphs of Functions and Curves. The *graph of a function* $y = f(x)$ for all $x \in D$, the domain of f, is the set of points in the coordinate plane given by $\Gamma = \{(x, f(x)) : x \in D\}$.

If f is a continuous function then its graph is a *curve*. In general, we think about a curve as the trajectory of a moving point.

If a continuous function $x \mapsto f(x)$, $x \in D$ is implicitly defined by an equation $F(x, y) = 0$, then its graph is part of the *curve* $C = \{(x, y) \in \mathbb{R}^2 : F(x, y) = 0\}$. We will also say that the curve C is defined by the equation $F(x, y) = 0$.

For example, the graph of the function $x \mapsto \sqrt{1 - x^2}$, $x \in [-1, 1]$ is the upper semicircle of the unit circle defined by $F(x, y) = x^2 + y^2 - 1 = 0$.

3.2 *Bank of functions*

Power Functions. If a and b are real numbers, then the function $y = f(x) = bx^a$, $x \in D \subseteq \mathbb{R}$ is called a *power function*.

We assume that $b \neq 0$ and distinguish several cases.

If $a = 0$, we define $f(0) = b$ to obtain $f(x) = b$ for all $x \in \mathbb{R}$. In this case, f is a *constant function*.

If a is a positive integer, then the domain of the function f is the set $\mathbb{R}$ and $f(x) = b \cdot \underbrace{x \cdot x \cdot \ldots \cdot x}_{a}$ for all $x \in \mathbb{R}$.

If a is a negative integer, then the domain of the function f is the set $\mathbb{R} \backslash \{0\}$ and $f(x) = \frac{b}{x^{-a}}$ for all $x \in \mathbb{R} \backslash \{0\}$.

For $m \in \mathbb{N} \backslash \{1\}$ and $x \in \mathbb{R}$, we define $x^{\frac{1}{m}}$ as the real number for which $\left(x^{\frac{1}{m}}\right)^m = x$ if such number exists. If there are two numbers with this property (this may happen if m is even) we choose one that is non-negative. The number $x^{\frac{1}{m}}$ is called the mth root of x and is commonly denoted as $\sqrt[m]{x}$.

If a is a rational number $a = \frac{n}{m}$ with $n \in \mathbb{Z} \backslash \{0\}$, $m \in \mathbb{N} \backslash \{1\}$, and m and n mutually prime, then

$$f(x) = b \sqrt[m]{x^n} = b \left(\sqrt[m]{x}\right)^n.$$

In the case that m is an odd integer the domain of the function f is $\mathbb{R}$ or $\mathbb{R}\backslash\{0\}$, depending if n is positive or negative. If m is an even integer, then the domain of the function f is the interval $[0, \infty)$ or $(0, \infty)$, depending if n is positive or negative.

If a is an irrational number, then the domain of f is the interval $[0, \infty)$ with $f(0) = 0$. To define $f(x) = bx^a$ for $x > 0$, we need calculus. Let $\{r_k\}$ be a sequence of rational numbers such that $\lim_{k \to \infty} r_k = a$. Then, by definition, $x^a = \lim_{k \to \infty} x^{r_k}$.[d]

Any power function is differentiable on the interior of its domain and $\frac{d}{dx}(bx^a) = abx^{a-1}$.

Polynomials. A *polynomial* is any function of the form $y = f(x) = a_0 + a_1 x + a_2 x^2 + \cdots + a_n x^n$, where n is a non-negative integer and $a_0, a_1, a_2, \ldots$, and a_n are given real numbers. The domain of any polynomial is the set $\mathbb{R}$. The derivative of f is the polynomial $f'(x) = a_1 + 2a_2 x + \cdots + na_n x^{n-1}$.

If $n \geq 1$ and $a_n \neq 0$, then the number n is called the *degree* of the polynomial f and is denoted by $\deg(f) = n$. Any non-zero constant function is a polynomial of degree 0.

A *linear function* $y = a_0 + a_1 x$, $a_1 \neq 0$ is a polynomial of degree 1.

A *quadratic function* $y = a_0 + a_1 x + a_2 x^2$, $a_2 \neq 0$ is a polynomial of degree 2.

If a is a non-negative integer and b is a non-zero real number, then the power function $f(x) = bx^a$ is a polynomial of degree a.

A polynomial of degree $n \geq 1$ can have at most n real roots (zeros).

Rational Functions. A *rational function* is a function of the form $f(x) = \frac{p(x)}{q(x)}$, where $p(x)$ and $q(x)$ are polynomials. The domain of a rational function f is the set $\{x \in \mathbb{R} : q(x) \neq 0\}$. Any polynomial is a rational function. Any rational function is differentiable on its domain and $\frac{d}{dx}\left(\frac{p(x)}{q(x)}\right) = \frac{p'(x)q(x) - p(x)q'(x)}{q^2(x)}$.

If a is a negative integer and b is a non-zero real number, then the power function $f(x) = bx^a$ is a rational function with domain $\mathbb{R}\backslash\{0\}$.

---

[d]We are using the fact that the sequence $\{r_k\}$ exists and that each such sequence yields the same limit.

Exponential and Logarithmic Functions. An *exponential function* is a function of the form $y = a^x$, where a is a positive real number not equal 1. The domain of an exponential function is the set $\mathbb{R}$ and the range is the interval $(0, \infty)$.

Observe the difference between an exponential function $x \mapsto a^x$ and a power function $x \mapsto x^a$ for some (fixed) number a.

If the base of the exponential function is the number e, Euler's number, then the exponential function $x \mapsto e^x$ is called the *natural exponential function*.

For any $0 < a \neq 1$, $f(x) = a^x$ is a one-to-one function. Its inverse is the *logarithmic function with base a*, $f^{-1}(x) = \log_a x$. Hence, if u and v are numbers such that $u = a^v$, then $v = \log_a u$. In particular, $\log_a 1 = 0$ and $\log_a a = 1$ for any $0 < a \neq 1$.

The domain of any logarithmic function is the interval $(0, \infty)$ and the range is the set $\mathbb{R}$.

The inverse of the natural exponential function is commonly denoted as $y = \ln x$ and is called the *natural logarithmic function*.

Exponential and logarithmic functions are differentiable on their domains: $\frac{d}{dx}\left(a^x\right) = a^x \ln a$ and $\frac{d}{dx}\left(\log_a x\right) = \frac{1}{x \ln a}$.

Trigonometric Functions and Inverse Trigonometric Functions. Let x be any real number. Let ℓ_x be the ray obtained by rotating the ray that corresponds to the positive direction of the x-axis by an angle of x radians.

Trigonometric functions *cosine*, $y = \cos x$, and *sine*, $y = \sin x$, are respectively defined as the abscissa (horizontal) coordinate and the ordinate (vertical) coordinate of the point at which the ray ℓ_x intersects the unit circle.

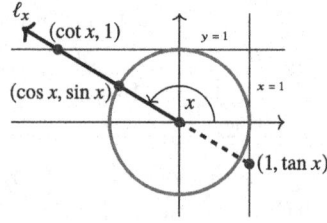

Functions cosine and sine are defined for all real numbers. Their range is the interval $[-1, 1]$.

Trigonometric functions *tangent, $y = \tan x$*, and *cotangent, $y = \cot x$*, are defined in the following way.

If x is not an odd multiple of $\frac{\pi}{2}$, the ray ℓ_x, extended to a line if necessary, intersects the line $x = 1$ at the point $(1, \tan x)$. The domain of the tangent function is the set $\left\{ x \in \mathbb{R} : x \neq \frac{(2k+1)\pi}{2},\ k \in \mathbb{Z} \right\}$. The range is the set $\mathbb{R}$.

If x is not a multiple of π, the ray ℓ_x, extended to a line if necessary, intersects the line $y = 1$ at the point $(\cot x, 1)$. The domain of the cotangent function is the set $\{ x \in \mathbb{R} : x \neq k\pi,\ k \in \mathbb{Z} \}$. The range is the set $\mathbb{R}$.

Each of the above listed trigonometric functions is a periodic function. Cosine and sine are periodic with period $T = 2\pi$ and tangent and cotangent are periodic with period $T = \pi$. The functions sine, tangent, and cotangent are odd. The cosine function is even.

Trigonometric functions are differentiable on their domains: $\frac{d}{dx}(\cos x) = -\sin x$, $\frac{d}{dx}(\sin x) = \cos x$, $\frac{d}{dx}(\tan x) = \frac{1}{\cos^2 x} = \sec^2 x$, and $\frac{d}{dx}(\cot x) = -\frac{1}{\sin^2 x} = -\csc^2 x$.

None of the trigonometric functions is one-to-one and, consequently, they do not have inverse functions. To obtain the so-called *inverse trigonometric functions*, we consider the appropriate restrictions of the corresponding trigonometric functions:

arcsin : $[-1, 1] \to \left[-\frac{\pi}{2}, \frac{\pi}{2} \right]$ is the inverse function of the function $x \mapsto \sin x$, $x \in \left[-\frac{\pi}{2}, \frac{\pi}{2} \right]$.[e]

arccos : $[-1, 1] \to [0, \pi]$ is the inverse function of the function $x \mapsto \cos x$, $x \in [0, \pi]$.

arctan : $\mathbb{R} \to \left(-\frac{\pi}{2}, \frac{\pi}{2} \right)$ is the inverse function of the function $x \mapsto \tan x$, $x \in \left(-\frac{\pi}{2}, \frac{\pi}{2} \right)$.

arccot : $\mathbb{R} \to (0, \pi)$ is the inverse function of the function $x \mapsto \cot x$, $x \in (0, \pi)$.

Hyperbolic Functions. We define *hyperbolic functions* in terms of the exponential function:

The *hyperbolic sine* is defined by $\sinh x = \frac{e^x - e^{-x}}{2}$, $x \in \mathbb{R}$.

---

[e]It is common to write $\arcsin(x) = \sin^{-1}(x)$.

The *hyperbolic cosine* is defined by $\cosh x = \frac{e^x + e^{-x}}{2}$, $x \in \mathbb{R}$.

The *hyperbolic tangent* is defined by $\tanh x = \frac{e^x - e^{-x}}{e^x + e^{-x}} = \frac{\sinh x}{\cosh x}$, $x \in \mathbb{R}$.

The *hyperbolic cotangent* is defined by $\coth x = \frac{e^x + e^{-x}}{e^x - e^{-x}} = \frac{\cosh x}{\sinh x}$, $x \in \mathbb{R}\backslash\{0\}$.

Functions hyperbolic sine, hyperbolic tangent, and hyperbolic cotangent are odd. The hyperbolic cosine function is even. Hyperbolic functions are differentiable on their domains: $\frac{d}{dx}\left(\cosh x\right) = \sinh x$, $\frac{d}{dx}\left(\sinh x\right) = \cosh x$, $\frac{d}{dx}\left(\tan x\right) = \frac{1}{\cosh^2 x}$, and $\frac{d}{dx}\left(\coth x\right) = -\frac{1}{\sinh^2 x}$.

3.3   *Graphs of commonly used functions*

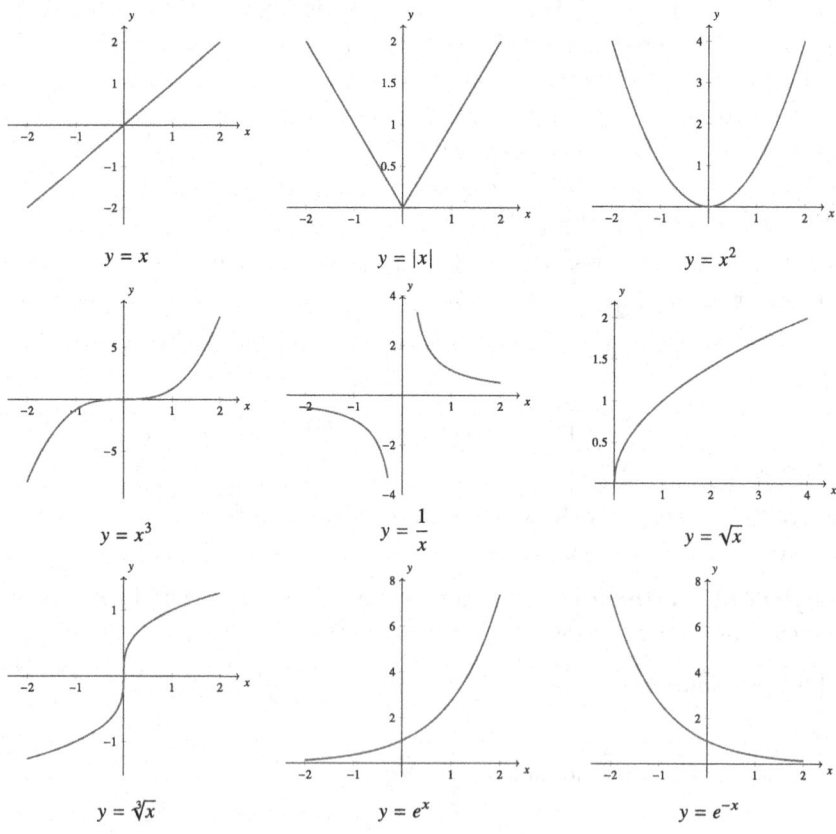

$y = x$ $y = |x|$ $y = x^2$

$y = x^3$ $y = \dfrac{1}{x}$ $y = \sqrt{x}$

$y = \sqrt[3]{x}$ $y = e^x$ $y = e^{-x}$

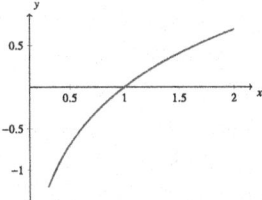

$$y = \ln x$$

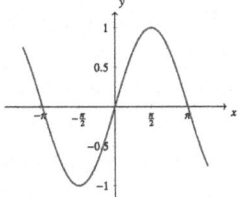

$$y = \sin x$$

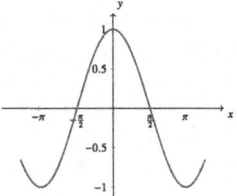

$$y = \cos x$$

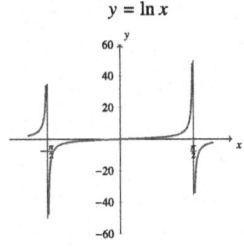

$$y = \tan x$$

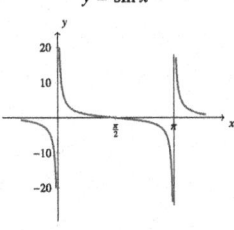

$$y = \cot x$$

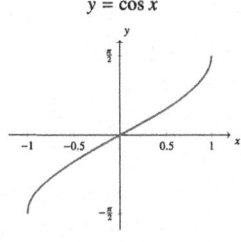

$$y = \arcsin x$$

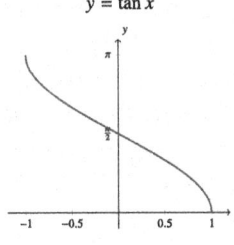

$$y = \arccos x$$

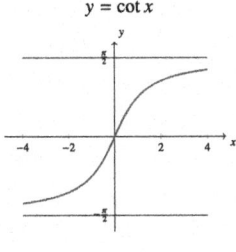

$$y = \arctan x$$

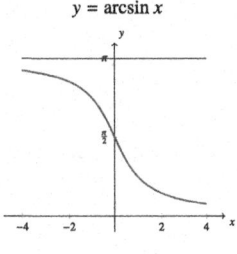

$$y = \operatorname{arccot} x$$

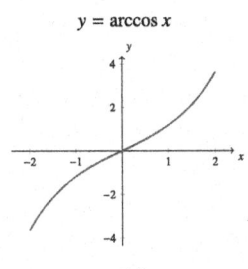

$$y = \sinh x$$

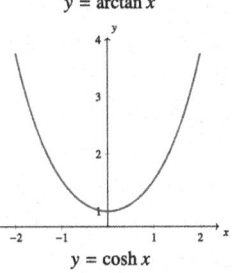

$$y = \cosh x$$

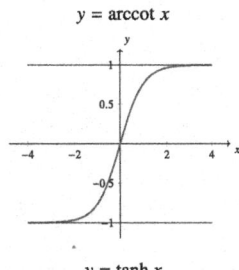

$$y = \tanh x$$

Chapter 1

Limits and Continuity

1.1 Introduction

Use the following definitions, theorems, and properties to solve the problems contained in this chapter.

Limit. We write $\lim_{x \to a} f(x) = L$ and say *the limit of $f(x)$ as x approaches a equals L* if the values of $f(x)$ are arbitrarily close to L for all $x \neq a$ that are sufficiently close to a.

$\varepsilon - \delta$ Limit Definition. Let f be a function defined on an open interval that contains a, except possibly at a itself. We say that *the limit of $f(x)$ as x approaches a equals L*, and we write $\lim_{x \to a} f(x) = L$ if for every number $\varepsilon > 0$ there is a $\delta = \delta(\varepsilon) > 0$ such that $|f(x) - L| < \varepsilon$ whenever $0 < |x - a| < \delta$.

Right-Hand and Left-Hand Limits. We write $\lim_{x \to a^+} f(x) = L$ and say *the right-hand limit of $f(x)$ as x approaches a equals L* if the values of $f(x)$ are arbitrarily close to L for all $x > a$ that are sufficiently close to a. Similarly, we write $\lim_{x \to a^-} f(x) = L$ and say *the left-hand limit of $f(x)$ as x approaches a equals L* if the values of $f(x)$ are arbitrarily close to L for all $x < a$ that are sufficiently close to a.

Limit Via Right-Hand and Left-Hand Limits. $\lim_{x \to a} f(x) = L$ if and only if $\lim_{x \to a^-} f(x) = L$ and $\lim_{x \to a^+} f(x) = L$.

Infinite Limit. Let f be a function defined on an open interval that contains a, except possibly at a itself. Then $\lim_{x \to a} f(x) = \infty$

1

means that the values of $f(x)$ are arbitrarily large for all x sufficiently close to a, but not equal to a. A similar definition holds for $\lim_{x \to a} f(x) = -\infty$.

Vertical Asymptote. The line $x = a$ is called a *vertical asymptote* of the graph of the function $y = f(x)$ if at least one of the following statements is true:

$$\lim_{x \to a} f(x) = \infty, \qquad \lim_{x \to a^-} f(x) = \infty, \qquad \lim_{x \to a^+} f(x) = \infty,$$

$$\lim_{x \to a} f(x) = -\infty, \qquad \lim_{x \to a^-} f(x) = -\infty, \qquad \lim_{x \to a^+} f(x) = -\infty.$$

Limit at Infinity. Let f be a function defined on (a, ∞). Then $\lim_{x \to \infty} f(x) = L$ means the values of $f(x)$ are arbitrarily close to L for all sufficiently large x. A similar definition holds for $\lim_{x \to -\infty} f(x) = L$.

Horizontal Asymptote. The line $y = a$ is called a *horizontal asymptote* of the graph of the function $y = f(x)$ if at least one of the following statements is true:

$$\lim_{x \to \infty} f(x) = a, \qquad \lim_{x \to -\infty} f(x) = a.$$

Slant Asymptote. The line $y = mx + b$, $m \neq 0$ is called a *slant asymptote* of the graph of the function $y = f(x)$ if at least one of the following statements is true:

$$\lim_{x \to \infty} [f(x) - (mx + b)] = 0, \qquad \lim_{x \to -\infty} [f(x) - (mx + b)] = 0.$$

Limit Laws. Let c be a constant and the limits $\lim_{x \to a} f(x)$ and $\lim_{x \to a} g(x)$ exist. Then

(1) $\lim_{x \to a} (f(x) \pm g(x)) = \lim_{x \to a} f(x) \pm \lim_{x \to a} g(x)$;

(2) $\lim_{x \to a} (c \cdot f(x)) = c \cdot \lim_{x \to a} f(x)$;

(3) $\lim_{x \to a} (f(x) \cdot g(x)) = \lim_{x \to a} f(x) \cdot \lim_{x \to a} g(x)$;

(4) $\lim_{x \to a} \dfrac{f(x)}{g(x)} = \dfrac{\lim_{x \to a} f(x)}{\lim_{x \to a} g(x)}$ if $\lim_{x \to a} g(x) \neq 0$.

Squeeze Theorem. If $f(x) \leq g(x) \leq h(x)$ for all x sufficiently close to a, except possibly at a, and $\lim_{x \to a} f(x) = \lim_{x \to a} h(x) = L$, then $\lim_{x \to a} g(x) = L$.

Trigonometric Limits. $\lim\limits_{x\to 0}\frac{\sin x}{x}=1$ and $\lim\limits_{x\to 0}\frac{\cos x-1}{x}=0$.

The Number e. $\lim\limits_{x\to 0}(1+x)^{\frac{1}{x}}=\lim\limits_{x\to\infty}\left(1+\frac{1}{x}\right)^{x}=e$.

L'Hospital's Rule. Suppose that f and g are differentiable and $g'(x)\neq 0$ on an open interval that contains a, except possibly at a. Suppose that $\lim\limits_{x\to a}\frac{f(x)}{g(x)}$ is an indeterminate form of type $\frac{0}{0}$ or $\pm\frac{\infty}{\infty}$. (In other words, $\lim\limits_{x\to a}f(x)=\lim\limits_{x\to a}g(x)=0$ or $\lim\limits_{x\to a}|f(x)|=\lim\limits_{x\to a}|g(x)|=\infty$.) Then $\lim\limits_{x\to a}\frac{f(x)}{g(x)}=\lim\limits_{x\to a}\frac{f'(x)}{g'(x)}$ if $\lim\limits_{x\to a}\frac{f'(x)}{g'(x)}$ exists or equals ∞ or $-\infty$.

Continuity. Let S be a union of intervals and let $f:S\to\mathbb{R}$. If a is an interior point of S, then we say that f *is continuous at a* if $\lim\limits_{x\to a}f(x)=f(a)$. This fact entails three pieces of information: (1) $f(a)$ is defined, (2) $\lim\limits_{x\to a}f(x)$ exists, and (3) the value $f(a)$ equals to the limit of $f(x)$ as x approaches a. If a is a boundary point of S, then we say that the function f *is continuous at a* if the value of the corresponding one-sided limit equals to $f(a)$. We say that f is a continuous function on the set $S\subseteq\mathbb{R}$ if it is continuous at any number $a\in S$. *Note*: It is common to interchangeably use the phrases "f is continuous at a point a" and "f is continuous at a number a." This is based on the fact that every point of a number line corresponds to a unique real number, and every real number corresponds to a point.

Limits and Continuity. If a function f is continuous at b and a function g is such that $\lim\limits_{x\to a}g(x)=b$, then $\lim\limits_{x\to a}f(g(x))=f(\lim\limits_{x\to a}g(x))=f(b)$.

Discontinuities. We say that a function f has a discontinuity at a number a if it is defined on an open interval containing a, except perhaps at a, and is not continuous at a. We distinguish four types of discontinuities: $x=a$ is a *removable discontinuity* if $\lim\limits_{x\to a}f(x)$ exists but it is not equal to $f(a)$ or $f(a)$ is not defined; $x=a$ is a *jump discontinuity* if both $\lim\limits_{x\to a^+}f(x)$ and $\lim\limits_{x\to a^-}f(x)$ exist but are not equal; $x=a$ is an *infinite discontinuity* if the line $x=a$ is a vertical asymptote of the graph of the function $y = f(x)$; and $x = a$ is an *essential discontinuity* if at least one of $\lim\limits_{x \to a^+} f(x)$ and $\lim\limits_{x \to a^-} f(x)$ does not exist but is not infinite.

Intermediate Value Theorem. Let f be a continuous function on the closed interval $[a, b]$ and $f(a) \neq f(b)$. For any number M between $f(a)$ and $f(b)$ there exists a number c in the open interval (a, b) such that $f(c) = M$.

1.2 Types of Limits

For all limit evaluations, use the limit properties, not the ε–δ limit definition, unless stated otherwise.

1.2.1 *Routine miscellaneous limits*

Evaluate the following limits. If a limit fails to exist, provide a reason.

Problem 1.1. $\lim\limits_{x \to 4} \left[\dfrac{1}{\sqrt{x} - 2} - \dfrac{4}{x - 4} \right]$.

Problem 1.2. $\lim\limits_{x \to 2\pi} \dfrac{1 - \cos x}{x^2}$.

Problem 1.3. $\lim\limits_{x \to 3} \dfrac{\sin x - x}{x^3}$.

Problem 1.4. $\lim\limits_{x \to 10} \dfrac{x^2 - 100}{x - 10}$.

Problem 1.5. $\lim\limits_{x \to -4} \dfrac{x^2 - 16}{x + 4} \ln |x|$.

Problem 1.6. $\lim\limits_{x \to 10} \dfrac{x^2 - 99}{x - 10}$.

Problem 1.7. $\lim\limits_{x \to 10} \dfrac{x^2 - 100}{x - 9}$.

Problem 1.8. $\lim\limits_{x \to 10} \sqrt{-x^2 + 20x - 100}$.

Problem 1.9. $\lim\limits_{x \to 1} \dfrac{\ln x}{x}$.

Problem 1.10. $\lim\limits_{x \to 1} \dfrac{\ln(1 + 3x)}{2x}$.

Problem 1.11. $\lim\limits_{x \to -2} \dfrac{2 - |x|}{2 + x}$.

Problem 1.12. $\lim\limits_{x \to 8} \dfrac{(x - 8)(x + 2)}{|x - 8|}$.

Problem 1.13. $\lim\limits_{x \to 2} \left(\dfrac{1}{x^2 + 5x + 6} - \dfrac{1}{x - 2} \right)$.

Problem 1.14. $\lim\limits_{x \to -1} \dfrac{x^2 - x - 2}{3x^2 - x - 1}$.

Problem 1.15. $\lim\limits_{x \to 16} \dfrac{\sqrt{x} - 4}{x - 16}$.

Problem 1.16. $\lim\limits_{x \to 8} \dfrac{\sqrt[3]{x} - 2}{x - 8}$.

Problem 1.17. $\lim\limits_{x \to 4} \dfrac{2 - \sqrt{x}}{4x - x^2}$.

Problem 1.18. $\lim\limits_{x \to 0} \dfrac{\sqrt{1 + 2x} - \sqrt{1 - 4x}}{x}$.

Problem 1.19. $\lim\limits_{x \to 5} e^{\frac{x - 5}{\sqrt{x - 1} - 2}}$.

Problem 1.20. $\lim\limits_{x \to 7} e^{\frac{\sqrt{x + 2} - 3}{x - 7}}$.

Problem 1.21. $\lim\limits_{t \to 0} \dfrac{\sqrt{\sin t + 1} - 1}{t}$.

Problem 1.22. $\lim\limits_{x \to 0} \dfrac{1 - e^{-x}}{1 - x}$.

Problem 1.23. $\lim\limits_{x \to 0} \dfrac{e^{-x^2} \cos(x^2)}{x^2}$.

Problem 1.24. $\lim\limits_{\theta \to \frac{3\pi}{2}} \dfrac{\cos \theta + 1}{\sin \theta}$.

1.2.2   *Not-so-routine miscellaneous limits*

Problem   1.25. Evaluate $\lim_{x \to a}[f(x) \cdot g(x)]$ given that $\lim_{x \to a}[f(x) + g(x)] = 3$ and $\lim_{x \to a}[f(x) - g(x)] = 4$.

Problem 1.26. What can be said about $\lim_{x \to 0} \dfrac{f(x)}{|x|}$ given that $\lim_{x \to 0} xf(x) = 3$?

Problem 1.27. Evaluate $\lim_{h \to 0} \dfrac{\sqrt[4]{16+h}-2}{2h}$.

Problem 1.28. Let $x_1 = 100$ and $x_{n+1} = \frac{1}{2}\left(x_n + \dfrac{100}{x_n}\right)$ for $n \geq 1$. Assume that $L = \lim_{n \to \infty} x_n$ exists. Evaluate L.

Problem 1.29. Convert the repeating decimal fraction $a = 0.21555\ldots$ into a common fraction.

Problem 1.30. Evaluate $\lim_{x \to 10} f(x)$, where $f(x) = x^2$ for all $x \neq 10$, but $f(10) = 99$.

Problem   1.31. Determine constants a and b such that $\lim_{x \to 0} \dfrac{\sqrt{ax+b}-2}{x} = 1$.

Problem 1.32. What will happen to the roots of the quadratic equation $ax^2 - 2x - 1 = 0$ with $a > 0$ if the coefficient a approaches zero?

Problem   1.33. Is there a number b such that $\lim_{x \to -2} \dfrac{bx^2+15x+15+b}{x^2+x-2}$ exists? If so, determine the value of b and the value of the limit.

Problem 1.34. Let $\lim_{x \to \infty} f(x) = 12$ and $\lim_{x \to \infty} \dfrac{f(x)}{g(x)+1} = 4$. Evaluate $\lim_{x \to \infty} g(x)$.

Problem 1.35. Sketch a graph of the function $f(x) = \lim_{n \to \infty} (\sin x)^{2n}$.

Problem 1.36. Sketch a graph of the function $f(x) = \lim_{n \to \infty} \dfrac{x+1}{1-x^{2n}}$. State $\lim_{x \to 1^+} f(x)$ and $\lim_{x \to 1^-} f(x)$.

Problem 1.37. Let $m, n \in \mathbb{N}$. Evaluate $\lim\limits_{x \to 1} \dfrac{x^m - 1}{x^n - 1}$.

Problem 1.38. Prove that any exponential function $x \mapsto a^x$, $a > 1$, grows faster than any polynomial function.

1.2.3 Left-hand and right-hand limits

Evaluate the following limits. If a limit fails to exist, provide a reason.

Problem 1.39. $\lim\limits_{x \to 1^-} \dfrac{x + 1}{x^2 - 1}$.

Problem 1.40. $\lim\limits_{x \to 1^+} \dfrac{\sqrt{x - 1}}{x^2 - 1}$.

Problem 1.41. Evaluate $\lim\limits_{x \to 1^-} f(x)$ if $f(x) = \begin{cases} \frac{x^2 - 1}{|x - 1|} & \text{if } x \neq 1 \\ 4 & \text{if } x = 1. \end{cases}$

Problem 1.42. Let $F(x) = \dfrac{2x^2 - 3x}{|2x - 3|}$. Evaluate $\lim\limits_{x \to 1.5^+} F(x)$ and $\lim\limits_{x \to 1.5^-} F(x)$. Does $\lim\limits_{x \to 1.5} F(x)$ exist?

Problem 1.43. $\lim\limits_{x \to 2^-} \dfrac{|x^2 - 4|}{10 - 5x}$.

Problem 1.44. $\lim\limits_{x \to -1^+} \arcsin x$.

Problem 1.45. $\lim\limits_{x \to 0^+} (x + \sin x)^{\frac{1}{x}}$.

Problem 1.46. $\lim\limits_{x \to e^+} (\ln x)^{\frac{1}{x}}$.

Problem 1.47. $\lim\limits_{x \to \frac{\pi}{2}^+} \dfrac{x}{\cot x}$.

Problem 1.48. $\lim\limits_{x \to \frac{\pi}{2}^+} \left(\dfrac{1}{\tan x} + \arctan \dfrac{1}{\frac{\pi}{2} - x} \right)$.

Problem 1.49. $\lim\limits_{x \to 0^-} \left(e^{\frac{1}{x}} + \dfrac{1}{\ln(-x)} \right)$.

1.2.4 *Limit at infinity*

Evaluate the following limits. If a limit fails to exist, provide a reason.

Problem 1.50. $\lim\limits_{x \to -\infty} \dfrac{\sqrt{x^2 + 4x}}{4x + 1}$.

Problem 1.51. $\lim\limits_{u \to \infty} \dfrac{u}{\sqrt{u^2 + 1}}$.

Problem 1.52. $\lim\limits_{x \to \infty} \dfrac{\sqrt{x^4 + 2}}{x^4 - 4}$.

Problem 1.53. $\lim\limits_{x \to \infty} \dfrac{\sqrt{4x^2 + 3x} - 7}{7 - 3x}$.

Problem 1.54. $\lim\limits_{x \to -\infty} \dfrac{\sqrt{x^6 - 3}}{\sqrt{x^6 + 5}}$.

Problem 1.55. $\lim\limits_{x \to \infty} \dfrac{1 + 3x}{\sqrt{2x^2 + x}}$.

Problem 1.56. $\lim\limits_{x \to -\infty} \dfrac{\sqrt[3]{x^3 - 9}}{2x - 1}$.

Problem 1.57. $\lim\limits_{x \to \infty} \dfrac{1.01^x}{x^{100}}$.

Problem 1.58. $\lim\limits_{x \to \infty} \dfrac{x^2}{e^x}$.

Problem 1.59. $\lim\limits_{x \to \infty} \dfrac{x^2}{e^{4x} - 1 - 4x}$.

Problem 1.60. $\lim\limits_{x \to \infty} \dfrac{\ln x}{\sqrt{x}}$.

Problem 1.61. $\lim\limits_{x \to \infty} \dfrac{\ln x^{100}}{x^2}$.

Problem 1.62. $\lim\limits_{x \to -\infty} \dfrac{3x^6 - 7x^5 + x}{5x^6 + 4x^5 - 3}$.

Problem 1.63. $\lim\limits_{x \to -\infty} \dfrac{5x^7 - 7x^5 + 1}{2x^8 + 6x^6 - 3}$.

Problem 1.64. $\lim\limits_{x\to-\infty} \dfrac{2x+3x^3}{x^2+2x-1}$.

Problem 1.65. $\lim\limits_{x\to\infty} \dfrac{(x+1)^7+(x+2)^7+\cdots+(x+7)^7}{x^7+7^7}$.

Problem 1.66. $\lim\limits_{x\to\infty} \dfrac{ax^{17}+bx}{cx^{17}-dx^3}$, $c\neq 0$.

Problem 1.67. $\lim\limits_{x\to\infty} \dfrac{3x+|1-3x|}{1-5x}$.

Problem 1.68. $\lim\limits_{x\to\infty} \left(\sqrt{x^2+x}-x\right)$.

Problem 1.69. $\lim\limits_{x\to-\infty} \left(\sqrt{x^2+5x}-\sqrt{x^2+2x}\right)$.

Problem 1.70. $\lim\limits_{x\to\infty} \left(\sqrt[3]{x^3+x^2}-x\right)$.

Problem 1.71. $\lim\limits_{x\to\infty} \left(\sqrt{x+\sqrt{x+\sqrt{x}}}-\sqrt{x}\right)$.

1.2.5 *Trigonometric limits*

Evaluate the following limits by using limits $\lim\limits_{x\to 0}\frac{\sin x}{x}=1$ and $\lim\limits_{x\to 0}\frac{1-\cos x}{x}=0$.

Problem 1.72. $\lim\limits_{x\to 0} \dfrac{\sin 6x}{2x}$.

Problem 1.73. $\lim\limits_{x\to 1} \dfrac{\sin(x-1)}{x^2+x-2}$.

Problem 1.74. $\lim\limits_{x\to 0} \dfrac{\sin 4x}{\sqrt{x+1}-1}$.

Problem 1.75. $\lim\limits_{x\to 0} \dfrac{1-\cos x}{x^2}$.

Problem 1.76. $\lim\limits_{x\to 0} \cot(3x)\sin(7x)$.

Problem 1.77. $\lim\limits_{x\to 0} \dfrac{(\sin x)^{100}}{x^{99} \sin 2x}$.

Problem 1.78. $\lim\limits_{x\to 0} \dfrac{x^{100} \sin 7x}{(\sin x)^{99}}$.

Problem 1.79. $\lim\limits_{x\to 0} \dfrac{x^{100} \sin 7x}{(\sin x)^{101}}$.

Problem 1.80. $\lim\limits_{x\to 0} \dfrac{\sin x}{\sqrt{x \sin 4x}}$.

Problem 1.81. $\lim\limits_{x\to 0} \dfrac{1 - \cos x}{x \sin x}$.

Problem 1.82. $\lim\limits_{x\to 0} \dfrac{\sin(\sin x)}{x}$.

Problem 1.83. $\lim\limits_{x\to \frac{\pi}{2}^+} \dfrac{\sqrt{1 + \cos 2x}}{\sqrt{\pi} - \sqrt{2x}}$.

Problem 1.84. $\lim\limits_{x\to \frac{\pi}{3}} \dfrac{\sin\left(x - \frac{\pi}{3}\right)}{1 - 2 \cos x}$.

1.2.6   *Limit definition of the number e*

Evaluate the following limits by using any of the limit definitions of the number e.

Problem 1.85. $\lim\limits_{x\to 0} (1 - 2x)^{\frac{1}{x}}$.

Problem 1.86. $\lim\limits_{x\to 0} \left(1 + \dfrac{x}{2}\right)^{\frac{3}{x}}$.

Problem 1.87. $\lim\limits_{x\to 0} (1 + \tan x)^{\cot x}$.

Problem 1.88. $\lim\limits_{x\to 0} \dfrac{\ln(1 + x)}{x}$.

Problem 1.89. $\lim\limits_{x\to 0} \dfrac{e^x - 1}{x}$.

Problem 1.90. $\lim\limits_{x\to 1}(2-x)^{\frac{1}{x-1}}$.

Problem 1.91. $\lim\limits_{x\to\infty}\left(1+\dfrac{1}{x}\right)^{2x}$.

Problem 1.92. $\lim\limits_{x\to\infty}\left(1+\dfrac{3}{x}\right)^{2x}$.

Problem 1.93. $\lim\limits_{x\to\infty}x(\ln(x+1)-\ln x)$.

Problem 1.94. $\lim\limits_{x\to\infty}\left(\dfrac{x^2+x+1}{x^2-x-1}\right)^{x}$.

Problem 1.95. $\lim\limits_{x\to\infty}\left(\dfrac{mx+n}{mx+p}\right)^{x}$, $m,n,p\in\mathbb{R}$, $m\neq 0$, and $n\neq p$.

1.2.7 Squeeze Theorem

Evaluate the following limits by applying the Squeeze Theorem.

Problem 1.96. Let I be an open interval such that $4\in I$, and let a function f be defined on a set $D=I\backslash\{4\}$. Evaluate $\lim\limits_{x\to 4}f(x)$, where $x+2\le f(x)\le x^2-10$ for all $x\in D$.

Problem 1.97. $\lim\limits_{x\to\infty}\frac{\sin x}{x}$.

Problem 1.98. $\lim\limits_{x\to 0}x\sin\left(\frac{1}{x}\right)$.

Problem 1.99. $\lim\limits_{x\to 0}e^{x\cos(1/x)}$.

Problem 1.100. $\lim\limits_{x\to 0}x^4\sin\left(\frac{1}{x}\right)$.

Problem 1.101. $\lim\limits_{x\to 0^+}\left(\sqrt{x}e^{\sin(1/x)}\right)$.

Problem 1.102. $\lim\limits_{x\to 0^+}(x^2+x)^{\frac{1}{3}}\sin\left(\frac{1}{x^2}\right)$.

Problem 1.103. $\lim\limits_{x\to 0}\dfrac{x^3\sin\left(\frac{1}{x^2}\right)}{\sin x}$.

Problem 1.104. $\lim\limits_{x\to\infty} \frac{5x^2-\sin 3x}{x^2+10}$.

Problem 1.105. Suppose that $\lim\limits_{x\to 0} f(x)$ exists and that $\frac{e^{\frac{2x}{3}}-1}{x} \leq \frac{f(x)}{x^2+2} \leq \frac{\sin 2x}{\sin 3x}$ for all $x \in (-1,0) \cup (0,1)$. Evaluate $\lim\limits_{x\to 0} f(x)$.

1.2.8 *L'Hospital's Rule*

Problem 1.106. Let $f(x) = x^2 \sin\left(\frac{1}{x}\right)$ for $x \neq 0$, and let $g(x) = \sin x$.

(1) Show that $\lim\limits_{x\to 0} f(x) = \lim\limits_{x\to 0} g(x) = 0$, i.e., show that $\frac{f(x)}{g(x)}$ is an indeterminate form of type $\frac{0}{0}$.
(2) Show that $\lim\limits_{x\to 0} \frac{f'(x)}{g'(x)}$ does not exist.
(3) Use the Squeeze Theorem to evaluate $\lim\limits_{x\to 0} \frac{f(x)}{g(x)}$.
(4) Explain why your result in (3) does not contradict L'Hospital's Rule.

Problem 1.107. Let $f(x) = 2x - \sin x$ and $g(x) = 2x + \sin x$.

(1) Show that $\lim\limits_{x\to\infty} f(x) = \lim\limits_{x\to\infty} g(x) = \infty$, i.e., show that $\frac{f(x)}{g(x)}$ is an indeterminate form of type $\frac{\infty}{\infty}$.
(2) Show that $\lim\limits_{x\to\infty} \frac{f'(x)}{g'(x)}$ does not exist.
(3) Evaluate $\lim\limits_{x\to\infty} \frac{f(x)}{g(x)}$.

Problem 1.108. Let $f(x) = e^x + e^{-x}$ and $g(x) = e^x - e^{-x}$.

(1) Show that $\lim\limits_{x\to\infty} f(x) = \lim\limits_{x\to\infty} g(x) = \infty$, i.e., show that $\frac{f(x)}{g(x)}$ is an indeterminate form of type $\frac{\infty}{\infty}$.
(2) Show that $\lim\limits_{x\to\infty} f'(x) = \lim\limits_{x\to\infty} g'(x) = \infty$, i.e., show that $\frac{f'(x)}{g'(x)}$ is an indeterminate form of type $\frac{\infty}{\infty}$.
(3) Show that $\frac{f(x)}{g(x)} = \frac{f''(x)}{g''(x)}$. Conclude that L'Hospital's Rule is not useful in this situation.
(4) Evaluate $\lim\limits_{x\to\infty} \frac{f(x)}{g(x)}$.

Each of the following limits is an indeterminate form of the following type: $\frac{0}{0}$, $\frac{\infty}{\infty}$, $0 \cdot \infty$, or $\infty - \infty$. Use L'Hospital's Rule to evaluate each limit. If a limit fails to exist, provide a reason.

Problem 1.109. $\displaystyle\lim_{x \to 0} \frac{\sinh 2x}{xe^x}$.

Problem 1.110. $\displaystyle\lim_{x \to 1} \frac{x^2 - 1}{e^{1-x^2} - 1}$.

Problem 1.111. $\displaystyle\lim_{x \to 1} \frac{\ln x}{\sin(\pi x)}$.

Problem 1.112. $\displaystyle\lim_{x \to 0} \frac{(2 + x)^{2023} - 2^{2023}}{x}$.

Problem 1.113. $\displaystyle\lim_{x \to 0} \frac{\ln(2 + 2x) - \ln 2}{x}$.

Problem 1.114. $\displaystyle\lim_{x \to 0} \frac{\ln(1 + 3x)}{2x}$.

Problem 1.115. $\displaystyle\lim_{\theta \to \frac{\pi}{2}^+} \frac{\ln(\sin \theta)}{\cos \theta}$.

Problem 1.116. $\displaystyle\lim_{x \to 1} \frac{1 - x + \ln x}{1 + \cos(\pi x)}$.

Problem 1.117. $\displaystyle\lim_{x \to 0} \frac{e^{2x} - 1 - 2x}{x^2}$.

Problem 1.118. $\displaystyle\lim_{x \to 1} \frac{x^2 - 1}{e^{1-x^7} - 1}$.

Problem 1.119. $\displaystyle\lim_{x \to 1} \frac{x^a - 1}{x^b - 1}$, $a, b \neq 0$.

Problem 1.120. $\displaystyle\lim_{x \to 0} \frac{\arcsin 3x}{\arcsin 5x}$.

Problem 1.121. $\displaystyle\lim_{x \to 0} \frac{\sin 3x}{\sin 5x}$.

Problem 1.122. $\displaystyle\lim_{x \to 0} \frac{x - \sin x}{x^3}$.

Problem 1.123. $\displaystyle\lim_{x\to\infty} \frac{(\ln x)^2}{x}$.

Problem 1.124. $\displaystyle\lim_{x\to\infty} \frac{\ln\sqrt{2x}}{\ln\sqrt[3]{3x}}$.

Problem 1.125. $\displaystyle\lim_{x\to 0^+} x^{0.01}\ln x$.

Problem 1.126. $\displaystyle\lim_{x\to 0^+} (\sin x)(\ln x)$.

Problem 1.127. $\displaystyle\lim_{x\to\infty} \left(x\cdot\ln\frac{x-1}{x+1}\right)$.

Problem 1.128. $\displaystyle\lim_{x\to\frac{\pi}{2}} \left(x-\frac{\pi}{2}\right)\tan x$.

Problem 1.129. $\displaystyle\lim_{x\to\infty} x\tan(1/x)$.

Problem 1.130. $\displaystyle\lim_{x\to 0^+} (\sin x)(\ln\sin x)$.

Problem 1.131. $\displaystyle\lim_{x\to 0} \left(\frac{1}{x^2}-\frac{1}{\tan x}\right)$.

Problem 1.132. $\displaystyle\lim_{x\to 0^+} \left(\frac{1}{x}-\frac{1}{e^x-1}\right)$.

Problem 1.133. $\displaystyle\lim_{x\to 0} \left(\frac{1}{\sin x}-\frac{1}{x}\right)$.

Problem 1.134. $\displaystyle\lim_{x\to 0} (\csc x-\cot x)$.

Problem 1.135. Suppose that f' is a continuous function. Use L'Hospital's Rule to show that

$$\lim_{h\to 0} \frac{f(x+h)-f(x-h)}{2h} = f'(x).$$

Problem 1.136. Suppose that $f''(x)$ exists. Use L'Hospital's Rule to show that

$$\lim_{h\to 0} \frac{f(x+h)+f(x-h)-2f(x)}{h^2} = f''(x).$$

1.2.9 *Indeterminate forms of types* 0^0, ∞^0, *and* 1^∞

Evaluate each of the following limits by first transforming the given indeterminate form type into type $0 \cdot \infty$ via identity $f(x)^{g(x)} = e^{g(x)\ln(f(x))}$, where $0 < f(x) \neq 1$. Use an additional transformation to obtain an indeterminate form of type $\frac{0}{0}$ or type $\frac{\infty}{\infty}$. Then use L'Hospital's Rule. If a limit fails to exist, provide a reason.

Problem 1.137. Let $a \in (0,1)$. Consider the two functions $f(x) = e^{x^{-2}\ln a}$ and $g(x) = \sin^2 x$.

(1) Evaluate $\lim\limits_{x\to 0} f(x)$, $\lim\limits_{x\to 0} g(x)$, and $\lim\limits_{x\to 0} f(x)^{g(x)}$.

(2) Conclude that 0^0 is an indeterminate form.

Problem 1.138. $\lim\limits_{x\to 0^+} x^x$.

Problem 1.139. $\lim\limits_{x\to 0^+} x^{\sqrt{x}}$.

Problem 1.140. $\lim\limits_{x\to 0^+} x^{\tan x}$.

Problem 1.141. $\lim\limits_{x\to 0^+} (\sin x)^{\tan x}$.

Problem 1.142. $\lim\limits_{x\to 0^+} (\tan x)^{\sin x}$.

Problem 1.143. $\lim\limits_{x\to 0^+} \left(\frac{x}{x+1}\right)^x$.

Problem 1.144. Let $a \in (1,\infty)$. Consider the two functions $f(x) = e^{x^{-2}\ln a}$ and $g(x) = \sin^2 x$.

(1) Evaluate $\lim\limits_{x\to 0} f(x)$, $\lim\limits_{x\to 0} g(x)$, and $\lim\limits_{x\to 0} f(x)^{g(x)}$.

(2) Conclude that ∞^0 is an indeterminate form.

Problem 1.145. $\lim\limits_{x\to\infty} x^{\frac{1}{x}}$.

Problem 1.146. $\lim\limits_{x\to\infty} (e^x + x)^{\frac{1}{x}}$.

Problem 1.147. $\lim\limits_{x\to\infty} (x + \sin x)^{\frac{1}{x}}$.

Problem 1.148. $\lim\limits_{x\to\infty} (\ln x)^{\frac{1}{x}}$.

Problem 1.149. $\lim\limits_{x\to\infty} (\ln x)^{e^{-x}}$.

Problem 1.150. Let $a \in (0,\infty)$. Consider the two functions $f(x) = e^{x^2 \ln a}$ and $g(x) = \frac{1}{\sin^2 x}$.

(1) Evaluate $\lim\limits_{x\to 0} f(x)$, $\lim\limits_{x\to 0} g(x)$, and $\lim\limits_{x\to 0} f(x)^{g(x)}$.

(2) Conclude that 1^∞ is an indeterminate form.

Problem 1.151. $\lim\limits_{x\to 0} (1 - 2x)^{\frac{1}{x}}$.

Problem 1.152. $\lim\limits_{x\to 0} (\cosh x)^{\frac{1}{x^2}}$.

Problem 1.153. $\lim\limits_{x\to 0^+} (\cos x)^{\frac{1}{x}}$.

Problem 1.154. $\lim\limits_{x\to 0} (1 + \sin x)^{\frac{1}{x}}$.

Problem 1.155. $\lim\limits_{x\to\infty} \left(1 + \sin \frac{3}{x}\right)^x$.

Problem 1.156. $\lim\limits_{x\to e^+} (\ln x)^{\frac{1}{x-e}}$.

1.2.10 *Asymptotes*

Solve the following asymptote problems.

Problem 1.157. Determine all asymptotes of the graph of the function $f(x) = \frac{x^2+x-6}{x+2}$.

Problem 1.158. Determine all asymptotes of the graph of the function $f(x) = \frac{x^2+x-6}{x^2-5x+6}$.

Problem 1.159. Determine the value of a so that the line $y = x+3$ is a slant asymptote of the graph of the function $f(x) = \frac{x^2+ax+5}{x+1}$.

Problem 1.160. Prove that the x-axis is a horizontal asymptote of the graph of the function $f(x) = \frac{\ln x}{x}$.

Determine all asymptotes of the graphs of the following functions:

Problem 1.161. $f(x) = \sin\left(\frac{1}{x}\right)$.

Problem 1.162. $g(x) = \frac{\sin x}{x}$.

Problem 1.163. $h(x) = x \sin \left(\frac{1}{x} \right)$.

Problem 1.164. $i(x) = \frac{1}{\sin x}$.

Problem 1.165. $j(x) = x + \sin \left(\frac{1}{x} \right)$.

1.2.11 ε–δ *limit definition*

Problem 1.166. Use the ε–δ limit definition to prove that $\lim\limits_{x \to 0} x^3 = 0$.

Problem 1.167.

(1) Sketch a graph of the function $f(x) = 2x^2$ on $[0, 2]$. Show the points $P = (1, 0)$ and $Q = (0, 2)$ on this graph. When using the ε–δ limit definition to establish that $\lim\limits_{x \to 1} f(x) = 2$, a number ε and another number δ, that depends on ε, are used. For $\varepsilon = \frac{1}{4}$, draw corresponding intervals on the y- and x-axis and highlight points on the graph that this value of ε determines. Recall that for a given $\varepsilon > 0$, a number $\delta > 0$ must be such that if $x \in (1 - \delta, 1 + \delta)$, then $f(x) \in (2 - \varepsilon, 2 + \varepsilon)$.
(2) Use the graph to determine a positive number δ so that whenever $|x - 1| < \delta$ it is always true that $|2x^2 - 2| < \frac{1}{4}$.
(3) State exactly what has to be proved to establish this limit property of the function f.

Problem 1.168. Use the ε–δ limit definition to show that $\lim\limits_{x \to 0} \sin \left(\frac{1}{x} \right)$ does not exist.

1.3 Continuity

Solve the following problems by using the appropriate definitions and theorems that are stated in the introduction to this chapter.

Problem 1.169. Give one example of a function f such that:

(1) f is continuous for all values of x except when $x = 3$, where f has a removable discontinuity;

(2) f is continuous for all values of x except when $x = 0$, where f has a jump discontinuity;

(3) f is continuous for all values of x except when $x = -1$, where f has an infinite discontinuity.

Problem 1.170. Sketch the graph of a function f such that:

(1) f has a removable discontinuity at $x = 2$ and an infinite discontinuity at $x = 7$, but f is continuous everywhere else.

(2) f has a jump discontinuity at $x = 3$ and a removable discontinuity at $x = 5$, but f is continuous everywhere else.

Problem 1.171. Determine all points where the function

$$f(x) = \begin{cases} x^2 - 4 & \text{if } x < -1 \\ -3 & \text{if } x = -1 \\ -x - 4 & \text{if } x > -1 \end{cases}$$

is continuous.

Problem 1.172. Determine all points where the function

$$f(x) = \begin{cases} \dfrac{x+3}{x^2 - 3x - 18} & \text{if } x \neq -3,\ x \neq 6 \\ 0 & \text{if } x = -3 \end{cases}$$

is discontinuous and identify the type of discontinuity.

Problem 1.173. Determine all points where the function

$$f(x) = \begin{cases} \dfrac{x^2 - 4}{x - 2} & \text{if } x \neq -2,\ x \neq 2 \\ 4 & \text{if } x = -2 \end{cases}$$

is discontinuous and identify the type of discontinuity.

Problem 1.174. You are given the function

$$f(x) = \begin{cases} c - x & \text{if } x \leq \pi \\ c \sin x & \text{if } x > \pi. \end{cases}$$

(1) Determine the constant c so that the function $f(x)$ is continuous on its domain.

(2) For the value of c found above, verify that the three conditions of continuity are satisfied for $x = \pi$.

(3) Draw a graph of $f(x)$ from $x = -\pi$ to $x = 3\pi$.

Problem 1.175. Determine the constant b so that the function

$$f(x) = \begin{cases} x^3 + bx - 7 & \text{if } x \le 2 \\ be^{x-2} & \text{if } x > 2 \end{cases}$$

is continuous everywhere.

Problem 1.176. Determine the constant a so that the function

$$f(x) = \begin{cases} ax^2 + 9 & \text{if } x > 0 \\ x^3 + a^3 & \text{if } x \ge 0 \end{cases}$$

is continuous everywhere.

Problem 1.177. Determine $a \in \mathbb{R}$ such that the function

$$f(x) = \begin{cases} \dfrac{\cos x}{2x - \pi} & \text{if } x > \frac{\pi}{2} \\ ax & \text{if } x \le \frac{\pi}{2} \end{cases}$$

is continuous everywhere.

Problem 1.178. Determine $a, b \in \mathbb{R}$ such that the function

$$f(x) = \begin{cases} 2 & \text{if } x \le -1 \\ ax + b & \text{if } -1 < x < 3 \\ -2 & \text{if } x \ge 3 \end{cases}$$

is continuous everywhere.

Investigate the following functions for continuity:

Problem 1.179. $\text{sign}(x) = \begin{cases} 1 & \text{if } x > 0 \\ 0 & \text{if } x = 0 \\ -1 & \text{if } x < 0. \end{cases}$

Problem 1.180. $f(x) = \begin{cases} \sin\left(\dfrac{1}{x}\right) & \text{if } x \ne 0 \\ 0 & \text{if } x = 0. \end{cases}$

Problem 1.181. $g(x) = \begin{cases} x\sin\left(\dfrac{1}{x}\right) & \text{if } x \neq 0 \\ 0 & \text{if } x = 0. \end{cases}$

Problem 1.182. $h(x) = \begin{cases} \arctan\left(\dfrac{1}{x}\right) & \text{if } x \neq 0 \\ 0 & \text{if } x = 0. \end{cases}$

Problem 1.183. $i(x) = \lim\limits_{n\to\infty} \arctan(nx)$.

Problem 1.184. Prove that the Dirichlet function

$$\chi(x) = \begin{cases} 1 & \text{if } x \text{ is rational} \\ 0 & \text{if } x \text{ is irrational} \end{cases}$$

has an essential discontinuity at every point.

1.3.1   *Intermediate Value Theorem*

Use the Intermediate Value Theorem to solve the following problems.

Problem 1.185. Prove that the polynomial $p(x) = x^3 - 7x - 5$ has at least three real roots.

Problem 1.186. Prove that the polynomial $p(x) = x^5 + x - 1$ has at least one real root.

Problem 1.187. Prove that any polynomial of odd degree has at least one real root.

Problem 1.188. Find an example of a polynomial with only two non-zero coefficients such that:

(1) The polynomial has no root;
(2) The polynomial has exactly one root;
(3) The polynomial has at least two roots.

Problem 1.189.

(1) Show that $2^x = \frac{10}{x}$ for some $x > 0$.
(2) Show that the equation $2^x = \frac{10}{x}$ has no solution for $x < 0$.

Problem 1.190. Let f be a continuous function defined on a closed interval $[a, b]$. Suppose that $c \in \mathbb{R}$ is such that $f(x) \neq c$ for every $x \in [a, b]$. Show that if $f(a) < c$, then $f(b) < c$.

Problem 1.191. Let f be a continuous function defined on $[0, 3]$. Suppose that f has no roots on $[0, 3]$ and that $f(0) = 1$. Show that $f(x) > 0$ for all $x \in [0, 3]$.

Problem 1.192. The function $f(x) = \frac{1}{x-1}$ never takes the value zero. Yet, $f(0) = -1$ is negative and $f(2) = 1$ is positive. Why is this not a counterexample to the Intermediate Value Theorem?

Problem 1.193. Prove that the equation $x^3 - 6x + 3 = 0$ has a root in the interval $(0.5, 1)$.

Problem 1.194. Prove that Kepler's equation $x = a \sin x + b$ with $a \in (0, 1)$ and $b \in (0, \infty)$ has a positive root that is not greater than $a + b$.

Problem 1.195. A function $h : I \to I$ is said to have a fixed point at $x = c \in I$ if $h(c) = c$. Suppose that the domain and range of a function f are both the interval $[0, 1]$ and that f is continuous on its domain with $f(0) \neq 0$ and $f(1) \neq 1$. Prove that f has at least one fixed point, i.e., prove that $f(c) = c$ for some $c \in (0, 1)$.

1.4 Answers, Hints, and Solutions

Solution 1.1. $\lim\limits_{x \to 4} \left[\frac{1}{\sqrt{x}-2} - \frac{4}{x-4} \right] = \lim\limits_{x \to 4} \frac{(\sqrt{x}+2)-4}{x-4} = \lim\limits_{x \to 4} \frac{\sqrt{x}-2}{x-4} = \lim\limits_{x \to 4} \frac{1}{\sqrt{x}+2} = \frac{1}{4}$.

Solution 1.2. $\lim\limits_{x \to 2\pi} \frac{1-\cos x}{x^2} = 0$. The function $x \mapsto f(x) = \frac{1-\cos x}{x^2}$ is continuous at $x = 2\pi$.

Solution 1.3. $\lim\limits_{x \to 3} \frac{\sin x - x}{x^3} = \frac{\sin 3 - 3}{27} \approx 0.109$.

Solution 1.4. $\lim\limits_{x\to 10}\dfrac{x^2-100}{x-10}=\lim\limits_{x\to 10}\dfrac{(x-10)(x+10)}{x-10}=\lim\limits_{x\to 10}(x+10)=20.$

Solution 1.5. $\lim\limits_{x\to -4}\dfrac{x^2-16}{x+4}\ln|x|=-16\ln 2.$ Factor the numerator.

Solution 1.6. $\lim\limits_{x\to 10}\dfrac{x^2-99}{x-10}$ does not exist, since $\lim\limits_{x\to 10+}\dfrac{x^2-99}{x-10}=\infty.$
Observe that $\lim\limits_{x\to 10-}\dfrac{x^2-99}{x-10}=-\infty.$

Solution 1.7. $\lim\limits_{x\to 10}\dfrac{x^2-100}{x-9}=0.$

Solution 1.8. $\lim\limits_{x\to 10}\sqrt{-x^2+20x-100}$ does not exist. Observe that the function $x\mapsto \sqrt{-x^2+20x-100}$ is defined only at $x=10$. The definition of a limit, as x approaches a, requires that the function is defined on an interval that contains a, except possibly at a itself.

Solution 1.9. $\lim\limits_{x\to 1}\dfrac{\ln x}{x}=0.$

Solution 1.10. $\lim\limits_{x\to 1}\dfrac{\ln(1+3x)}{2x}=\dfrac{\ln 4}{2}=\ln\sqrt{4}=\ln 2.$

Solution 1.11. $\lim\limits_{x\to -2}\dfrac{2-|x|}{2+x}=1.$ Consider how $|x|$ can be rewritten when $x<0$.

Solution 1.12. $\lim\limits_{x\to 8}\dfrac{(x-8)(x+2)}{|x-8|}$ does not exist. Use the left-hand and the right-hand limits. Consider how $|x-8|$ can be rewritten for $x>8$ and $x<8$.

Solution 1.13. $\lim\limits_{x\to 2}\left(\dfrac{1}{x^2+5x+6}-\dfrac{1}{x-2}\right)$ does not exist. Note that the limit of the first rational function exists. Observe that $\lim\limits_{x\to 2-}\left(\dfrac{1}{x^2+5x+6}-\dfrac{1}{x-2}\right)=\infty$ and $\lim\limits_{x\to 2+}\left(\dfrac{1}{x^2+5x+6}-\dfrac{1}{x-2}\right)=-\infty.$

Solution 1.14. $\lim\limits_{x\to -1}\dfrac{x^2-x-2}{3x^2-x-1}=0.$

Solution 1.15. $\lim\limits_{x\to 16}\dfrac{\sqrt{x}-4}{x-16}=\dfrac{1}{8}.$ Rationalize the numerator or factor the denominator.

Solution 1.16 $\lim\limits_{x\to 8}\dfrac{\sqrt[3]{x}-2}{x-8}=\dfrac{1}{12}.$ Recall that $x-8=(\sqrt[3]{x}-2)(\sqrt[3]{x^2}+2\sqrt[3]{x}+4).$

Solution 1.17. $\lim\limits_{x\to 4}\frac{2-\sqrt{x}}{4x-x^2}=\frac{1}{16}$. Rationalize the numerator or factor the denominator.

Solution 1.18. $\lim\limits_{x\to 0}\frac{\sqrt{1+2x}-\sqrt{1-4x}}{x}=3$. Rationalize the numerator.

Solution 1.19. $\lim\limits_{x\to 5} e^{\frac{x-5}{\sqrt{x-1}-2}}=e^4$. Rationalize the denominator. Next, use the fact that the function $x\mapsto e^x$ is continuous everywhere.

Solution 1.20. $\lim\limits_{x\to 7} e^{\frac{\sqrt{x+2}-3}{x-7}}=e^{\frac16}$. Rationalize the numerator.

Solution 1.21. $\lim\limits_{t\to 0}\frac{\sqrt{\sin t+1}-1}{t}=\frac12$. Rationalize the numerator.

Solution 1.22. $\lim\limits_{x\to 0}\frac{1-e^{-x}}{1-x}=0$. Note that the function $x\mapsto\frac{1-e^{-x}}{1-x}$ is continuous at $x=0$.

Solution 1.23. $\lim\limits_{x\to 0}\frac{e^{-x^2}\cos(x^2)}{x^2}=\infty$. Note that $\lim\limits_{x\to 0}e^{-x^2}\cos(x^2)=1$ and $\lim\limits_{x\to 0}x^2=0^+$.

Solution 1.24. $\lim\limits_{\theta\to\frac{3\pi}{2}}\frac{\cos\theta+1}{\sin\theta}=-1$. Note that the function $\theta\mapsto\frac{\cos\theta+1}{\sin\theta}$ is continuous at $\theta=\frac{3\pi}{2}$.

Solution 1.25. From $4(f(x)\cdot g(x))=(f(x)+g(x))^2-(f(x)-g(x))^2$, it follows that $\lim\limits_{x\to a}[f(x)\cdot g(x)]=\frac{3^2-4^2}{4}=-\frac74$.

Solution 1.26. $\lim\limits_{x\to 0}\frac{f(x)}{|x|}$ does not exist. Consider $\lim\limits_{x\to 0}\frac{xf(x)}{x|x|}$. Observe that $x|x|<0$ if $x<0$ and $x|x|>0$ if $x>0$.

Solution 1.27. $\lim\limits_{h\to 0}\frac{\sqrt[4]{16+h}-2}{2h}=\frac{1}{64}$. *Solution 1*: Use the fact that $a^4-b^4=(a-b)(a+b)(a^2+b^2)$ to obtain $\lim\limits_{h\to 0}\frac{\sqrt[4]{16+h}-2}{h}=$
$\lim\limits_{h\to 0}\frac{\sqrt[4]{(16+h)^4}-2^4}{h(\sqrt[4]{16+h}+2)(\sqrt{16+h}+4)}=\lim\limits_{h\to 0}\frac{1}{(\sqrt[4]{16+h}+2)(\sqrt{16+h}+4)}$. *Solution 2*: The given limit is one-half of the derivative of the function $f(x)=\sqrt[4]{x}$ at $x=16$. Recall that $f'(x)=\frac14 x^{-\frac34}$.

Solution 1.28. $L=10$. Use the fact that $L=\lim\limits_{n\to\infty}x_n$ to conclude $L^2=100$. Can L be negative?

Solution 1.29. We use the fact that $1 - a^n = (1 - a)(1 + a + a^2 + \cdots a^{n-1})$. From $0.21\underbrace{55\ldots5}_{n} = \frac{21}{100} + \frac{5}{1000} \cdot \left(1 + \frac{1}{10} + \cdots + \frac{1}{10^{n-1}}\right) =$

$\frac{21}{100} + \frac{5}{1000} \cdot \frac{1 - \frac{1}{10^n}}{1 - \frac{1}{10}}$ and $\lim_{n\to\infty} \frac{1}{10^n} = 0$, we conclude that $0.2155\ldots = \frac{194}{900}$.

Solution 1.30. $\lim_{x\to 10} f(x) = \lim_{x\to 10} x^2 = 100$. Recall the definition of a limit.

Solution 1.31. $a = b = 4$. Rationalize the numerator. Choose the value of b so that x becomes a factor in the numerator.

Solution 1.32. By the Quadratic Formula, the solutions of the given equation are $x_1 = \frac{1+\sqrt{1+a}}{a}$ and $x_2 = \frac{1-\sqrt{1+a}}{a}$. It follows that $\lim_{a\to 0} x_1 = \infty$ and $\lim_{a\to 0} x_2 = -\frac{1}{2}$.

Solution 1.33. Since the denominator approaches 0 as $x \to -2$, the necessary condition for this limit to exist is that the numerator approaches 0 as $x \to -2$. We solve $4b - 30 + 15 + b = 0$ to obtain $b = 3$. Therefore, $\lim_{x\to -2} \frac{3x^2+15x+18}{x^2+x-2} = -1$.

Solution 1.34. From $\lim_{x\to\infty} \frac{1}{g(x)+1} = \lim_{x\to\infty} \frac{1}{f(x)} \cdot \frac{f(x)}{g(x)+1} = \frac{1}{3}$, it follows that $\lim_{x\to\infty} g(x) = 2$.

Solution 1.35. If $x = \frac{\pi}{2} + k\pi$ for some $k \in \mathbb{Z}$, then $\sin^2 x = 1$ and $f(x) = 1$. If $x \neq \frac{\pi}{2} + k\pi$ for $k \in \mathbb{Z}$, then $0 \leq \sin^2 x < 1$ and $f(x) = 0$.

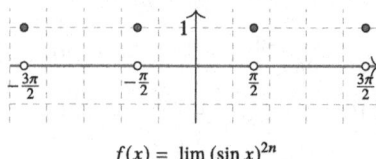

$$f(x) = \lim_{n\to\infty} (\sin x)^{2n}$$

Solution 1.36. The function f is not defined at $x = \pm 1$.
If $x \in (-1, 1)$, then $\lim_{n\to\infty} x^{2n} = 0$ and $f(x) = x + 1$. If $x \in (-\infty, -1) \cup (1, \infty)$, then $\lim_{n\to\infty} x^{2n} = \infty$ and $f(x) = 0$. Hence, $\lim_{x\to 1^-} f(x) = 2$ and $\lim_{x\to 1^+} f(x) = 0$.

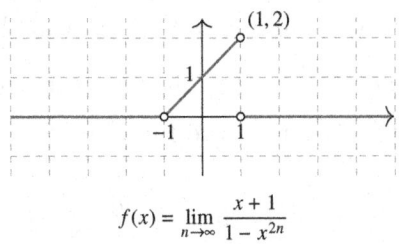

$$f(x) = \lim_{n \to \infty} \frac{x+1}{1-x^{2n}}$$

Solution 1.37. $\lim\limits_{x \to 1} \frac{x^m - 1}{x^n - 1} = \frac{m}{n}$. Use the identity $x^k - 1 = (x - 1)$
$(x^{k-1} + x^{k-2} + \cdots + 1)$, $k \in \mathbb{N}$.

Solution 1.38. The question is to prove that for any $n \in \mathbb{N}$,
$\lim\limits_{x \to \infty} \frac{x^n}{a^x} = 0$. From $\frac{x^n}{a^x} = \left(\frac{x}{(\sqrt[n]{a})^x} \right)^n$, it follows that it is enough
to prove that $\lim\limits_{x \to \infty} \frac{x}{a^x} = 0$ for any $a > 1$.
Let $b > 0$ be such that $a = 1+b$. Let $x > 2$ and let $c = c(x) \in [0, 1)$
be such that $x - c \in \mathbb{N}$. Then, by the Binomial Theorem, $a^x \geq$
$(1 + b)^{x-c} \geq 1 + b(x - c) + \frac{b^2(x-c)(x-c-1)}{2} > \frac{b^2(x-1)(x-2)}{2}$. It follows
that $0 < \frac{x}{a^x} < \frac{2x}{b^2(x-1)(x-2)}$. By the Squeeze Theorem, $\lim\limits_{x \to \infty} \frac{x}{a^x} = 0$.

Solution 1.39. $\lim\limits_{x \to 1^-} \frac{x+1}{x^2 - 1} = \lim\limits_{x \to 1^-} \frac{x+1}{(x+1)(x-1)} = \lim\limits_{x \to 1^-} \frac{1}{x-1} = -\infty$.

Solution 1.40. $\lim\limits_{x \to 1^+} \frac{\sqrt{x-1}}{x^2 - 1} = \infty$. Note that $x^2 - 1 = (x - 1)(x + 1)$.

Solution 1.41. $\lim\limits_{x \to 1^-} f(x) = -2$. Consider how $|x - 1|$ can be rewritten when $x < 1$.

Solution 1.42. $\lim\limits_{x \to 1.5^+} F(x) = 1.5$ and $\lim\limits_{x \to 1.5^-} F(x) = -1.5$. Since
the left-hand limit and the right-hand limit are not equal, $\lim\limits_{x \to 1.5} F(x)$
does not exist.

Solution 1.43. $\lim\limits_{x \to 2^-} \frac{|x^2 - 4|}{10 - 5x} = \frac{4}{5}$. Consider how $|x^2 - 4|$ can be rewritten when $x < 2$, then factor each polynomial and simplify.

Solution 1.44. $\lim\limits_{x \to -1^+} \arcsin x = -\frac{\pi}{2}$. Recall the definition of the
function $x \mapsto \arcsin x$.

Solution 1.45. $\lim\limits_{x\to 0^+} (x+\sin x)^{\frac{1}{x}} = 0$. If x is a small positive number then $(x+\sin x)^{\frac{1}{x}} = e^{\frac{1}{x}\cdot \ln(x+\sin x)}$. This, together with $\lim\limits_{x\to 0^+} \frac{1}{x}\cdot \ln(x+\sin x) = -\infty$ and the fact that the function $x \mapsto e^x$ is continuous, implies the result.

Solution 1.46. $\lim\limits_{x\to e^+} (\ln x)^{\frac{1}{x}} = \lim\limits_{x\to e^+} e^{\frac{1}{x}\ln(\ln x)} = e^0 = 1$.

Solution 1.47. $\lim\limits_{x\to \frac{\pi}{2}^+} \frac{x}{\cot x} = \lim\limits_{x\to \frac{\pi}{2}^+} \frac{x\sin x}{\cos x} = -\infty$.

Solution 1.48. $\lim\limits_{x\to \frac{\pi}{2}^+} \left(\frac{1}{\tan x} + \arctan \frac{1}{\frac{\pi}{2}-x} \right) = 0 - \frac{\pi}{2} = -\frac{\pi}{2}$.

Solution 1.49. $\lim\limits_{x\to 0^-} \left(e^{\frac{1}{x}} + \frac{1}{\ln(-x)} \right) = 0 + 0 = 0$.

Solution 1.50. $\lim\limits_{x\to -\infty} \frac{\sqrt{x^2+4x}}{4x+1} = \lim\limits_{x\to -\infty} \frac{\sqrt{x^2}\cdot\sqrt{1-\frac{4}{x^2}}}{x\left(4+\frac{1}{x}\right)} = -\frac{1}{4}$. Note that $x < 0$ implies that $\sqrt{x^2} = -x$.

Solution 1.51. $\lim\limits_{u\to \infty} \frac{u}{\sqrt{u^2+1}} = \lim\limits_{u\to \infty} \frac{1}{\sqrt{1+\frac{1}{u^2}}} = 1$. Note that $u > 0$ implies that $u = \sqrt{u^2}$.

Solution 1.52.

Solution 1: $\lim\limits_{x\to \infty} \frac{\sqrt{x^4+2}}{x^4-4} = \lim\limits_{x\to \infty} \left(\frac{1}{x^2}\cdot \frac{\sqrt{1+\frac{2}{x^4}}}{1-\frac{4}{x^4}} \right) = 0$.

Solution 2: For large values of x we have that $\sqrt{x^4+2} \approx \sqrt{x^4} = x^2$.

Solution 1.53. $\lim\limits_{x\to \infty} \frac{\sqrt{4x^2+3x}-7}{7-3x} = \lim\limits_{x\to \infty} \frac{\sqrt{x^2}\left(\sqrt{4+\frac{3}{x}} - \frac{7}{\sqrt{x^2}}\right)}{x\left(\frac{7}{x}-3\right)} = -\frac{2}{3}$.

Note that $x > 0$.

Solution 1.54. $\lim\limits_{x\to -\infty} \frac{\sqrt{x^6-3}}{\sqrt{x^6+5}} = 1$. You may use the fact that $x^6-3 \approx x^6 + 5$ for large values of $|x|$.

Solution 1.55. $\lim\limits_{x\to \infty} \frac{1+3x}{\sqrt{2x^2+x}} = \frac{3}{\sqrt{2}}$. You may use the fact that $\sqrt{2x^2+x} \approx \sqrt{2x^2} = |x|\sqrt{2}$ for large values of $|x|$.

Solution 1.56. $\lim\limits_{x\to -\infty} \frac{\sqrt[3]{x^3-9}}{2x-1} = \frac{1}{2}$. You may use the fact that $\sqrt[3]{x^3-9} \approx \sqrt[3]{x^3} = x$ for large values of $|x|$.

Solution 1.57. $\lim\limits_{x\to\infty} \frac{1.01^x}{x^{100}} = \infty$. Use the fact that the exponential function $x \mapsto a^x$ with $a > 1$ grows faster than any polynomial function.

Solution 1.58. $\lim\limits_{x\to\infty} \frac{x^2}{e^x} = 0$. Use the fact that the polynomial function grows slower than the exponential function.

Solution 1.59. $\lim\limits_{x\to\infty} \frac{x^2}{e^{4x}-1-4x} = 0$. Use the fact that the polynomial function grows slower than the exponential function.

Solution 1.60. $\lim\limits_{x\to\infty} \frac{\ln x}{\sqrt{x}} = 0$. Use the fact that the logarithmic function grows slower than the square root function.

Solution 1.61. $\lim\limits_{x\to\infty} \frac{\ln x^{100}}{x^2} = \lim\limits_{x\to\infty} \frac{100\ln x}{x^2} = 0$. Use the fact that the logarithmic function grows slower than the square function.

Solution 1.62. $\lim\limits_{x\to-\infty} \frac{3x^6-7x^5+x}{5x^6+4x^5-3} = \frac{3}{5}$. *Solution 1*: Divide the numerator and denominator by the highest power. *Solution 2*: Recall that, for large values of $|x|$ and $a_n \neq 0$, $a_n x^n + a_{n-1}x^{n-1} + \cdots + a_0 \approx a_n x^n$. Keep in mind that the meaning of "$\approx$" in this case is that $\lim\limits_{x\to\pm\infty} \frac{a_n x^n + a_{n-1}x^{n-1} + \cdots + a_0}{a_n x^n} = 1$, **not** that $\lim\limits_{x\to\pm\infty} ((a_n x^n + a_{n-1}x^{n-1} + \cdots + a_0) - a_n x^n) = 0$. The meaning of "$a_n x^n + a_{n-1}x^{n-1} + \cdots + a_0 \approx a_n x^n$ as $x \to \pm\infty$" is that, for large values of $|x|$, the difference $(a_n x^n + a_{n-1}x^{n-1} + \cdots + a_0) - a_n x^n$ is *small* compared to the size of x. For example, take the polynomial $p(x) = x+2$. Here, $\lim\limits_{x\to\infty} \frac{x+2}{x} = 1$, i.e., for large values of x, $p(x) = x + 2 \approx x$. For very large values of x, say $x = 10^{10}$, the difference $p(x) - x = p(10^{10}) - 10^{10} = 2$ is very small compared to the size of x.

Solution 1.63. $\lim\limits_{x\to-\infty} \frac{5x^7-7x^5+1}{2x^8+6x^6-3} = \lim\limits_{x\to-\infty} \frac{5x^7}{2x^8} = 0$.

Solution 1.64. $\lim\limits_{x\to-\infty} \frac{2x+3x^3}{x^2+2x-1} = \lim\limits_{x\to-\infty} \frac{3x^3}{x^2} = -\infty$.

Solution 1.65. $\lim\limits_{x\to\infty} \frac{(x+1)^7+(x+2)^7+\cdots+(x+7)^7}{x^7+7^7} = 7$. For large values of x and any $a \in \mathbb{R}$, $(x+a)^7 \approx x^7$.

Solution 1.66. $\lim\limits_{x\to\infty} \frac{ax^{17}+bx}{cx^{17}-dx^3} = \frac{a}{c}$, where $c \neq 0$.

Solution 1.67. $\lim\limits_{x\to\infty} \frac{3x+|1-3x|}{1-5x} = -\frac{6}{5}$. Observe that, for $x > \frac{1}{3}$, $3x + |1 - 3x| = 6x - 1$.

Solution 1.68. $\lim\limits_{x\to\infty}\left(\sqrt{x^2+x}-x\right) = \lim\limits_{x\to\infty}\left(\sqrt{x^2+x}-x\right) \cdot \frac{\sqrt{x^2+x}+x}{\sqrt{x^2+x}+x} = \lim\limits_{x\to\infty} \frac{x}{\sqrt{x^2+x}+x} = \frac{1}{2}$. Use the fact that, for large positive values of x, $\sqrt{x^2+x} \approx \sqrt{x^2} = x$. *Note:* The expression $\sqrt{x^2+x} \approx x$ means that $\lim\limits_{x\to\infty} \frac{\sqrt{x^2+x}}{x} = 1$, **not** that $\lim\limits_{x\to\infty}\left(\sqrt{x^2+x}-x\right) = 0$. The meaning of "$\sqrt{x^2+x} \approx x$ as $x \to \infty$" is that, for large values of x, the difference $\sqrt{x^2+x} - x$ is *small* compared to the size of x. For example, for $x = 1000$, $\sqrt{1000^2+1000} - 1000 \approx 0.4998$. Keep in mind that "$\infty - \infty$" is a type of indeterminate form.

Solution 1.69. $\lim\limits_{x\to-\infty}\left(\sqrt{x^2+5x}-\sqrt{x^2+2x}\right) = -\frac{3}{2}$. Rationalize the numerator. Note that $x \to -\infty$, so $x < 0$ and, therefore, $\sqrt{x^2} = |x| = -x$.

Solution 1.70. Rationalize the numerator by using the identity $a^3 - b^3 = (a-b)(a^2 + ab + b^2)$ to obtain $\sqrt[3]{x^3+x^2} - x = \frac{\left(\sqrt[3]{x^3+x^2}\right)^3 - x^3}{\sqrt[3]{(x^3+x^2)^2}+x\sqrt[3]{x^3+x^2}+x^2} = \frac{x^2}{\sqrt[3]{(x^3+x^2)^2}+x\sqrt[3]{x^3+x^2}+x^2}$, $x \neq 0$. Use the fact that, for large values of x, $\sqrt[3]{x^3+x^2} \approx \sqrt[3]{x^3} = x$ to conclude $\lim\limits_{x\to\infty}\left(\sqrt[3]{x^3+x^2}-x\right) = \frac{1}{3}$.

Solution 1.71.
$$\lim\limits_{x\to\infty}\left(\sqrt{x+\sqrt{x+\sqrt{x}}}-\sqrt{x}\right) = \lim\limits_{x\to\infty} \frac{\sqrt{x+\sqrt{x}}}{\sqrt{x+\sqrt{x+\sqrt{x}}}+\sqrt{x}} = \frac{1}{2}.$$

Solution 1.72. $\lim\limits_{x\to0} \frac{\sin 6x}{2x} = \lim\limits_{x\to0} \frac{3\cdot\sin 6x}{6x} = \left|\begin{matrix}\text{Substitution:}\\ 6x = \theta\end{matrix}\right| = 3\lim\limits_{\theta\to0} \frac{\sin\theta}{\theta} = 3\cdot1 = 3.$

Solution 1.73. $\lim\limits_{x\to1} \frac{\sin(x-1)}{x^2+x-2} = \lim\limits_{x\to1} \frac{\sin(x-1)}{x-1} \cdot \frac{1}{x+2} = 1\cdot\frac{1}{3} = \frac{1}{3}.$

Solution 1.74. $\lim\limits_{x\to0} \frac{\sin 4x}{\sqrt{x+1}-1} = \lim\limits_{x\to0} \frac{\sin 4x}{\sqrt{x+1}-1} \cdot \frac{\sqrt{x+1}+1}{\sqrt{x+1}+1} = \lim\limits_{x\to0} \frac{\sin 4x}{x} \cdot \left(\sqrt{x+1}+1\right) = 4\cdot2 = 8.$

Solution 1.75.
$$\lim\limits_{x\to0} \frac{1-\cos x}{x^2} = \lim\limits_{x\to0} \frac{2\sin^2 \frac{x}{2}}{x^2} = \frac{1}{2}\lim\limits_{x\to0}\left(\frac{\sin\frac{x}{2}}{\frac{x}{2}}\right)^2 = \frac{1}{2}.$$

Solution 1.76. $\lim\limits_{x\to 0} \cot(3x)\sin(7x) = \lim\limits_{x\to 0}\dfrac{\cos(3x)\sin(7x)}{\sin(3x)} =$
$\lim\limits_{x\to 0} \dfrac{7}{3}\cos(3x)\cdot\dfrac{3x}{\sin(3x)}\cdot\dfrac{\sin(7x)}{7x} = \dfrac{7}{3}.$

Solution 1.77. $\lim\limits_{x\to 0}\dfrac{(\sin x)^{100}}{x^{99}\sin(2x)} = \lim\limits_{x\to 0}\left(\dfrac{(\sin x)^{99}}{x^{99}}\cdot\dfrac{\sin x}{\sin(2x)}\right) =$
$\lim\limits_{x\to 0}\left(\dfrac{\sin x}{x}\right)^{99}\cdot\lim\limits_{x\to 0}\dfrac{1}{2\cos x} = \dfrac{1}{2}.$

Solution 1.78. $\lim\limits_{x\to 0}\dfrac{x^{100}\sin 7x}{(\sin x)^{99}} = 0.$ Rewrite $\dfrac{x^{100}\sin 7x}{(\sin x)^{99}} = x\sin 7x \cdot$
$\left(\dfrac{x}{\sin x}\right)^{99}.$

Solution 1.79. $\lim\limits_{x\to 0}\dfrac{x^{100}\sin 7x}{(\sin x)^{101}} = 7.$ Rewrite $\dfrac{x^{100}\sin 7x}{(\sin x)^{101}} = 7\cdot\left(\dfrac{x}{\sin x}\right)^{101}\cdot$
$\dfrac{\sin 7x}{7x}.$

Solution 1.80. $\lim\limits_{x\to 0}\dfrac{\sin x}{\sqrt{x\sin 4x}}$ does not exist. For x close but not equal
to 0, rewrite $\dfrac{\sin x}{\sqrt{x\sin 4x}} = \dfrac{\sin x}{2|x|}\cdot\dfrac{1}{\sqrt{\frac{\sin 4x}{4x}}}.$ Note that the left-hand and the
right-hand limits as $x\to 0$ are not equal.

Solution 1.81. $\lim\limits_{x\to 0}\dfrac{1-\cos x}{x\sin x} = \dfrac{1}{2}.$ For x close but not equal to 0,
rewrite $\dfrac{1-\cos x}{x\sin x} = \dfrac{2\sin^2\frac{x}{2}}{2x\sin\frac{x}{2}\cos\frac{x}{2}} = \dfrac{1}{2\cos\frac{x}{2}}\cdot\dfrac{\sin\frac{x}{2}}{\frac{x}{2}}.$

Solution 1.82. $\lim\limits_{x\to 0}\dfrac{\sin(\sin x)}{x} = \lim\limits_{x\to 0}\dfrac{\sin(\sin x)}{\sin x}\cdot\dfrac{\sin x}{x} = 1\cdot 1 = 1.$

Solution 1.83. $\lim\limits_{x\to\frac{\pi}{2}^+}\dfrac{\sqrt{1+\cos 2x}}{\sqrt{\pi}-\sqrt{2x}} = -\sqrt{2\pi}.$ Observe that $x\to\frac{\pi}{2}^+$
implies that $\cos x < 0.$ For x close but not equal to $\frac{\pi}{2}$, rewrite
$\dfrac{\sqrt{1+\cos 2x}}{\sqrt{\pi}-\sqrt{2x}} = \dfrac{-\sqrt{2}\cos x}{\pi-2x}\cdot\left(\sqrt{\pi}+\sqrt{2x}\right) = -\dfrac{\sqrt{2}}{2}\cdot\dfrac{\sin\left(\frac{\pi}{2}-x\right)}{\frac{\pi}{2}-x}\cdot\left(\sqrt{\pi}+\sqrt{2x}\right).$

Solution 1.84. $\lim\limits_{x\to\frac{\pi}{3}}\dfrac{\sin\left(x-\frac{\pi}{3}\right)}{1-2\cos x} = \dfrac{1}{\sqrt{3}}.$ Observe that, via the substitution $x = \frac{\pi}{3}+t$, the given limit becomes $\lim\limits_{t\to 0}\dfrac{\sin t}{1-2\cos\left(\frac{\pi}{3}+t\right)} =$
$\lim\limits_{t\to 0}\dfrac{\sin t}{1-\cos t+\sqrt{3}\sin t} = \lim\limits_{t\to 0}\dfrac{\frac{\sin t}{t}}{\frac{1-\cos t}{t}+\sqrt{3}\cdot\frac{\sin t}{t}}.$

Solution 1.85. $\lim\limits_{x\to 0}(1-2x)^{\frac{1}{x}} = \left|\begin{array}{c}\text{Substitution:}\\ x = -\frac{u}{2} \\ x\to 0 \Leftrightarrow u\to 0\end{array}\right| = \lim\limits_{u\to 0}(1+u)^{\frac{-2}{u}} =$
$\left(\lim\limits_{u\to 0}(1+u)^{\frac{1}{u}}\right)^{-2} = e^{-2}.$

Solution 1.86. $\lim\limits_{x\to 0}\left(1+\frac{x}{2}\right)^{\frac{3}{x}} = \begin{vmatrix} \text{Substitution:} \\ x=2u \\ x\to 0 \Leftrightarrow u\to 0 \end{vmatrix} = \lim\limits_{u\to 0}(1+u)^{\frac{3}{2u}} =$

$\left(\lim\limits_{u\to 0}(1+u)^{\frac{1}{u}}\right)^{\frac{3}{2}} = e^{\frac{3}{2}}.$

Solution 1.87. $\lim\limits_{x\to 0}(1+\tan x)^{\cot x} = e$. Use the substitution $\tan x = u$.

Solution 1.88. $\lim\limits_{x\to 0}\frac{\ln(1+x)}{x} = \lim\limits_{x\to 0}\ln(1+x)^{\frac{1}{x}} = \ln e = 1.$

Solution 1.89. $\lim\limits_{x\to 0}\frac{e^x-1}{x} = \begin{vmatrix} e^x-1=u \Leftrightarrow x=\ln(u+1) \\ x\to 0 \Leftrightarrow u\to 0 \end{vmatrix} =$

$\lim\limits_{u\to 0}\frac{u}{\ln(u+1)} = 1.$

Solution 1.90. $\lim\limits_{x\to 1}(2-x)^{\frac{1}{x-1}} = e^{-1}$. Use the substitution $u=1-x$.

Solution 1.91. $\lim\limits_{x\to\infty}\left(1+\frac{1}{x}\right)^{2x} = \lim\limits_{x\to\infty}\left(\left(1+\frac{1}{x}\right)^x\right)^2 = e^2.$

Solution 1.92. $\lim\limits_{x\to\infty}\left(1+\frac{3}{x}\right)^{2x} = \begin{vmatrix} \text{Substitution:} \\ x=3u \\ x\to\infty \Leftrightarrow u\to\infty \end{vmatrix} =$

$\lim\limits_{u\to\infty}\left(1+\frac{1}{u}\right)^{2\cdot 3u} = e^6.$

Solution 1.93. $\lim\limits_{x\to\infty}x(\ln(x+1)-\ln x) = \lim\limits_{x\to\infty}\ln\left(\frac{x+1}{x}\right)^x =$

$\lim\limits_{x\to\infty}\ln\left(1+\frac{1}{x}\right)^x = 1.$

Solution 1.94. $\lim\limits_{x\to\infty}\left(\frac{x^2+x+1}{x^2-x-1}\right)^x = e^2$. Observe that, for $x\neq\frac{1\pm\sqrt5}{2}$,

$$\left(\frac{x^2+x+1}{x^2-x-1}\right)^x = \left(1+\frac{2x+2}{x^2-x-1}\right)^x$$

$$= \left(\left(1+\frac{2x+2}{x^2-x-1}\right)^{\frac{x^2-x-1}{2x+2}}\right)^{\frac{2x^2+2x}{x^2-x-1}}.$$

Note that $\lim\limits_{x\to\infty}\frac{2x^2+2x}{x^2-x-1} = 2$. Via the substitution $\frac{x^2-x-1}{2x+2} = u$, it follows that $\lim\limits_{x\to\infty}\left(1+\frac{2x+2}{x^2-x-1}\right)^{\frac{x^2-x-1}{2x+2}} = e$. *Note:* Suppose that

$\lim\limits_{x\to\infty} f(x) = a > 0$, $a \neq 1$, and $\lim\limits_{x\to\infty} g(x) = b$ and that $h(x) = f(x)^{g(x)}$. Then $\ln(h(x)) = g(x)\ln(f(x))$ and $\lim\limits_{x\to\infty} \ln(h(x)) = b\ln a = \ln a^b$. Hence, $\lim\limits_{x\to\infty} h(x) = a^b$ and our result follows.

Solution 1.95. $\lim\limits_{x\to\infty} \left(\dfrac{mx+n}{mx+p}\right)^x = e^{\frac{n-p}{m}}$. Observe that, for $x \neq -\dfrac{p}{n}$,

$$\left(\frac{mx+n}{mx+p}\right)^x = \left(1 + \frac{n-p}{mx+p}\right)^x = \left(\left(1 + \frac{n-p}{mx+p}\right)^{\frac{mx+p}{n-p}}\right)^{\frac{(n-p)x}{mx+p}}.$$

Note that $\lim\limits_{x\to\infty} \dfrac{(n-p)x}{mx+p} = \dfrac{n-p}{m}$ and $\lim\limits_{x\to\infty} \left(1 + \dfrac{n-p}{mx+p}\right)^{\frac{mx+p}{n-p}} = e$.

Solution 1.96. From $\lim\limits_{x\to 4}(x+2) = 6$ and $\lim\limits_{x\to 4}(x^2 - 10) = 6$, by the Squeeze Theorem, it follows $\lim\limits_{x\to 4} f(x) = 6$.

Solution 1.97. $\lim\limits_{x\to\infty} \dfrac{\sin x}{x} = 0$. Recall that $\lim\limits_{x\to\infty} \sin x$ does not exist. From $-1 \leq \sin x \leq 1$ it follows that, for $x > 0$, $-\dfrac{1}{x} \leq \dfrac{\sin x}{x} \leq \dfrac{1}{x}$. Observe that $\lim\limits_{x\to\infty} \dfrac{-1}{x} = \lim\limits_{x\to\infty} \dfrac{1}{x} = 0$ and apply the Squeeze Theorem.

Solution 1.98. $\lim\limits_{x\to 0} x\sin\left(\dfrac{1}{x}\right) = 0$. Recall that $\lim\limits_{x\to 0} \sin\left(\dfrac{1}{x}\right)$ does not exist. Also, recall that, for $x \neq 0$, $-1 \leq \sin\left(\dfrac{1}{x}\right) \leq 1$. For $x > 0$, we have $-x \leq x\sin\left(\dfrac{1}{x}\right) \leq x$. Since $\lim\limits_{x\to 0^+}(-x) = \lim\limits_{x\to 0^+} x = 0$, by the Squeeze Theorem, it follows that $\lim\limits_{x\to 0^+} x\sin\left(\dfrac{1}{x}\right) = 0$. Apply a similar reasoning for $x < 0$.

Solution 1.99. Recall that $-1 \leq \cos\left(\dfrac{1}{x}\right) \leq 1$ for $x \neq 0$. For $x < 0$, we have $x \leq x\cos\left(\dfrac{1}{x}\right) \leq -x$, which implies $e^x \leq e^{x\cos\left(\frac{1}{x}\right)} \leq e^{-x}$ since $t \mapsto e^t$ is a monotone increasing function. From $\lim\limits_{x\to 0^-} e^x = 1$ and $\lim\limits_{x\to 0^-} e^{-x} = 1$, by the Squeeze Theorem, it follows that $\lim\limits_{x\to 0^-} e^{x\cos(1/x)} = 1$. Apply a similar reasoning for $x > 0$ to conclude that $\lim\limits_{x\to 0} e^{x\cos(1/x)} = 1$.

Solution 1.100. Note that $x^4 \geq 0$. So, for $x \neq 0$, $-1 \leq \sin\left(\dfrac{1}{x}\right) \leq 1$ implies $-x^4 \leq x^4 \sin\left(\dfrac{1}{x}\right) \leq x^4$. Since $\lim\limits_{x\to 0}(-x^4) = \lim\limits_{x\to 0} x^4 = 0$, by the Squeeze Theorem, it follows that $\lim\limits_{x\to 0} x^4 \sin\left(\dfrac{1}{x}\right) = 0$.

Solution 1.101. Let $x > 0$. Note that $-1 \leq \sin(1/x) \leq 1$ implies $\sqrt{x}e^{-1} \leq \sqrt{x}e^{\sin(1/x)} \leq \sqrt{x}e$. Use the Squeeze Theorem to evaluate $\lim\limits_{x \to 0^+} \sqrt{x}e^{\sin\left(\frac{1}{x}\right)} = 0$.

Solution 1.102. $\lim\limits_{x \to 0^+} (x^2 + x)^{\frac{1}{3}} \sin\left(\frac{1}{x^2}\right) = 0$.

Solution 1.103. $\lim\limits_{x \to 0} \dfrac{x^3 \sin\left(\frac{1}{x^2}\right)}{\sin x} = 0$. Let x be close, but not equal, to zero. Then $\dfrac{x^3}{\sin x} = \dfrac{x^2}{\frac{\sin x}{x}} > 0$ which, together with $-1 \leq \sin\left(\frac{1}{x^2}\right) \leq 1$, implies $-\dfrac{x^3}{\sin x} \leq \dfrac{x^3 \sin\left(\frac{1}{x^2}\right)}{\sin x} \leq \dfrac{x^3}{\sin x}$. Observe that $\lim\limits_{x \to 0} \dfrac{x^3}{\sin x} = 0$ and apply the Squeeze Theorem.

Solution 1.104. $\lim\limits_{x \to \infty} \dfrac{5x^2 - \sin 3x}{x^2 + 10} = 5$. *Solution 1*: Note that $5x^2 - 1 \leq 5x^2 - \sin 3x \leq 5x^2 + 1$, $x \in \mathbb{R}$, implies $\dfrac{5x^2 - 1}{x^2 + 10} \leq \dfrac{5x^2 - \sin 3x}{x^2 + 10} \leq \dfrac{5x^2 + 1}{x^2 + 10}$. Apply the Squeeze Theorem. *Solution 2*: Observe that $\dfrac{5x^2 - \sin 3x}{x^2 + 10} = \dfrac{5 - \frac{\sin 3x}{x^2}}{1 + \frac{10}{x^2}}$ and then use the Squeeze Theorem to conclude that $\lim\limits_{x \to \infty} \dfrac{\sin 3x}{x^2} = 0$.

Solution 1.105. $\lim\limits_{x \to 0} f(x) = \frac{4}{3}$. From $\lim\limits_{x \to 0} \dfrac{\sin 2x}{\sin 3x} = \lim\limits_{x \to 0} \dfrac{2 \cdot \frac{\sin 2x}{2x}}{3 \cdot \frac{\sin 3x}{3x}} = \frac{2}{3}$ and $\lim\limits_{x \to 0} \dfrac{e^{\frac{2x}{3}} - 1}{x} = \lim\limits_{x \to 0} \dfrac{2}{3} \cdot \dfrac{e^{\frac{2x}{3}} - 1}{\frac{2x}{3}} = \frac{2}{3}$, by the Squeeze Theorem, it follows that $\lim\limits_{x \to 0} \dfrac{f(x)}{x^2 + 2} = \frac{2}{3}$.

Solution 1.106. (1) By the Squeeze Theorem, $\lim\limits_{x \to 0} f(x) = \lim\limits_{x \to 0} x^2 \sin\left(\frac{1}{x}\right) = 0$.

Observe from the following image that, for $0 < x < \frac{\pi}{2}$, the length of the line segment $\overline{AB}$ is smaller than the length of the arc $\overset{\frown}{AC}$, i.e., that $0 < \sin x < x$. Use the Squeeze Theorem to conclude $\lim\limits_{x \to 0} g(x) = \lim\limits_{x \to 0} \sin x = 0$.

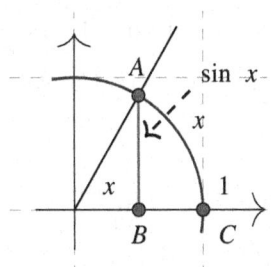

(2) For x close, but not equal, to zero, we have $\frac{f'(x)}{g'(x)} =$ $\frac{2x\sin\left(\frac{1}{x}\right)-\cos\left(\frac{1}{x}\right)}{\cos x}$. Recall that $\lim\limits_{x\to 0} 2x\sin\left(\frac{1}{x}\right) = 0$, $\lim\limits_{x\to 0}\cos x = 1$ and that $\lim\limits_{x\to 0}\cos\left(\frac{1}{x}\right)$ does not exist. Therefore, $\lim\limits_{x\to 0}\frac{f'(x)}{g'(x)}$ does not exist. (3) Let x be a small positive number. Then $-\frac{x^2}{\sin x} \le \frac{f(x)}{g(x)} = \frac{x^2}{\sin x}\cdot\sin\left(\frac{1}{x}\right) \le \frac{x^2}{\sin x}$. Use the fact that $\lim\limits_{x\to 0}\frac{x^2}{\sin x} = \lim\limits_{x\to 0}\frac{x}{\frac{\sin x}{x}} = 0$ and the Squeeze Theorem to conclude $\lim\limits_{x\to 0}\frac{f(x)}{g(x)} = 0$. (4) We can apply L'Hospital's Rule only if the limit $\lim\limits_{x\to a}\frac{f'(x)}{g'(x)}$ exists. This example shows that there exist functions f and g such that $\lim\limits_{x\to a}\frac{f(x)}{g(x)}$ is an indeterminate form of type $\frac{0}{0}$ and that $\lim\limits_{x\to a}\frac{f(x)}{g(x)}$ exists, but that $\lim\limits_{x\to a}\frac{f'(x)}{g'(x)}$ does not exist. No contradiction!

Solution 1.107. (1) Observe that, for any $x \in \mathbb{R}$, $f(x) \ge 2x - 1$ and $g(x) \ge 2x - 1$. Since $\lim\limits_{x\to\infty}(2x - 1) = \infty$, by the Squeeze Theorem, it follows that $\lim\limits_{x\to\infty} f(x) = \lim\limits_{x\to\infty} g(x) = \infty$. (2) From $\frac{f'(x)}{g'(x)} = \frac{2-\cos x}{2+\cos x} = \frac{4}{2+\cos x} - 1$ and the fact that $\lim\limits_{x\to\infty}\cos x$ does not exist, it follows that $\lim\limits_{x\to\infty}\frac{f'(x)}{g'(x)}$ does not exist. (3) $\lim\limits_{x\to\infty}\frac{f(x)}{g(x)} = 1$. Use the fact that $\lim\limits_{x\to\infty}\frac{\sin x}{x} = 0$.

Solution 1.108. (1) From $\lim\limits_{x\to\infty} e^x = \infty$ and $\lim\limits_{x\to\infty} e^{-x} = 0$ it follows that $\lim\limits_{x\to\infty} f(x) = \lim\limits_{x\to\infty} g(x) = \infty$. (2) Note that $f'(x) = g(x)$ and $g'(x) = f(x)$. Hence, $\frac{f'(x)}{g'(x)} = \frac{g(x)}{f(x)}$ is an indeterminate form of type $\frac{\infty}{\infty}$. (3) Note that $f''(x) = f(x)$ and $g''(x) = g(x)$. Hence, L'Hospital's

Rule is not useful in this situation. (4) $\lim\limits_{x\to\infty} \frac{f(x)}{g(x)} = \lim\limits_{x\to\infty} \frac{e^x - e^{-x}}{e^x + e^{-x}} =$
$\lim\limits_{x\to\infty} \frac{1-e^{-2x}}{1+e^{-2x}} = 1$. Note that, in accordance with L'Hospital's Rule,
$\lim\limits_{x\to\infty} \frac{f'(x)}{g'(x)}$ is also 1.

Solution 1.109. Type $\frac{0}{0}$. $\lim\limits_{x\to0} \frac{\sinh 2x}{xe^x} = \lim\limits_{x\to0} \frac{e^{2x}-e^{-2x}}{2xe^x} =$
$\lim\limits_{x\to0} \frac{e^x - e^{-3x}}{2x} \overset{L'H}{=} \lim\limits_{x\to0} \frac{e^x + 3e^{-3x}}{2} = 2$.

Solution 1.110. Type $\frac{0}{0}$. $\lim\limits_{x\to1} \frac{x^2-1}{e^{1-x^2}-1} \overset{L'H}{=} \lim\limits_{x\to1} \frac{2x}{-2xe^{1-x^2}} =$
$- \lim\limits_{x\to1} \frac{1}{e^{1-x^2}} = -1$.

Solution 1.111. Type $\frac{0}{0}$. $\lim\limits_{x\to1} \frac{\ln x}{\sin(\pi x)} \overset{L'H}{=} \lim\limits_{x\to1} \frac{1}{\pi x \cos(\pi x)} = -\frac{1}{\pi}$.

Solution 1.112. Type $\frac{0}{0}$. $\lim\limits_{x\to0} \frac{(2+x)^{2023}-2^{2023}}{x} \overset{L'H}{=}$
$\lim\limits_{x\to0} \frac{2023(2+x)^{2022}}{1} = 2023 \cdot 2^{2022}$.

Solution 1.113. Type $\frac{0}{0}$. $\lim\limits_{x\to0} \frac{\ln(2+2x)-\ln 2}{x} = 1$.

Solution 1.114. Type $\frac{0}{0}$. $\lim\limits_{x\to0} \frac{\ln(1+3x)}{2x} = \frac{3}{2}$.

Solution 1.115. Type $\frac{0}{0}$. $\lim\limits_{\theta\to\frac{\pi}{2}+} \frac{\ln(\sin\theta)}{\cos\theta} = 0$.

Solution 1.116. Type $\frac{0}{0}$. $\lim\limits_{x\to1} \frac{1-x+\ln x}{1+\cos(\pi x)} = -\frac{1}{\pi^2}$. Apply L'Hospital's Rule twice.

Solution 1.117. Type $\frac{0}{0}$. $\lim\limits_{x\to0} \frac{e^{2x}-1-2x}{x^2} = 2$. Apply L'Hospital's Rule twice.

Solution 1.118. Type $\frac{0}{0}$. $\lim\limits_{x\to1} \frac{x^2-1}{e^{1-x^7}-1} = -\frac{2}{7}$.

Solution 1.119. Type $\frac{0}{0}$. $\lim\limits_{x\to1} \frac{x^a-1}{x^b-1} = \frac{a}{b}$, $a,b \neq 0$.

Solution 1.120. Type $\frac{0}{0}$. $\lim\limits_{x\to0} \frac{\arcsin 3x}{\arcsin 5x} = \frac{3}{5}$. Observe that we can evaluate this limit without using L'Hospital's Rule:
$\lim\limits_{x\to0} \frac{\arcsin 3x}{\arcsin 5x} = \frac{3}{5} \lim\limits_{x\to0} \frac{\frac{\arcsin 3x}{3x}}{\frac{\arcsin 5x}{5x}} = \frac{3}{5} \lim\limits_{x\to0} \frac{\frac{\arcsin 3x}{\sin(\arcsin 3x)}}{\frac{\arcsin 5x}{\sin(\arcsin 5x)}}$. Now, from

$\lim\limits_{x\to 0} \frac{\arcsin nx}{\sin(\arcsin nx)} = |t = \arcsin nx| = \lim\limits_{t\to 0} \frac{t}{\sin t} = 1$, the result follows.

Solution 1.121. Type $\frac{0}{0}$. $\lim\limits_{x\to 0} \frac{\sin 3x}{\sin 5x} = \frac{3}{5}$. Observe that we can evaluate this limit without using L'Hospital's Rule: $\lim\limits_{x\to 0} \frac{\sin 3x}{\sin 5x} = \lim\limits_{x\to 0} \frac{3\cdot\frac{\sin 3x}{3x}}{5\cdot\frac{\sin 5x}{5x}} = \frac{3}{5}$.

Solution 1.122. Type $\frac{0}{0}$. $\lim\limits_{x\to 0} \frac{x-\sin x}{x^3} = \frac{1}{6}$. Apply L'Hospital's Rule three times.

Solution 1.123. Type $\frac{\infty}{\infty}$. $\lim\limits_{x\to\infty} \frac{(\ln x)^2}{x} \overset{L'H}{=} \lim\limits_{x\to\infty} \frac{2\ln x}{x} = 0$.

Solution 1.124. Type $\frac{\infty}{\infty}$. $\lim\limits_{x\to\infty} \frac{\ln((2x)^{\frac{1}{2}})}{\ln((3x)^{\frac{1}{3}})} = \frac{3}{2}$. Use properties of logarithms first.

Solution 1.125. Type $0 \cdot \infty$. Transform to type $\frac{\infty}{\infty}$ and apply L'Hospital's Rule: $\lim\limits_{x\to 0^+} x^{0.01}\ln x = \lim\limits_{x\to 0^+} \frac{\ln x}{x^{-0.01}} \overset{L'H}{=}$ $\lim\limits_{x\to 0^+} \frac{1}{-0.01x^{-1.001}\cdot x} = \lim\limits_{x\to 0^+} \frac{1}{-0.01x^{-0.01}} = 0$.

Solution 1.126. Type $0 \cdot \infty$. Transform to type $\frac{\infty}{\infty}$ and apply L'Hospital's Rule: $\lim\limits_{x\to 0^+} (\sin x)(\ln x) = \lim\limits_{x\to 0^+} \frac{\ln x}{\csc x} \overset{L'H}{=}$ $\lim\limits_{x\to 0^+} \frac{1}{x(-\csc x \cot x)} = -\lim\limits_{x\to 0^+} \frac{\sin x \tan x}{x} = -1\cdot 0 = 0$.

Solution 1.127. Type $0 \cdot \infty$. Transform to type $\frac{0}{0}$, $\lim\limits_{x\to\infty} \left(x\cdot\ln\frac{x-1}{x+1}\right) = \lim\limits_{x\to\infty} \frac{\ln\frac{x-1}{x+1}}{\frac{1}{x}}$ and apply L'Hospital's Rule to obtain $\lim\limits_{x\to\infty} \left(x\cdot\ln\frac{x-1}{x+1}\right) = -2$.

Solution 1.128. Type $0 \cdot \infty$. Transform to type $\frac{0}{0}$. $\lim\limits_{x\to\frac{\pi}{2}} \left(x-\frac{\pi}{2}\right)\tan x = -1$.

Solution 1.129. Type $0 \cdot \infty$. Transform to type $\frac{0}{0}$. $\lim\limits_{x\to\infty} x\tan(1/x) = 1$. Observe that we can evaluate this limit without using L'Hospital's Rule: $\lim\limits_{x\to\infty} x\tan(1/x) = \left|t = \frac{1}{x}\right| = \lim\limits_{t\to 0} \frac{\sin t}{t\cos t}$ and the result follows.

Solution 1.130. Type $0 \cdot \infty$. Transform to type $\frac{\infty}{\infty}$. $\lim\limits_{x \to 0^+} (\sin x)(\ln \sin x) = 0$.

Solution 1.131. Type $\infty - \infty$. Transform to type $\frac{0}{0}$, $\frac{1}{x^2} - \frac{1}{\tan x} = \frac{\sin x - x^2 \cos x}{x^2 \sin x}$ and apply L'Hospital's Rule to obtain $\lim\limits_{x \to 0} \left(\frac{1}{x^2} - \frac{1}{\tan x} \right) = \infty$.

Solution 1.132. Type $\infty - \infty$. Transform to type $\frac{0}{0}$, $\frac{1}{x} - \frac{1}{e^x - 1} = \frac{e^x - 1 - x}{x\left(e^x - 1\right)}$ and apply L'Hospital's Rule twice to obtain

$\lim\limits_{x \to 0^+} \left(\frac{1}{x} - \frac{1}{e^x - 1} \right) = \frac{1}{2}$.

Solution 1.133. Type $\infty - \infty$. Transform to type $\frac{0}{0}$, $\frac{1}{\sin x} - \frac{1}{x} = \frac{x - \sin x}{x \sin x}$ and apply L'Hospital's Rule to obtain $\lim\limits_{x \to 0} \left(\frac{1}{\sin x} - \frac{1}{x} \right) = 0$.

Solution 1.134. Type $\infty - \infty$. Transform to type $\frac{0}{0}$, $\csc x - \cot x = \frac{1 - \cos x}{\sin x}$ and apply L'Hospital's Rule to obtain $\lim\limits_{x \to 0}(\csc x - \cot x) = 0$. Observe that we can evaluate this limit without using L'Hospital's Rule: $\lim\limits_{x \to 0} \frac{1 - \cos x}{\sin x} = \lim\limits_{x \to 0} \frac{\frac{1 - \cos x}{x}}{\frac{\sin x}{x}}$ and the result follows.

Solution 1.135. Observe that $\frac{f(x+h) - f(x-h)}{2h}$ is an indeterminate form of type $\frac{0}{0}$. (Keep in mind that "x" is a constant and that "h" denotes the variable in this case.) L'Hospital's Rule gives $\lim\limits_{h \to 0} \frac{f(x+h) - f(x-h)}{2h} \overset{L'H}{=} \lim\limits_{h \to 0} \frac{f'(x+h) + f'(x-h)}{2}$. Since f' is continuous, it follows $\lim\limits_{h \to 0} f'(x+h) = \lim\limits_{h \to 0} f'(x-h) = f'(x)$.

Solution 1.136. Observe that $\frac{f(x+h) + f(x-h) - 2f(x)}{h^2}$ is an indeterminate form of type $\frac{0}{0}$. By L'Hospital's Rule, $\lim\limits_{h \to 0} \frac{f(x+h) + f(x-h) - 2f(x)}{h^2}$

$\overset{L'H}{=} \lim\limits_{h \to 0} \frac{f'(x+h) - f'(x-h)}{2h} = \frac{1}{2} \cdot \lim\limits_{h \to 0} \left(\frac{f'(x+h) - f'(x)}{h} + \frac{f'(x-h) - f'(x)}{-h} \right) = f''(x)$.

Note: In evaluating the following limits we use the fact that $x \mapsto e^x$ and $x \mapsto \ln x$ are continuous functions on their domains. In other words, if $\lim\limits_{x \to a} f(x) = b$, then $\lim\limits_{x \to a} e^{f(x)} = e^{\lim\limits_{x \to a} f(x)} = e^b$, and if $b > 0$, then $\lim\limits_{x \to a} \ln f(x) = \ln(\lim\limits_{x \to a} f(x)) = \ln b$.

Solution 1.137. (1) Since $a \in (0,1)$ implies that $\frac{\ln a}{x^2} < 0$ for all $x \neq 0$, it follows that $\lim_{x \to 0} \frac{\ln a}{x^2} = -\infty$. Therefore, $\lim_{x \to 0} f(x) = \lim_{x \to 0} e^{\frac{\ln a}{x^2}} = 0$. Clearly, $\lim_{x \to 0} g(x) = \lim_{x \to 0} \sin^2 x = 0$. Observe that $\lim_{x \to 0} f(x)^{g(x)}$ is of the type 0^0. From $f(x)^{g(x)} = e^{\ln a \cdot \frac{\sin^2 x}{x^2}}$ and the fact that $\lim_{x \to 0} \frac{\sin^2 x}{x^2} = 1$, it follows that $\lim_{x \to 0} f(x)^{g(x)} = e^{\ln a} = a$. (2) From (1) it follows that a limit of the type 0^0 may take any value between 0 and 1. Hence, 0^0 is an indeterminate form.

Solution 1.138. Type 0^0. $\lim_{x \to 0^+} x^x = e^{\ln\left(\lim_{x \to 0^+} x^x\right)} = e^{\lim_{x \to 0^+} \ln x^x} = e^{\lim_{x \to 0^+} x \ln x} = e^{\lim_{x \to 0^+} \frac{\ln x}{\frac{1}{x}}} \stackrel{L'H}{=} e^{\lim_{x \to 0^+} \frac{\frac{1}{x}}{-\frac{1}{x^2}}} = e^0 = 1$.

Solution 1.139. Type 0^0. $\lim_{x \to 0^+} x^{\sqrt{x}} = e^{\lim_{x \to 0^+} \sqrt{x} \cdot \ln x} = 1$.

Solution 1.140. Type 0^0. $\lim_{x \to 0^+} x^{\tan x} = e^{\lim_{x \to 0^+} \tan x \cdot \ln x} = 1$.

Solution 1.141. Type 0^0. $\lim_{x \to 0^+} (\sin x)^{\tan x} = e^{\lim_{x \to 0^+} \tan x \cdot \ln(\sin x)} = 1$.

Solution 1.142. Type 0^0. $\lim_{x \to 0^+} (\tan x)^{\sin x} = e^{\lim_{x \to 0^+} \sin x \cdot \ln(\tan x)} = 1$.

Solution 1.143. Type 0^0. $\lim_{x \to 0^+} \left(\frac{x}{x+1}\right)^x = e^{\lim_{x \to 0^+} x \cdot \ln\left(\frac{x}{x+1}\right)} = 1$.

Solution 1.144. (1) Since $a \in (1,\infty)$ implies that $\frac{\ln a}{x^2} > 0$ for all $x \neq 0$, it follows that $\lim_{x \to 0} \frac{\ln a}{x^2} = \infty$. Therefore, $\lim_{x \to 0} f(x) = \lim_{x \to 0} e^{\frac{\ln a}{x^2}} = \infty$. Clearly, $\lim_{x \to 0} g(x) = \lim_{x \to 0} \sin^2 x = 0$. Observe that $\lim_{x \to 0} f(x)^{g(x)}$ is of type ∞^0. From $f(x)^{g(x)} = e^{\ln a \cdot \frac{\sin^2 x}{x^2}}$ and the fact that $\lim_{x \to 0} \frac{\sin^2 x}{x^2} = 1$ it follows that $\lim_{x \to 0} f(x)^{g(x)} = e^{\ln a} = a$. (2) From (1) it follows that a limit of type ∞^0 may take any value $a > 1$. Hence, ∞^0 is an indeterminate form.

Solution 1.145. Type ∞^0. $\displaystyle\lim_{x\to\infty} x^{\frac{1}{x}} = e^{\ln\left(\lim_{x\to\infty} x^{\frac{1}{x}}\right)} = e^{\lim_{x\to\infty}\ln x^{\frac{1}{x}}} =$
$e^{\lim_{x\to\infty}\frac{\ln x}{x}} \overset{L'H}{=} e^{\lim_{x\to\infty}\frac{1}{x}} = e^0 = 1.$

Solution 1.146. Type ∞^0. $\displaystyle\lim_{x\to\infty}(e^x + x)^{\frac{1}{x}} = e^{\lim_{x\to\infty}\frac{\ln(e^x+x)}{x}}$. Now,
apply L'Hospital's Rule three times to obtain $\displaystyle\lim_{x\to\infty}(e^x+x)^{\frac{1}{x}} = e$.

Solution 1.147. Type ∞^0. $\displaystyle\lim_{x\to\infty}(x + \sin x)^{\frac{1}{x}} = e^{\lim_{x\to\infty}\frac{\ln(x+\sin x)}{x}} = 1$.

Solution 1.148. Type ∞^0. $\displaystyle\lim_{x\to\infty}(\ln x)^{\frac{1}{x}} = e^{\lim_{x\to\infty}\frac{\ln(\ln x)}{x}} = 1$.

Solution 1.149. Type ∞^0. $\displaystyle\lim_{x\to\infty}(\ln x)^{e^{-x}} = e^{\lim_{x\to\infty}\frac{\ln(\ln x)}{e^x}} = 1$.

Solution 1.150. (1) Clearly, $\displaystyle\lim_{x\to 0} e^{x^2\ln a} = 1$ and $\displaystyle\lim_{x\to 0}\frac{1}{\sin^2 x} = \infty$.
Observe that $\displaystyle\lim_{x\to 0} f(x)^{g(x)}$ is of type 1^∞. From $f(x)^{g(x)} = e^{\ln a\cdot\frac{x^2}{\sin^2 x}}$
and the fact that $\displaystyle\lim_{x\to 0}\frac{x^2}{\sin^2 x} = 1$ it follows that $\displaystyle\lim_{x\to 0} f(x)^{g(x)} = e^{\ln a} =$
a.   (2) From (1) it follows that a limit of type 1^∞ may take any
positive value. Hence, 1^∞ is an indeterminate form.

Solution 1.151. Type 1^∞. $\displaystyle\lim_{x\to 0}(1 - 2x)^{\frac{1}{x}} = e^{\ln\left(\lim_{x\to 0}(1-2x)^{\frac{1}{x}}\right)} =$
$e^{\lim_{x\to 0}\ln(1-2x)^{\frac{1}{x}}} = e^{\lim_{x\to 0}\frac{\ln(1-2x)}{x}}$. Now, apply L'Hospital's Rule to obtain
$\displaystyle\lim_{x\to 0}(1 - 2x)^{\frac{1}{x}} = e^{-2}$.

Solution 1.152. Type 1^∞. $\displaystyle\lim_{x\to 0}(\cosh x)^{\frac{1}{x^2}} = e^{\lim_{x\to 0}\frac{\ln(\cosh x)}{x^2}} = e^{\frac{1}{2}}$.

Solution 1.153. Type 1^∞. $\displaystyle\lim_{x\to 0^+}(\cos x)^{\frac{1}{x}} = e^{\lim_{x\to 0^+}\frac{\ln(\cos x)}{x}} = 1$.

Solution 1.154. Type 1^∞. $\displaystyle\lim_{x\to 0}(1 + \sin x)^{\frac{1}{x}} = e^{\lim_{x\to 0}\frac{\ln(1+\sin x)}{x}} = e$.

Solution 1.155. Type 1^∞. $\displaystyle\lim_{x\to\infty}\left(1+\sin\frac{3}{x}\right)^x = e^{\lim_{x\to\infty} x\ln\left(1+\sin\frac{3}{x}\right)} = e^3$.

Solution 1.156. Type 1^∞. $\displaystyle\lim_{x\to e^+}(\ln x)^{\frac{1}{x-e}} = e^{\lim_{x\to e^+}\frac{\ln(\ln x)}{x-e}} = e^{\frac{1}{e}}$.

Solution 1.157. From $f(x) = \frac{x^2+x-6}{x+2} = \frac{x(x+2)-(x+2)-4}{x+2} = x-1-\frac{4}{x+2}$
for $x \neq -2$, we obtain that $\lim\limits_{x\to\infty} (f(x) - (x-1)) = \lim\limits_{x\to\infty} \left(-\frac{4}{x+2}\right) =$
0. Conclude that the line $y = x - 1$ is a slant asymptote of the graph
of the function $y = f(x)$. Observe that f is a rational function that is
not defined at $x = -2$. From $\lim\limits_{x\to-2^+} f(x) = \lim\limits_{x\to-2^+} \frac{(x-2)(x+3)}{x+2} = -\infty,$
we conclude that the line $x = -2$ is a vertical asymptote of the graph
of the function $y = f(x)$.

Solution 1.158. Observe that the domain of the function f is given
by $\{x : x^2 - 5x + 6 \neq 0\} = \mathbb{R}\backslash\{2,3\}$. From $f(x) = \frac{(x-2)(x+3)}{(x-2)(x-3)} =$
$\frac{x+3}{x-3}$ for $x \neq 2$ and $x \neq 3$, we conclude that $\lim\limits_{x\to2} f(x) = -5$ and
$\lim\limits_{x\to3^+} f(x) = \infty$. The line $x = 3$ is the only vertical asymptote of the
graph of the function $y = f(x)$. From $\lim\limits_{x\to\infty} f(x) = 1$, we conclude
that the line $y = 1$ is a horizontal asymptote of the graph of the
function.

Solution 1.159. $a = 4$. The question is to determine a real number
a so that $\lim\limits_{x\to\infty} \left(\frac{x^2+ax+5}{x+1} - (x+3)\right) = \lim\limits_{x\to\infty} \frac{(a-4)x+2}{x+1} = 0.$

Solution 1.160. $\lim\limits_{x\to\infty} \frac{\ln x}{x} = 0.$

Solution 1.161. The function f is continuous on its domain $\mathbb{R}\backslash\{0\}$.
Since $\lim\limits_{x\to0} \sin\left(\frac{1}{x}\right)$ does not exist and since the function is bounded,
the graph of the function $y = f(x)$ has no vertical asymptote.

From $\lim\limits_{x\to\infty} \sin\left(\frac{1}{x}\right) = 0$, it follows that the line $y = 0$ is a horizontal
asymptote of the graph of the function.

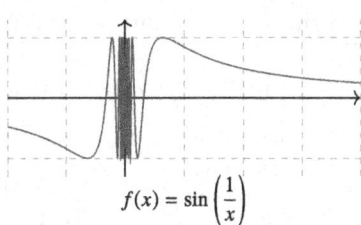

$$f(x) = \sin\left(\frac{1}{x}\right)$$

Solution 1.162. The function g is continuous on its domain $\mathbb{R}\backslash\{0\}$. Since $\lim\limits_{x\to 0} \frac{\sin x}{x} = 1$, the graph of the function $y = g(x)$ has no vertical asymptote. From $\lim\limits_{x\to\pm\infty} \frac{\sin x}{x} = 0$, it follows that the line $y = 0$ is a horizontal asymptote of the graph of the function.

Observe that, for each $k \in \mathbb{Z}$, $g(k\pi) = 0$. Hence, this curve intersects its horizontal asymptote an infinite number of times.

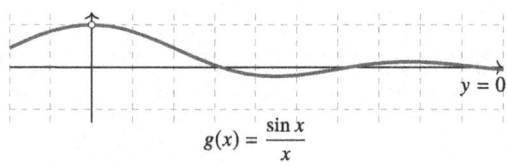

$$g(x) = \frac{\sin x}{x}$$

Solution 1.163. The function h is continuous on its domain $\mathbb{R}\backslash\{0\}$. Since $\lim\limits_{x\to 0} x \sin\left(\frac{1}{x}\right) = 0$, the graph of the function $y = h(x)$ has no vertical asymptote.

From $\lim\limits_{x\to\pm\infty} x \sin\left(\frac{1}{x}\right) = \lim\limits_{x\to\pm\infty} \frac{\sin\left(\frac{1}{x}\right)}{\frac{1}{x}} = 1$, it follows that the line $y = 1$ is a horizontal asymptote of the graph of the function.

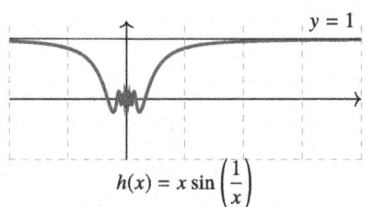

$$h(x) = x \sin\left(\frac{1}{x}\right)$$

Solution 1.164. The function i is continuous on its domain $\mathbb{R}\backslash\{k\pi : k \in \mathbb{Z}\}$.

Since $\lim\limits_{x\to\pm\infty} \frac{1}{\sin x}$ do not exist, the graph of the function $y = i(x)$ has no horizontal asymptote. Observe that i is an odd periodic function with a period 2π. From $\lim\limits_{x\to 0^+} \frac{1}{\sin x} = \lim\limits_{x\to\pi^-} \frac{1}{\sin x} = \infty$, it follows that, for any $k \in \mathbb{Z}$, the line $x = k\pi$ is a vertical asymptote of the graph of the function $y = i(x)$.

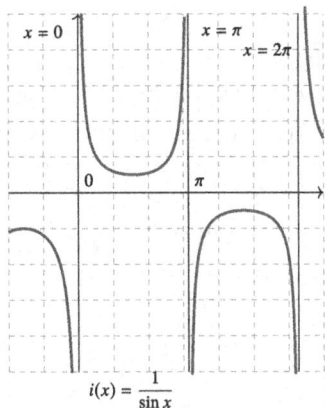

$$i(x) = \frac{1}{\sin x}$$

Solution 1.165. The function j is continuous on its domain $\mathbb{R}\backslash\{0\}$. Since $\lim\limits_{x\to 0}\left(x+\sin\left(\frac{1}{x}\right)\right)$ does not exist and since the function is bounded on the set $\left(-\frac{2}{\pi},0\right)\cup\left(0,\frac{2}{\pi}\right)$, the graph of the function $y = j(x)$ has no vertical asymptote. From $\lim\limits_{x\to\infty}\left(j(x)-x\right) = \lim\limits_{x\to\infty}\sin\left(\frac{1}{x}\right) = 0$, it follows that the line $y = x$ is a slant asymptote of the graph of the function.

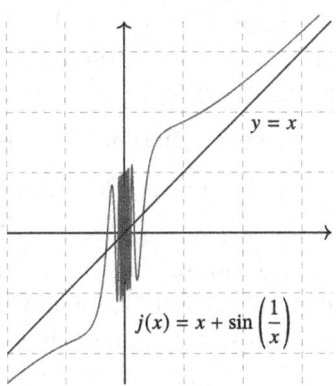

$$j(x) = x + \sin\left(\frac{1}{x}\right)$$

Solution 1.166. Let $\varepsilon > 0$ be given. We need to find $\delta = \delta(\varepsilon) > 0$ such that $|x - 0| < \delta$ implies $|x^3 - 0| < \varepsilon$, which is the same as $|x| < \delta$ implies $|x^3| < \varepsilon$. Clearly, we can take $\delta = \sqrt[3]{\varepsilon}$: $|x| < \sqrt[3]{\varepsilon}$ implies $|x|^3 = |x^3| < \varepsilon$ and, by definition, $\lim\limits_{x\to 0} x^3 = 0$. *Note:* The notation $\delta = \delta(\varepsilon)$ means that δ is a function of ε, i.e., that the value of δ depends on the value of ε.

Solution 1.167.

(1) See the following figure.

(2) Using the graph, we work backwards from $2 - \frac{1}{4} = \frac{7}{4}$ and $2 + \frac{1}{4} = \frac{9}{4}$ to obtain their corresponding x-values via the function $f(x) = 2x^2$. Since $\lim\limits_{x \to 1} f(x) = 2$, we reject negative roots to obtain $x = \sqrt{\frac{1}{2} \cdot \frac{7}{4}} \approx 0.93$ and $x = \sqrt{\frac{9}{8}} \approx 1.06$.

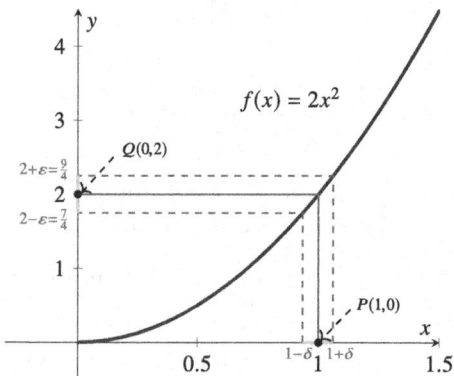

Since $|0.93 - 1| = 0.07$ and $|1.06 - 1| = 0.06$, if we, for example, choose $\delta = 0.05$, then for any x such that $|x - 1| < \delta$, we will have $|f(x) - 2| < \varepsilon = \frac{1}{4}$.   (3) For any $\varepsilon > 0$ there exists $\delta = \delta(\varepsilon) > 0$ such that $|x - 1| < \delta$ implies $|2x^2 - 2| < \varepsilon$.

Solution 1.168. Proof by contradiction: Suppose that $\lim\limits_{x \to 0} \sin\left(\frac{1}{x}\right) = L$, i.e., suppose that $\lim\limits_{x \to 0} \sin\left(\frac{1}{x}\right)$ exists. Recall that this means that for *any* $\varepsilon > 0$, we can find *at least one* $\delta > 0$ such that $|x - 0| = |x| < \delta$ implies $\left|\sin\left(\frac{1}{x}\right) - L\right| < \varepsilon$. Suppose that $L \neq 0$ and that $0 < \varepsilon < |L|$. Let δ be any positive number, and let a natural number k be so large that $k > \frac{1}{\pi\delta}$. Then, for $x = \frac{1}{k\pi}$, we have that $|x| = \frac{1}{k\pi} < \delta$ and $\left|\sin\left(\frac{1}{x}\right) - L\right| = |\sin(k\pi) - L| = |L| > \varepsilon$. Hence, by definition, $L \neq 0$ cannot be the value of the limit. Suppose that $L = 0$ and that $0 < \varepsilon < 1$. Let δ be any positive number, and let a natural number k be so large that $2k + 1 > \frac{2}{\pi\delta}$. Then, for $x = \frac{2}{(2k+1)\pi}$, we have that $|x| = \frac{2}{(2k+1)\pi} < \delta$ and $\left|\sin\left(\frac{1}{x}\right) - L\right| = \left|\sin\frac{(2k+1)\pi}{2} - 0\right| = 1 > \varepsilon$. Hence, by definition, $L = 0$ cannot be the value of the limit. Since our assumption that the limit exists leads to a contradiction, we conclude that $\lim\limits_{x \to 0} \sin\left(\frac{1}{x}\right)$ does not exists.

Solution 1.169. There are many functions that satisfy the required conditions.

(1) For example, $f(x) = \frac{x^2-9}{x-3}$ for $x \neq 3$ and $f(3) = 0$. The graph of f is a line $y = x + 3$ with a hole in the line at $x = 3$. By letting $f(3) = 6$ instead of 0, the hole is removed.

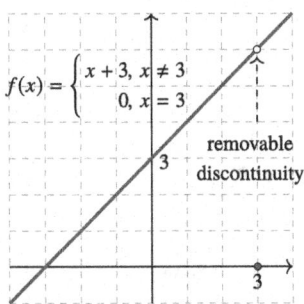

(2) For example, the Heaviside step function, which is one for positive arguments and zero otherwise, is discontinuous at $x = 0$. Investigating the left and right limits tells us that there is a jump discontinuity at $x = 0$.

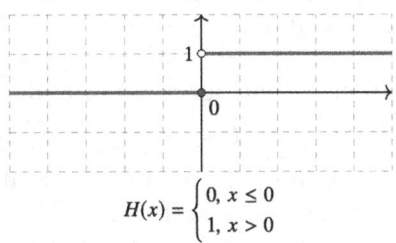

(3) For example, $f(x) = \frac{2}{x+1}$ for $x \neq -1$. The domain of f is the set $\mathbb{R}\backslash\{-1\}$. Hence, f is discontinuous at $x = -1$. Investigating the left- and right-hand limits tells us that this discontinuity is infinite.

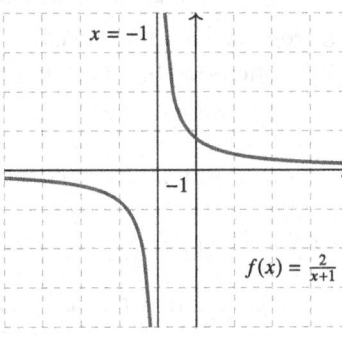

Solution 1.170. See the following figures: (1) a function with a removable and an infinite discontinuity (left); and (2) a function with a removable and a jump discontinuity (right).

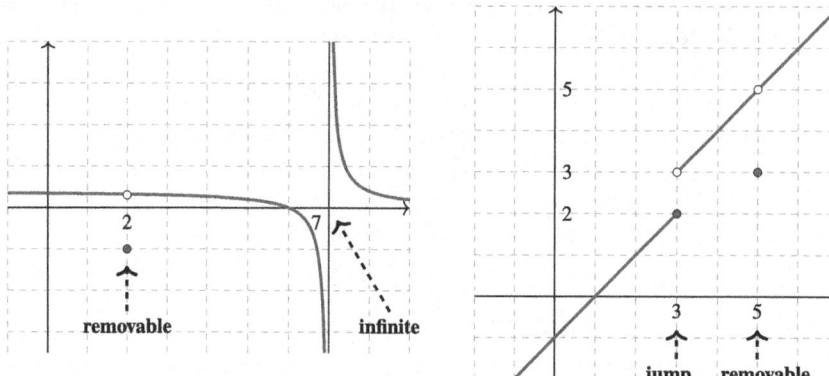

Solution 1.171. Note that each function piece is a polynomial and therefore continuous on its domain. Investigate the left- and the right-hand limits as $x \to -1$ and the function value at $x = -1$ to show that f is also continuous at $x = -1$.

Solution 1.172. Observe that, for $x \neq -3$ and $x \neq 6$, $\frac{x+3}{x^2-3x-18} = \frac{x+3}{(x+3)(x-6)} = \frac{1}{x-6}$. From $\lim_{x \to -3} f(x) = -\frac{1}{9} \neq f(-3)$, conclude that the function f has a removable discontinuity at $x = -3$. From $\lim_{x \to 6^+} f(x) = \infty$, conclude that the function f has an infinite discontinuity at $x = 6$.

Solution 1.173. Observe that, for $x \neq 2$, $\frac{x^2-4}{x-2} = x + 2$. From $\lim_{x \to 2} f(x) = 4$ and the fact that $f(2)$ is not defined, conclude that the function f has a removable discontinuity at $x = 2$. From $\lim_{x \to -2} f(x) = 0 \neq f(-2)$, conclude that the function f has a removable discontinuity at $x = -2$.

Solution 1.174. (1) $c = \pi$. Solve $\lim_{x \to \pi^-} f(x) = \lim_{x \to \pi^-} f(x)$ for c.

(2) If $f(x) = \begin{cases} \pi - x, \ x \leq \pi \\ \pi \sin x, \ x > \pi \end{cases}$ then: (a) $f(\pi) = \pi - \pi = 0$ and so f is defined. (b) $\lim_{x \to \pi^-} f(x) = \lim_{x \to \pi^-} (\pi - x) = 0$ and $\lim_{x \to \pi^+} f(x) =$

$\lim\limits_{x \to \pi^+} \pi \sin x = 0$. Therefore, $\lim\limits_{x \to \pi} f(x) = 0$ and so exists. Lastly, (c) the function value and limit of f at $x = \pi$ are equal, $f(\pi) = 0 = \lim\limits_{x \to \pi} f(x)$.

(3)

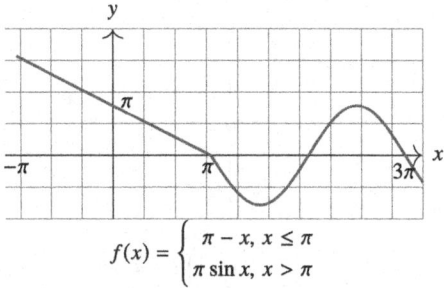

$$f(x) = \begin{cases} \pi - x, & x \leq \pi \\ \pi \sin x, & x > \pi \end{cases}$$

Solution 1.175. $b = -1$. Investigate the left- and the right-hand limits as $x \to 2$ and the function value at $x = 2$.

Solution 1.176. $a = \sqrt[3]{9}$. Investigate the left- and the right-hand limits as $x \to 0$ and the function value at $x = 0$.

Solution 1.177. $a = -\frac{1}{\pi}$. Investigate the left- and the right-hand limits as $x \to \frac{\pi}{2}$ and the function value at $x = \frac{\pi}{2}$.

Solution 1.178. $a = -1$ and $b = 1$. Solve the system of equations $\lim\limits_{x \to -1^-} 2 = \lim\limits_{x \to -1^+} (ax + b)$ and $\lim\limits_{x \to 3^-} (ax + b) = \lim\limits_{x \to 3^+} -2$.

Solution 1.179. From $\lim\limits_{x \to 0^+} \text{sign}(x) = 1$ and $\lim\limits_{x \to 0^-} \text{sign}(x) = -1$, conclude that the function $x \mapsto \text{sign}(x)$ has a jump discontinuity at $x = 0$.

Solution 1.180. As a composition of two continuous functions, the function f is continuous at any $x \neq 0$. Recall that $\lim\limits_{x \to 0} \sin\left(\frac{1}{x}\right)$ does not exist. Conclude that the function f is not continuous at $x = 0$. This also means that $x = 0$ is not a removable discontinuity. Since $-1 \leq f(x) \leq 1$ for any $x \in \mathbb{R}$, $x = 0$ cannot be an infinite discontinuity. Since neither $\lim\limits_{x \to 0^+} \sin\left(\frac{1}{x}\right)$ nor $\lim\limits_{x \to 0^-} \sin\left(\frac{1}{x}\right)$ exists, $x = 0$ is an essential discontinuity.

Solution 1.181. Use the Squeeze Theorem to show $\lim\limits_{x \to 0} x \sin\left(\frac{1}{x}\right) = 0$. Conclude that g is a continuous function on its domain.

Solution 1.182. As a composition of two continuous functions, the function h is continuous at any $x \neq 0$. Since $\lim\limits_{x \to 0^+} \arctan\left(\frac{1}{x}\right) = \lim\limits_{t \to \infty} \arctan t = \frac{\pi}{2}$ and $\lim\limits_{x \to 0^-} \arctan\left(\frac{1}{x}\right) = -\frac{\pi}{2}$, the function h has a jump discontinuity at $x = 0$.

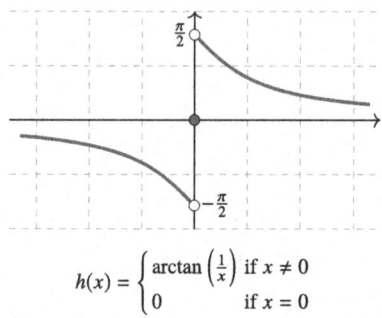

$$h(x) = \begin{cases} \arctan\left(\frac{1}{x}\right) & \text{if } x \neq 0 \\ 0 & \text{if } x = 0 \end{cases}$$

Solution 1.183. Notice that, for $x > 0$, $\lim\limits_{n \to \infty} nx = \infty$. Hence, if $x > 0$, then $i(x) = \frac{\pi}{2}$. Similarly, if $x < 0$, then $i(x) = -\frac{\pi}{2}$. Finally, $i(0) = \lim\limits_{n \to \infty} \arctan(n \cdot 0) = 0$. The function i is continuous at any $x \neq 0$. At $x = 0$ it has a jump discontinuity.

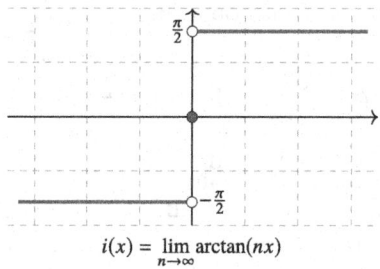

$$i(x) = \lim_{n \to \infty} \arctan(nx)$$

Solution 1.184. Let $a \in \mathbb{R}$. To show that the Dirichlet function χ is not continuous at a number $a \in \mathbb{R}$, it is enough to show that $\lim\limits_{x \to a} \chi(x)$ does not exist. Let $\varepsilon = \frac{1}{2}$ and $\delta > 0$.

Suppose that a is a rational number. Let a natural number k be so large that $k > \frac{\pi}{\delta}$. Then the number $b = a + \frac{\pi}{k}$ is irrational (otherwise $\pi = k(b-a)$ would be rational). Now, $|a - b| = \frac{\pi}{k} < \delta$ and $|\chi(a) - \chi(b)| = |1 - 0| = 1 > \varepsilon = \frac{1}{2}$. Hence, $\lim\limits_{x \to a^+} f(x)$ does not exist.

Suppose that a is an irrational number given in a decimal form $a = d.d_1 d_2 d_3 \ldots$. This means that $d \in \{0, 1, 2, \ldots\}$ and, for each $i \in \mathbb{N}$, $d_i \in \{0, 1, 2, \ldots, 9\}$ are such that $a = d + \frac{d_1}{10} + \frac{d_2}{10^2} + \frac{d_3}{10^3} + \cdots$.

Let a natural number k be so large that $10^k > \frac{1}{\delta}$. Consider the rational number $b = d.d_1d_2d_3\cdots d_k = d + \frac{d_1}{10} + \frac{d_2}{10^2} + \cdots + \frac{d_k}{10^k}$. Now, $|a - b| = \frac{d_{k+1}}{10^{k+1}} + \frac{d_{k+2}}{10^{k+2}} + \cdots < \frac{1}{10^k} < \delta$ and $|\chi(a) - \chi(b)| = |0 - 1| = 1 > \varepsilon = \frac{1}{2}$. Hence, $\lim\limits_{x \to a^-} f(x)$ does not exist.

Therefore, the Dirichlet function is discontinuous at every $a \in \mathbb{R}$ and has an essential discontinuity there.

Solution 1.185. Recall that, as a polynomial, p is a continuous function on $\mathbb{R}$. Note that $p(-3) < 0$, $p(-1) > 0$, $p(0) < 0$ and $p(3) > 0$. Apply the Intermediate Value Theorem on each of the closed intervals $[-3, -1], [-1, 0]$, and $[0, 3]$ to conclude that the polynomial p has a root in each of the corresponding open intervals. *Note*: A cubic polynomial cannot have more than three roots.

Solution 1.186. Note that $p(0) = -1 < 0$ and $p(1) = 1 > 0$. By the Intermediate Value Theorem, there is at least one $x \in (-1, 1)$ such that $p(x) = 0$. *Note*: From $p'(x) = 5x^4 + 1 > 0$ it follows that p is an increasing function. The polynomial p has only one real root.

Solution 1.187. Let n be an odd natural number and $p(x) = a_n x^n + a_{n-1}x^{n-1} + \cdots + a_0$ with $a_n > 0$. Recall that $\lim\limits_{x \to \pm\infty} \frac{p(x)}{a_n x^n} = 1$ implies that $\lim\limits_{x \to -\infty} p(x) = \lim\limits_{x \to -\infty} a_n x^n = -\infty$ (since n is odd we have that $a_n x^n < 0$ for $x < 0$) and $\lim\limits_{x \to \infty} p(x) = \lim\limits_{x \to \infty} a_n x^n = \infty$. Hence, there are $a < 0$ and $b > 0$ such that $p(a) < 0$ and $p(b) > 0$. By the Intermediate Value Theorem, there is at least one $x \in (a, b)$ such that $p(x) = 0$.

Solution 1.188. (1) For example, $f(x) = x^2 + 1$. Note that $f(x) > 0$ for all $x \in \mathbb{R}$. (2) For example, the linear function $f(x) = x + 1$ has $x = -1$ as its only root. (3) For example, the polynomial $f(x) = x^5 - x = x(x - 1)(x + 1)(x^2 + 1)$ has three roots.

Solution 1.189. Let $f(x) = 2^x - \frac{10}{x}$. Note that the domain of f is the set $\mathbb{R}\backslash\{0\}$ and that on its domain, as a sum of two continuous functions, f is a continuous function. (1) Since $f(1) = 2 - 10 = -8$ and $f(3) = 8 - \frac{10}{3} = \frac{14}{3} > 0$, by the Intermediate Value Theorem, there is $x \in (1, 3)$ such that $f(x) = 2^x - \frac{10}{x} = 0$. (2) For all $x \in (-\infty, 0)$, we have that $\frac{10}{x} < 0$, which implies that for all $x \in (-\infty, 0)$, we have that $f(x) > 0$.

Solution 1.190. Since f is a continuous function, the function $g(x) = f(x) - c$ also must be continuous on $[a, b]$. From $f(a) < c$, it follows that $g(a) = f(a) - c < 0$. Suppose that $f(b) > c$. Then $g(b) = f(b) - c > 0$. By the Intermediate Value Theorem, there is $k \in (a, b)$ such that $g(k) = 0$. But this means that $f(k) = c$, which is a contradiction. Hence, $f(b) < c$.

Solution 1.191. Suppose that $f(a) < 0$ for some $a \in (0, 3]$. Since f is continuous on the closed interval $[0, a]$ and $f(0) = 1 > 0$, by the Intermediate Value Theorem, there must be $c \in (0, a) \subseteq (0, 3)$ such that $f(c) = 0$. But this is a contradiction. Hence, $f(x) > 0$ for all $x \in [0, 3]$.

Solution 1.192. The Intermediate Value Theorem requires that the function is continuous on the closed interval in question. Here, the function f is not defined for $x = 1 \in [0, 2]$, so the Intermediate Value Theorem does not apply on $[0, 2]$.

Solution 1.193. Take $p(x) = x^3 - 6x + 3$. Note that $p(0.5) = 0.125 > 0$ and $p(1) = -2 < 0$. Apply the Intermediate Value Theorem.

Solution 1.194. Observe that if $a + b = \frac{\pi}{2} + 2(k - 1)\pi$ for $k \in \mathbb{N}$ and $x = a + b$, then we have $a \sin x + b = a \sin \left(\frac{\pi}{2} + 2(k - 1)\pi \right) + b = a + b = x$. Hence, in this case, $x = a + b$ is a solution of Kepler's equation. Next, suppose that $a + b \neq \frac{\pi}{2} + 2(k - 1)\pi$ for $k \in \mathbb{N}$. Let $f(x) = x - a \sin x - b$. Note that $f(0) = -b < 0$ and $f(a + b) = a(1 - \sin(a + b)) > 0$. Apply the Intermediate Value Theorem on the interval $[0, a + b]$.

Solution 1.195. Note that the function $g(x) = f(x) - x$ is continuous on the closed interval $[0, 1]$. Since $[0, 1]$ is the range of the function f and since $f(0) \neq 0$ and $f(1) \neq 1$, there are $a, b \in (0, 1)$ such that $f(a) = 0$ and $f(b) = 1$. This implies $g(a) < 0$ and $g(b) > 0$. By the Intermediate Value Theorem, there is $c \in (a, b)$ such that $g(c) = f(c) - c = 0$. Hence, $f(c) = c$ and $x = c$ is a fixed point of the function f.

Chapter 2

Derivatives

2.1 Introduction

Use the following definitions, techniques, properties, and algorithms to solve the problems contained in this chapter.

Derivative. The *derivative of a function f at a number a* is defined as $f'(a) = \lim\limits_{h \to 0} \dfrac{f(a+h)-f(a)}{h} = \lim\limits_{x \to a} \dfrac{f(x)-f(a)}{x-a}$ if this limit exists. If $f'(a)$ exists, we say that the *function f is differentiable at a*.

Notation. Depending on the context, to denote the first derivative of a function $y = f(x)$ we will use both $y' = f'(x)$, a notation attributed to Joseph-Louis Lagrange (1736–1813), and the so-called Leibniz's notation $\dfrac{dy}{dx}$, named in honor of Gottfried Wilhelm Leibniz (1646–1716).

Continuity and Differentiation. If a function $y = f(x)$ is differentiable at a number a, then it is continuous at the number a.

Tangent Line. If a function $y = f(x)$, $x \in D$, is differentiable at a number $a \in D$, then we say that the line determined by the equation $y = f(a) + f'(a)(x - a)$ is the *tangent line to the graph of f at the point $(a, f(a))$*.

Product Rule. If functions f and g are both differentiable at x, then $(f(x)g(x))' = f(x)g'(x) + g(x)f'(x)$.

Quotient Rule. If functions f and g are both differentiable at x and $g(x) \neq 0$, then $\left(\dfrac{f(x)}{g(x)}\right)' = \dfrac{g(x)f'(x)-f(x)g'(x)}{(g(x))^2}$.

Chain Rule. If functions f and g are both differentiable on their domains and $F = f \circ g$ is the composite function defined by $F(x) = f(g(x))$, then F is differentiable and $F'(x) = f'(g(x)) \cdot g'(x)$. If we write $y = f(x)$ and $z = F(x) = g(f(x)) = g(y)$, then the chain rule may also be expressed in Leibniz's notation as $\dfrac{dz}{dx} = \dfrac{dz}{dy} \cdot \dfrac{dy}{dx}$.

Implicit Differentiation. Let a function $y = y(x)$ be implicitly defined by $F(x, y) = c$, where c is a constant. Proceed as follows to determine the derivative $y' = \dfrac{dy}{dx}$:

(1) Use the Chain Rule to differentiate both sides of the expression $F(x, y) = c$ by taking x as the independent variable and y as a function of x.

(2) Solve the resulting equation for y'.

Method of Logarithmic Differentiation. If a function is of the form $y = f(x) = g(x)^{h(x)}$, or if the expression for $y = f(x)$ involves multiple factors, quotients, and/or integer and rational powers, then using logarithms may help us to determine $y' = \dfrac{dy}{dx}$ in a relatively straightforward way. Proceed as follows:

(1) Take natural logarithms of both sides of the equation $y = f(x)$.

(2) Use the properties of logarithms to simplify the expression $\ln f(x)$.

(3) Differentiate implicitly with respect to x.

(4) Solve the resulting equation for y'.

(5) Express y' in terms of x by using the initial equation $y = f(x)$.

Derivative of Inverse Function. Given a differentiable one-to-one function f, the derivative of its inverse function f^{-1} is given by $\left(f^{-1}\right)'(x) = \dfrac{1}{f'(f^{-1}(x))}$, for all x in the range of the function f such that $f'\left(f^{-1}(x)\right) \neq 0$.

Method of Related Rates. If two variables are related by an equation and both are functions of an independent third variable (e.g., time), then their rates of change are related as well with respect to this third variable. If we know the value of one rate of change, we can compute the other. We proceed as follows:

(1) Identify the independent variable on which the other quantities depend and assign it a symbol, such as t. Also, assign symbols to the variable quantities that depend on t.

(2) Determine an equation that relates the dependent variables.

(3) Differentiate both sides of the equation with respect to t. Keep in mind that the dependent variables are functions of t, so the Chain Rule applies.

(4) Substitute the given values into the related rates equation and solve for the unknown rate.

2.2 Definition of Derivative and Differentiation Rules

For all differentiation problems use the appropriate definition and rules.

2.2.1 *Definition of derivative*

Use the definition of derivative to solve the following problems.

Problem 2.1. Show that if $f(x) = \dfrac{1}{2x-1}$, then $f'(3) = -0.08$.

Problem 2.2. Use the definition of derivative to evaluate $f'(2)$ when $f(x) = x + \frac{1}{x}$.

Problem 2.3. Evaluate $f'(1)$ when $f(x) = 3x^2 - 4x + 1$.

Problem 2.4. Evaluate $f'(4)$ when $f(x) = \sqrt{5-x}$.

Problem 2.5. Determine $f'(x)$ when $f(x) = \sqrt{x}$.

Problem 2.6. Determine $f'(x)$ when $f(x) = \dfrac{x}{x-2}$. Then check your work by using the Quotient Rule.

Problem 2.7. Let f be a function that is continuous everywhere and

$$F(x) = \begin{cases} \dfrac{f(x)\sin^2 x}{x} & \text{if } x \neq 0, \\ 0 & \text{if } x = 0. \end{cases}$$

Evaluate $F'(0)$.

Problem 2.8. Evaluate $\lim\limits_{h\to 0} \dfrac{\sin^7\left(\dfrac{\pi}{6}+\dfrac{h}{2}\right)-\dfrac{1}{2^7}}{h}$.

Problem 2.9. If $g(x) = 2x^3 + \ln x$ is the derivative of $f(x)$, evaluate $\lim\limits_{x\to 0} \dfrac{f(1+x)-f(1)}{x}$.

Problem 2.10. Evaluate $\lim\limits_{x\to 0} \dfrac{\sqrt{1+x}+(1+x)^7-2}{x}$.

Problem 2.11. Determine a function f and a number a such that $\lim\limits_{h\to 0} \dfrac{(2+h)^6-64}{h} = f'(a)$.

2.2.2   Continuity and differentiation

Solve the following problems by using definitions of continuity and differentiation, as well as their connection.

Problem 2.12. State two properties that a continuous function f can have, either of which guarantees that the function is not differentiable at $x = a$. Sketch an example for each.

Problem 2.13. Sketch the graph of the function

$$f(x) = \begin{cases} 2 - x^2 & \text{if } 0 \le x < 1 \\[2mm] \dfrac{5}{2} & \text{if } x = 1 \\[2mm] |2 - x| & \text{if } 1 < x \le 3 \\[2mm] \dfrac{1}{x-3} & \text{if } 3 < x \le 5 \\[2mm] 2 + \sin(2\pi x) & \text{if } 5 < x \le 6 \\[2mm] 2 & \text{if } x > 6. \end{cases}$$

Answer the following questions with *Yes* or *No*.

(1) Is f continuous at
    (a) $x = 1$?   (b) $x = 6$?
(2) Do the following limits exist?
    (a) $\lim\limits_{x\to 1} f(x)$ (b) $\lim\limits_{x\to 3^-} f(x)$

(3) Is f differentiable
    (a) at $x = 1$?   (b) on the interval $(1, 3)$?

Problem 2.14. You are given that

$$f(x) = \begin{cases} 2 + \sqrt{x}, & \text{if } x \geq 1 \\ \dfrac{x}{2} + \dfrac{5}{2}, & \text{if } x < 1. \end{cases}$$

(1) Determine whether or not the function f is continuous at $x = 1$.
(2) Use the definition of the derivative to evaluate $f'(1)$.

Problem 2.15. Explain why the function

$$f(x) = \begin{cases} x^2 + 2x + 1, & \text{if } x \leq 0 \\ 1 + \sin x, & \text{if } x > 0 \end{cases}$$

is continuous but not differentiable on the interval $(-1, 1)$.

Problem 2.16. Justify whether the following statement is true or false: If a function f is defined and continuous at each number in the interval $(0, 2)$, then f must be differentiable at $x = 1$.

Problem 2.17. Justify whether the following statement is true or false: If a function g is differentiable at a number x_0, then g must be continuous at x_0.

2.2.3 *Differentiation rules*

Use differentiation rules to determine the derivative of the following functions. When requested, evaluate the derivative at a given number. Assume that the independent variable belongs to an interval where the given function is differentiable.

Problem 2.18. Evaluate $f'(2)$ for $f(x) = 3x^4 + \ln x$.

Problem 2.19. Evaluate $f'(0)$ for $f(x) = \sin^{-1}(x^2 + x) + 5^x$.

Problem 2.20. $y = \ln(\tan(7^{1-5x}))$.

Problem 2.21. $y = \tan(\cos^{-1}(e^{4x}))$.

Problem 2.22. $y = e^{4\cosh\sqrt{x}}$.

Problem 2.23. $y = e^{\cos x^2}$.

Problem 2.24. $y = x^{20} \arctan x$.

Problem 2.25. $y = e^{3 \ln(2x+1)}$.

Problem 2.26. $y = \sec(\sinh x)$.

Problem 2.27. $f(x) = \dfrac{3x^2 + 1}{e^x}$.

Problem 2.28. $g(z) = \sin \sqrt{z^2 + 1}$.

Problem 2.29. $h(y) = \sqrt{\dfrac{\cos y}{y}}$.

Problem 2.30. $f(x) = \dfrac{1}{x + \frac{1}{x}}$.

Problem 2.31. $f(x) = \arctan(\sqrt{x})$.

Problem 2.32. $f(x) = \cosh(5 \ln x)$.

Problem 2.33. $f(x) = 10^{3x}$.

Problem 2.34. $f(x) = x^{10} \tanh x$.

Problem 2.35. $y = \dfrac{e^{x^2+1}}{x \sin x}$.

Problem 2.36. $f(x) = \ln(\cos 3x)$.

Problem 2.37. $f(x) = 2^{2x} - (x^2 + 1)^{\frac{2}{3}}$.

Problem 2.38. $f(x) = \tan^2(x^2)$.

Problem 2.39. $f(x) = 5x + x^5 + 5^x + \sqrt[5]{x} + \ln 5$.

Problem 2.40. $f(x) = \ln(\sinh x)$.

Problem 2.41. $f(x) = e^{x \cos x}$.

Problem 2.42. $f(x) = \dfrac{\sin x}{1 + \cos x}$.

Problem 2.43. $f(x) = x^2 \sin^2(2x^2)$.

Problem 2.44. $y = \sec \sqrt{x^2 + 1}$.

Problem 2.45. $y = e^{-5x} \cosh 3x$.

Problem 2.46. $y = \arctan \sqrt{x^2 - 1}$.

Problem 2.47. $f(x) = \cos(e^{3x-4})$.

Problem 2.48. $y = (\sec^2 x - \tan^2 x)^{45}$.

Problem 2.49. $h(t) = e^{-\tan\left(\frac{t}{3}\right)}$.

Problem 2.50. $f(y) = 3^{\log_7(\arcsin y)}$.

Problem 2.51. $y = \dfrac{\ln(x^2 - 3x + 8)}{\sec(x^2 + 7x)}$.

Problem 2.52. $g(x) = \cosh\left(\dfrac{\sqrt{x+1}}{x^2 - 3}\right)$.

Problem 2.53. $f(x) = \dfrac{3^{\cos x}}{e^{2x}}$.

Problem 2.54. $f(x) = \dfrac{5^{\cos x}}{\sin x}$.

Problem 2.55. $f(x) = \dfrac{xe^x}{\cos(x^2)}$.

Problem 2.56. $f(x) = \dfrac{x \ln x}{\sin(2x + 3)}$.

Problem 2.57. $f(x) = \dfrac{e^{\cos x}}{x^2 + x}$.

Problem 2.58. $f(x) = \dfrac{5^{\log_2(\pi)} e^{\cos x}}{\cos x}$.

Problem 2.59. $f(x) = x^6 e^x + 5e^{2x}$.

Problem 2.60. $f(x) = \cos(\sin(x^3))$.

Problem 2.61. $g(z) = \sqrt{\log(|2z + 1|)}$.

Problem 2.62. $y = \dfrac{x \sec x}{5 \ln(x^2)}$.

Problem 2.63. $y = \sinh(7^{2x} - \sqrt{x})$.

2.2.4   *Higher-order derivatives*

Use the differentiation rules to solve the following problems.

Problem 2.64. Let f be a function differentiable on $\mathbb{R}$ and such that $f(x) = \frac{x^4-16}{x-2}$ for all $x \neq 2$. Evaluate $f^{(4)}(2)$.

Problem 2.65. Given $g(t) = \sqrt{4t^2 + 3}$, determine $g''(t)$.

Problem 2.66. Given $g(\theta) = \cos\left(\frac{\theta}{2}\right)$, determine $g^{(11)}(\theta)$.

Problem 2.67. Given $y = \frac{1}{x} + \cos 2x$, determine $\frac{d^5y}{dx^5}$.

Problem 2.68. Determine $\frac{d^2y}{dx^2}$ if $y = \arctan(x^2)$.

Problem 2.69. Determine y'' if $y = e^{e^x}$.

Problem 2.70. Determine $f'''(x)$ where $f(x) = \sinh(2x)$.

Problem 2.71. Given $y = \frac{-4}{x+2}$, determine y''.

2.2.5   *Not-so-routine differentiation*

Use your knowledge of limits, the Squeeze Theorem, the definition of derivative, and the differentiation rules to solve the following problems.

Problem 2.72. If a function $y = g(x)$ is differentiable on its domain and $f(x) = (\cos x)e^{g(x)}$, what is $f'(x)$?

Problem 2.73. If a function $y = g(x)$ is positive and differentiable on its domain and $f(x) = (\sin x)\ln g(x)$, what is $f'(x)$?

Problem 2.74. Differentiate the function $y = f(x) = g(x^3)$ if $g(x) = \frac{1}{x^2}$.

Problem 2.75. Suppose that functions f and g are differentiable and such that $f(g(x)) = x$ and $f'(x) = 1 + (f(x))^2$. Show that $g'(x) = \frac{1}{1+x^2}$.

Problem 2.76. If $y = \frac{\sqrt{x^2+1}-\sqrt{x^2-1}}{\sqrt{x^2+1}+\sqrt{x^2-1}}$, show that $\frac{dy}{dx} = 2x - \frac{2x^3}{\sqrt{x^4-1}}$.

Problem 2.77. If f and g are two functions for which $f'(x) = g(x)$ and $g'(x) = f(x)$ for all x, prove that $f^2 - g^2$ must be a constant function.

Problem 2.78. Show that if f and g are twice differentiable functions, then $(f(x)g(x))'' = f''(x)g(x) + 2f'(x)g'(x) + f(x)g''(x)$.

Problem 2.79. Suppose that f and g are differentiable functions on $\mathbb{R}$ and that a function h is defined by $h = f \cdot g$. You are given that $h(1) = 24$, $g(1) = 6$, $f'(1) = -2$, and $h'(1) = 20$, evaluate $g'(1)$.

Problem 2.80. Given $F(x) = (f(g(x)))^2$, $g(1) = 2$, $g'(1) = 3$, $f(2) = 4$, and $f'(2) = 5$, evaluate $F'(1)$.

Problem 2.81. Suppose that functions F and G are such that $F(3) = 2$, $G(3) = 4$, $G(0) = 3$, $F'(3) = -1$, $G'(3) = 0$, and $G'(0) = 0$.

(1) If $S(x) = \dfrac{F(x)}{G(x)}$, evaluate $S'(3)$.

(2) If $T(x) = F(G(x))$, evaluate $T'(0)$.

(3) If $U(x) = \ln(F(x))$, evaluate $U'(3)$.

Problem 2.82. Suppose that functions f and g are such that $f(2) = 3$, $g(2) = 4$, $g(0) = 2$, $f'(2) = -1$, $g'(2) = 0$, and $g'(0) = 3$.

(1) If $h(x) = f(x) \cdot g(x)$, evaluate $h'(2)$.

(2) If $k(x) = \dfrac{f(x)}{g(x)}$, evaluate $k'(2)$.

(3) If $p(x) = 2f(x) - 3g(x)$, evaluate $p'(2)$.

(4) If $r(x) = f(g(x))$, evaluate $r'(0)$.

Problem 2.83. If a function g is continuous but not differentiable at $x = 0$, $g(0) = 8$, and a function f is defined by $f(x) = xg(x)$, evaluate $f'(0)$.

Problem 2.84. Let $f(x) = \log_a(3x^2 - 2)$. For what value of a is $f'(1) = 3$?

Problem 2.85. Let $f(x) = e^{a(x^2-1)}$. For what value of a is $f'(1) = 4$?

Problem 2.86. Let $f(x) = \ln(x^2 + 1)^a$. For what value of a is $f'(2) = 2$?

Problem 2.87. The function

$$f(x) = \begin{cases} e^x, & \text{if } x \leq 1 \\ mx + b, & \text{if } x > 1 \end{cases}$$

is differentiable at $x = 1$. Determine the values for the constants m and b.

Problem 2.88. Determine the values of A and B that make

$$f(x) = \begin{cases} x^2 + 1, & \text{if } x \geq 0 \\ A \sin x + B \cos x, & \text{if } x < 0 \end{cases}$$

differentiable at $x = 0$.

Problem 2.89.

(1) Write down the formula for the derivative of the function $f(x) = \tan x$, $x \in \left(-\frac{\pi}{2}, \frac{\pi}{2}\right)$. State how you could use the derivative of the sine and cosine functions to derive this formula.
(2) Use the formula from part (1) to derive the formula for the derivative of the function $g(x) = \arctan x$.
(3) Use formulas indicated in parts (1) and (2) to evaluate the derivative of $h(x) = \tan(x^2) + \arctan(x^2)$ at $x = \frac{\sqrt{\pi}}{2}$.

Problem 2.90. Let $f(x) = x^2 \sin\left(\frac{1}{x}\right)$ if $x \neq 0$, and $f(0) = 0$. Evaluate $f'(0)$, or state why it does not exist.

Problem 2.91. Let I be a bounded function on $\mathbb{R}$ and $f(x) = x^2 I(x)$. Show that f is differentiable at $x = 0$.

Problem 2.92. Let $f(x) = 2x + \cos x$. Explain why $f(x)$ is an increasing function for all x. Then conclude that $g(x) = f^{-1}(x)$ exists. Evaluate $g'(0)$.

2.3   Implicit and Logarithmic Differentiation

The methods of implicit and logarithmic differentiation of finding a derivative of a function are based on the Chain Rule.

2.3.1 *Implicit differentiation*

Use implicit differentiation and differentiation rules to determine the derivative $\frac{dy}{dx}$ of the following implicitly given functions. When requested, evaluate the derivative at a given number.

Problem 2.93. $x^y = y^x$.

Problem 2.94. $x^2 + 2xy^2 = 3y + 4$.

Problem 2.95. $\ln(x + y) = xy - y^3$.

Problem 2.96. $e^x + e^y = x^e + y^e + e^3$.

Problem 2.97. $x^5 + y^5 = 5xy$.

Problem 2.98. $x^2 + x^3 y - xy^2 = 1$.

Problem 2.99. $e^y - 3^x = x \sinh y$.

Problem 2.100. $\sinh x - \cos y = x^2 y$.

Problem 2.101. $\ln y + x = x^2 + x \cos y$.

Problem 2.102. $x \sin y + y \cos x = 1$.

Problem 2.103. $\sin y + x = x^2 + x \cos y$.

Problem 2.104. $\cos x + e^y = x^2 + \tan^{-1} y$.

Problem 2.105. $y \cos(x^2) = x \sin(y^2)$.

Problem 2.106. $2y^{\frac{2}{3}} = 4y^2 \ln x$, $y > 0$.

Problem 2.107. Evaluate $h'(0)$ if $h(x) + x \cos(h(x)) = x^2 + 3x$.

Problem 2.108. Use the derivative of the exponential function $y = e^x$ to show that $\frac{d}{dx}(\ln x) = \frac{1}{x}$.

Problem 2.109. Use the derivative of the sine function $y = \sin x$ to show that $\frac{d}{dx}\left(\sin^{-1} x\right) = \frac{1}{\sqrt{1-x^2}}$.

2.3.2 *Logarithmic differentiation*

Use the method of logarithmic differentiation and differentiation rules to determine the derivative of the following functions. When requested, evaluate the derivative at a given number.

Problem 2.110. Determine $y'(u)$ as a function of u alone, where
$$y(u) = \left(\frac{(u+1)(u+2)}{(u^2+1)(u^2+2)} \right)^{1/3}.$$

Problem 2.111. $y = \dfrac{(x+2)^{3\ln x}}{(x^2+1)^{1/2}}.$

Problem 2.112. $y = \dfrac{e^{2x}}{(x^2+1)^3(1+\sin x)^5}.$

Problem 2.113. $f(x) = \dfrac{(x-1)^2}{(x+1)^3}.$

Problem 2.114. $y = \dfrac{x^5 e^{x^3} \sqrt[3]{x^2+1}}{(x+1)^4}.$

Problem 2.115. $g(x) = \dfrac{(2+\cos(3x^2))e^{\pi x}}{3\sqrt{x}}.$

Problem 2.116. $g(x) = \ln\left(\sqrt{x^2+1}\,\sin^4 x \right).$

Problem 2.117. If $y = \dfrac{\sqrt{1+2x}\,\sqrt[4]{1+4x}\,\sqrt[6]{1+6x}\ldots\sqrt[100]{1+100x}}{\sqrt[3]{1+3x}\,\sqrt[5]{1+5x}\,\sqrt[7]{1+7x}\ldots\sqrt[101]{1+101x}}$, determine $y'(0)$.

Problem 2.118. $f(x) = x^x.$

Problem 2.119. $y = x^{3x}.$

Problem 2.120. $y = x^{\sqrt{x}}.$

Problem 2.121. $y = x^{\ln x}.$

Problem 2.122. $y = x^{\sinh x}.$

Problem 2.123. $f(x) = x^{x^2}$.

Problem 2.124. $f(x) = x^{\arctan x}$.

Problem 2.125. $y = (\ln x)^{\cos x}$.

Problem 2.126. $f(x) = (x+2)^x$.

Problem 2.127. $y = x^{e^x}$.

Problem 2.128. $y = x^{\arcsin x}$.

Problem 2.129. $y = \cos^x(x)$.

Problem 2.130. $f(x) = (\tan x)^{\ln x + x^2}$.

2.4 Tangent Lines

Use your knowledge of the geometric interpretation of the derivative, differentiation rules, and differentiation techniques to solve the following tangent line related problems.

2.4.1 *Tangent lines to explicitly defined functions*

Use differentiation rules to solve the following tangent line problems that involve a function.

Problem 2.131.

(1) Evaluate $\lim\limits_{h \to 0} \dfrac{f(1+h)-f(1)}{h}$ where $f(x) = \dfrac{3x+1}{x-2}$.
(2) What does the result in (1) tell you about the tangent line to the graph of the function $y = f(x)$ at $x = 1$?
(3) Determine an equation of the tangent line to the graph of the function $y = f(x)$ at $x = 1$.

Problem 2.132. Suppose that functions f and g are such that $f(2) = 3$, $g(2) = 4$, $g(0) = 2$, $f'(2) = -1$, $g'(2) = 0$, and $g'(0) = 3$. Determine an equation of the tangent line to the graph of the function $F = f \circ g$ at the point $(0, F(0))$.

Problem 2.133. Determine an equation of the tangent line to the graph of the function $y = 4 + 3x + \cosh x$ at the point $(0, 5)$.

Problem 2.134. At what point on the graph of the function $y = \sinh x$ does the tangent line have a slope of 1?

Problem 2.135. Find the point(s) on the graph of the function $y = x^3$ where a line through the point $(4, 0)$ is tangent to the graph.

Problem 2.136. Determine an equation of the tangent line to the graph of the function $y = x^2 - 1$ that has a slope equal to -4.

Problem 2.137. Determine all tangent lines to the graph of the function $y = x^2$ passing through the point $(0, -3)$.

Problem 2.138. Determine an equation of the tangent line to the graph of the function $y = \arcsin x$ when $x = -\dfrac{1}{\sqrt{2}}$.

Problem 2.139. Determine an equation of the line that is tangent to the graph of the function $y = \sqrt{x} - \dfrac{1}{\sqrt{x}}$ at $x = 1$.

Problem 2.140. Determine an equation of the line that is tangent to the graph of the function $y = e^{x^2}$ at $x = 2$.

2.4.2 *Tangent lines to implicitly defined functions*

Use implicit differentiation to solve the following tangent line problems that involve implicitly defined functions.

Problem 2.141. Evaluate the slope of the tangent line to the curve defined by the equation $y + x \ln y - 2x = 0$ at the point $(0.5, 1)$.

Problem 2.142. Evaluate the slope of the tangent line to the curve defined by the equation $xy = 6e^{2x-3y}$ at the point $(3, 2)$.

Problem 2.143. Evaluate the slope of the tangent line to the curve defined by the equation $e^y \ln(x + y) + 1 = \cos(xy)$ at the point $(1, 0)$.

Problem 2.144. The curve defined by the equation $x \sin y + y \sin x = \pi$ passes through the point $P = \left(\dfrac{\pi}{2}, \dfrac{\pi}{2}\right)$.

(1) Evaluate the slope of the tangent line through P.
(2) Write the equation of the tangent line through P.

Problem 2.145. A special case of the so-called *devil's curve* is defined by the equation $y^2(y^2 - 4) = x^2(x^2 - 5)$.

(1) Determine the y-intercepts of the given devil's curve.
(2) Determine $\frac{dy}{dx}$ at the point (x, y).
(3) Determine an equation for the tangent line to the curve at $(\sqrt{5}, 0)$.

Problem 2.146. Let C denote the circle whose equation is $(x-5)^2 + y^2 = 25$. Notice that the point $(8, -4)$ lies on the circle C. Determine an equation of the line that is tangent to C at the point $(8, -4)$.

Problem 2.147. Determine an equation of the tangent line to the curve defined by the equation $\sin(x + y) = y^2 \cos x$ at $(0, 0)$.

Problem 2.148. Write the equation of the line tangent to the curve defined by the equation $\sin(x + y) = xe^{x+y}$ at the origin $(0, 0)$.

Problem 2.149. Write the equation of the line tangent to the curve defined by the equation $\sin(x + y) = 2x - 2y$ at the point (π, π).

Problem 2.150. A curve is defined by the equation $x^2 - xy + y^2 = 3$. Show that the tangent lines are parallel to each other at the points where the curve crosses the x-axis.

2.4.3 *Not-so-routine tangent line problems*

Use your knowledge of differentiation to solve the following tangent line problems.

Problem 2.151. Consider a tangent line to the graph of the function $y = 17 - 2x^2$ that passes through the point $(3, 1)$.

(1) Provide a diagram of this situation. Can you draw two tangent lines to the curve passing through the point $(3, 1)$?
(2) Evaluate the slopes of those tangent lines.
(3) Determine equations of those tangent lines.

Problem 2.152. The following figure depicts a circle with radius 1 inscribed in the parabola $y = x^2$. Determine the centre of the circle.

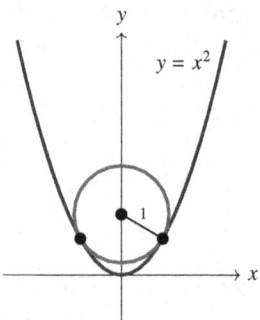

Problem 2.153. Determine all values of c such that the line $y = \frac{3x}{2} + 6$ is tangent to the graph of the function $y = c\sqrt{x}$.

Problem 2.154. Determine all values of a such that the tangent line to the graph of the function $f(x) = \frac{2x^2}{1+x^2}$ at the point $(a, f(a))$ is parallel to the tangent line to the graph of the function $g(x) = \tan^{-1}(x)$ at $(a, g(a))$.

Problem 2.155. Consider the function $y = 3x^2 + 2x - 10$ on the interval $[1, 5]$. Reason whether this function has a tangent line with a slope of 20 anywhere on this interval.

Problem 2.156. Let C be the graph of the function $y = (x - 1)^3$ and L be the line $3y + x = 0$.

(1) Determine equations of all lines that are tangent to C and perpendicular to L.
(2) Draw a labelled diagram showing the graph C, the line L, and the line(s) of your solution to part (1). For each line of your solution, mark on the diagram the point where it is tangent to C and the point where it is perpendicular to L.

Problem 2.157. Determine the x-coordinates of all points on the curve defined by the equation $x^3 + y^3 = 6xy$ where the tangent line to the curve is vertical.

Problem 2.158. A curve is defined by the equation $x^2y + ay^2 = b$, where a and b are constants.

(1) Determine $\frac{dy}{dx}$.

(2) Determine constants a and b if the point $(1, 1)$ is on the curve and the tangent line to the curve at $(1, 1)$ is $4x + 3y = 7$.

Problem 2.159. Let L be any tangent to the curve $\sqrt{x} + \sqrt{y} = \sqrt{k}$, $k > 0$. Show that the sum of the x-intercept and the y-intercept of L is k.

Problem 2.160. Consider the set of all line segments that are tangent to the curve $x^{\frac{2}{3}} + y^{\frac{2}{3}} = k^{\frac{2}{3}}$, $k \neq 0$, and cut off by the coordinate axes. Show that the length of all such segments is constant. What is this length?

2.5 Related Rates

Use the method of related rates to solve the following problems. Show all your work, do not forget to use appropriate units, and clearly explain your reasoning.

2.5.1 *Pythagorean relationship*

In this section, quantities are related by the Pythagorean Theorem.

Problem 2.161. A ladder $12\,$m long leans against a wall. The foot of the ladder is pulled away from the wall at the rate of $0.5\,$m/min. At what rate is the top of the ladder falling when the foot of the ladder is $4\,$m from the wall?

Problem 2.162. A rocket R is launched vertically and it is tracked from a radar station S that is 4 miles away from the launch site at the same height above sea level. At a certain instant after launch, R is 5 miles away from S and the distance from R to S is increasing at a rate of 3600 miles/h. Compute the vertical speed v of the rocket at this instant.

Problem 2.163. A boat is pulled into a dock by means of a rope attached to a pulley on the dock, as shown in the following figure. The rope is attached to the bow of the boat at a point $1\,$m below the pulley. If the rope is pulled through the pulley at a rate of $1\,$m/s, at what rate will the boat be approaching the dock when $10\,$m of rope is out.

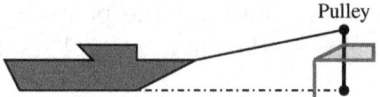

Pulley

Problem 2.164. An airplane is flying horizontally at an altitude of $y = 3\,\text{km}$ and at a speed of $480\,\text{km/h}$ passes directly above an observer on the ground. How fast is the distance D, from the observer to the airplane, increasing $30\,\text{s}$ later?

Problem 2.165. A kite is rising vertically at a constant speed of $2\,\text{m/s}$ from a location at ground level that is $8\,\text{m}$ away from the person handling the string of the kite.

(1) Let z be the distance from the kite to the person. Evaluate the rate of change of z with respect to time t when $z = 10\,\text{m}$.
(2) Let x be the angle in radians the string makes with the horizontal line parallel to the ground. Evaluate the rate of change of x with respect to time t when the kite is $y = 6\,\text{m}$ above ground.

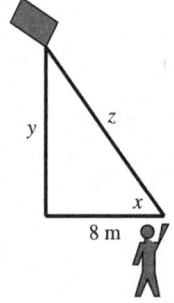

Problem 2.166. A girl flying a kite holds the string $5\,\text{ft}$ above the ground level and lets the string out at a rate of $2\,\text{ft/s}$ as the kite moves horizontally at an altitude of $105\,\text{ft}$.

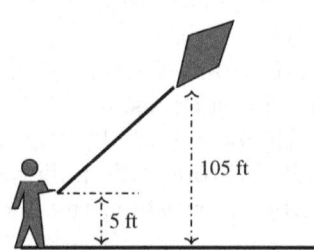

Assuming there is no sag in the string, evaluate the rate at which the kite is moving when 125 ft of string has been let out.

Problem 2.167. A balloon is rising at a constant speed 4 m/s. A boy is cycling along a straight road at a speed of 8 m/s. When he passes under the balloon, it is 36 m above him. How fast is the distance between the boy and the balloon increasing 3 s later?

Problem 2.168. A boy is standing on a road holding a balloon and a girl is running towards him at 3 m/s. At $t = 0$, the girl is 10 m away and the boy releases the balloon that now rises vertically at a speed of 2 m/s. How fast is the distance from the girl to the balloon changing 2 s later?

Problem 2.169. At what rate is the diagonal of a cube changing if its edges are decreasing at a rate of 3 cm/s?

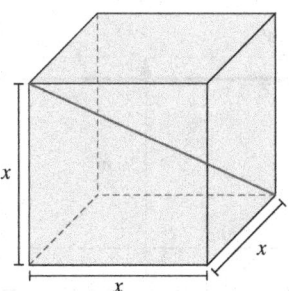

Problem 2.170. A boy starts walking north at a speed of 1.5 m/s and a girl starts running west at a speed of 2 m/s from the same point P at the same time. At what rate is the distance between the boy and the girl increasing 6 s later?

Problem 2.171. At noon of a certain day, ship A is 60 miles due north of ship B. If ship A sails east at a speed of 15 miles/h and ship B sails north at a speed of 12.25 miles/h, determine how rapidly the distance between them is changing 4 h later.

Problem 2.172. A ball is about to land at a point A, while two puppies, Koosen and Tilley, are watching. As soon as the ball lands, Koosen starts off 15 m north of A, running at 3 m/s in the direction of A, and Tilley starts off 12 m east of A, running at 2 m/s in the direction of A. At what rate is the distance between the puppies changing when they are 5 m apart?

Problem 2.173. Approaching a right-angled intersection from the north, a police car is chasing a speeding SUV that has turned the corner and is now moving straight east. When the police car is 0.6 km north of the intersection and the SUV is 0.8 kilometers east of the intersection, the police determine with radar that the distance between them and the SUV is increasing at 20 km/h. If the police car is moving at 60 km/h at the instant of measurement, what is the speed of the SUV?

Problem 2.174. A person is running east along a river bank path at a speed of 2 m/s. A cyclist is on a path on the opposite bank cycling west at a speed of 5 m. The cyclist is initially 500 m east of the runner.

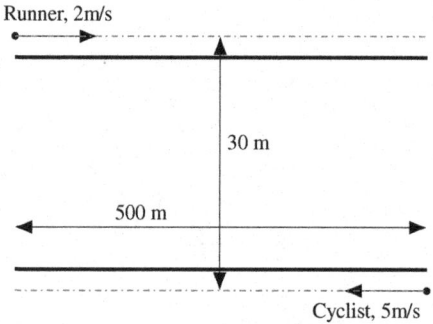

If the paths are 30 m apart, how fast is the distance between the runner and cyclist changing after one minute?

2.5.2 *Trigonometric relationship*

In this section, quantities are related by trigonometric ratios.

Problem 2.175. A particle is moving along the parabola $y = x^2 - 4x + 8$. As a function of time t in seconds, its x-coordinate is $x(t) = -2t^3 + 5$. Let l be the line joining the origin $(0,0)$ to the particle. Determine how quickly the angle between the x-axis and the line l is changing when $x = 3$.

Problem 2.176. A ladder 15 ft long rests against a vertical wall. Its top slides down the wall while its bottom moves away along the level ground at a speed of 2 ft/s. How fast is the angle between the top of the ladder and the wall changing when the angle is $\frac{\pi}{3}$ radians?

Problem 2.177. A person A, situated at the edge of the 10 m wide river, observes the passage of a speed boat going downstream. The boat travels exactly through the middle of the river.

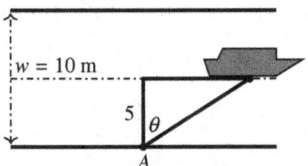

When the boat is at $\theta = 60°$, the observer measures the rate of change of the angle θ to be 2 radians/s. What is the speed v of the speed boat at that instant?

Problem 2.178. A high speed train is traveling at 3 km/min along a straight track. The train is moving away from a movie camera that is located 0.5 km from the track.

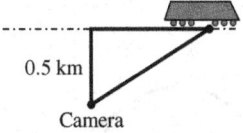

The camera keeps turning so as to always point at the front of the train. How fast, in radians per minute, is the camera rotating when the train is 1 kilometer from the camera?

Problem 2.179. An airplane flying horizontally at a constant height of 1000 m above a fixed radar station. At a certain instant the angle of elevation θ from the station is $\frac{\pi}{4}$ radians and decreasing at a rate of 0.1 radians/s. What is the speed of the aircraft at this moment?

Problem 2.180. A helicopter takes off from a point 80 m away from an observer located on the ground and rises vertically at 2 m/s. At what rate is the elevation angle of the observer's line of sight to the helicopter changing when the helicopter is 60 m above the ground?

Problem 2.181. A lighthouse is located on a small island 3 km off-shore from the nearest point P on a straight shoreline. The light of the lighthouse makes 4 revolutions/min. How fast is the light beam moving along the shoreline when it is shining on a point 1 km along the shoreline from P?

Problem 2.182. You are riding on a Ferris wheel of diameter 20 m. The wheel is rotating at 1 revolution/m. How fast are you rising when you are at the point P in the following figure, that is if you are 6 m horizontally away from the vertical line passing the centre of the wheel?

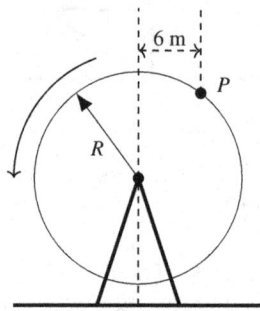

Problem 2.183. The following figure shows a rotating wheel with radius 40 cm and a connecting rod AP with length 1.2 m.

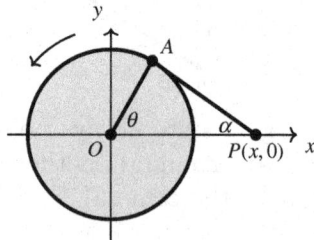

The pin P slides back and forth along the x-axis as the wheel rotates counterclockwise at a rate of 360 revolutions/min.

(1) Evaluate the angular velocity of the connecting rod $\dfrac{d\alpha}{dt}$ in radians per second when $\theta = \dfrac{\pi}{3}$.
(2) Express the distance $x = |OP|$ in terms of θ.
(3) Determine an expression for the velocity of the pin P in terms of θ.

2.5.3 *Similar triangles*

In this section, use similar triangles to establish the relationship between quantities.

Problem 2.184. A child 1.5 m tall walks towards a lamppost on the level ground at the rate of 0.25 m/s. The lamppost is 10 m high. How fast is the length of the child's shadow decreasing when the child is 4 m from the post?

Problem 2.185. A light moving at 2 m/s approaches a 2 m tall man standing 4 m from a wall. The light is 1 m above the ground.

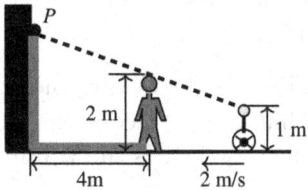

How fast is the tip P of the man's shadow moving up the wall when the light is 8 m from the wall?

Problem 2.186. A coffee filter has the shape of an inverted cone with a fixed top radius R and height H. Water drains out of the filter at a rate of $10\,\text{cm}^3/\text{min}$. When the depth h of the water is 8 cm, the depth is decreasing at a rate of 2 cm/min.

(1) Express the volume of water in the cone as a function of the depth of the water only.
(2) What is the ratio $\dfrac{R}{H}$ at the instant when $h = 8\,\text{cm}$?

2.5.4 *Area and volume*

In this section, use the common area or volume formulas to establish the relationship between quantities.

Problem 2.187. An oil slick on a lake is surrounded by a floating circular containment boom. As the boom is pulled in, the circular containment area shrinks, all the while maintaining the shape of a circle. If the boom is pulled in at the rate of 5 m/min, at what rate is the containment area shrinking when it has a diameter of 100 m?

Problem 2.188. A rectangle is inscribed in the unit circle so that its sides are parallel to the coordinate axes. Let $\theta \in \left(0, \frac{\pi}{2}\right)$ be the angle between the positive ray of the x-axis and the ray with the initial point at the origin and passing through the top-right vertex P of the rectangle. Suppose that the angle θ is increasing at the rate of 2 radians/s. Suppose also that all lengths are measured in centimeters. At what rate is the area of the rectangle changing when $\theta = \frac{\pi}{3}$? Is the area increasing or decreasing at that moment? Why?

Problem 2.189. Consider a cube of variable size, i.e., its edge length is increasing. Assume that the volume of the cube is increasing at the rate of $10\,\mathrm{cm}^3/\mathrm{min}$. How fast is the surface area increasing when the edge length is $8\,\mathrm{cm}$?

Problem 2.190. Consider a cube of variable size, i.e., its edge length is increasing. Assume that the surface area of the cube is increasing at the rate of $6\,\mathrm{cm}^2/\mathrm{min}$. How fast is the volume increasing when the edge length is $5\,\mathrm{cm}$?

Problem 2.191. The volume of an ice cube is decreasing at a rate of $5\,\mathrm{m}^3/\mathrm{s}$. What is the rate of change of the side length at the instant when the side lengths are $2\,\mathrm{m}$?

Problem 2.192. The height of a rectangular box is increasing at a rate of $2\,\mathrm{m}/\mathrm{s}$, while the volume is decreasing at a rate of $5\,\mathrm{m}^3/\mathrm{s}$. If the base of the box is a square, at what rate is one of the sides of the base decreasing at the moment when the base area is $64\,\mathrm{m}^2$ and the height is $8\,\mathrm{m}$?

Problem 2.193. Sand is pouring out of a tube at $1\,\mathrm{cm}^3/\mathrm{s}$ and is forming a pile in the shape of a cone. The height of the cone is equal to the radius of the circle at its base. How fast is the sand pile rising when it is $2\,\mathrm{m}$ high?

Problem 2.194. A water tank is in the shape of a cone with its apex pointing downwards. The tank has an upper radius of $3\,\mathrm{m}$ and is $5\,\mathrm{m}$ high. To begin with, the tank is full of water, but at time $t = 0\,\mathrm{s}$, a small hole at the apex is opened and the water begins to drain. When the height of water in the tank has dropped to $3\,\mathrm{m}$, the water is flowing out at $2\,\mathrm{cm}^3/\mathrm{s}$. At what rate is the water level dropping then?

Problem 2.195. A conical tank with an upper radius of 4 m and a height of 5 m drains into a cylindrical tank with radius of 4 m and a height of 5 m. The water level in the conical tank is dropping at a rate of 0.5 m/min when the water level of the conical tank is 3 m.

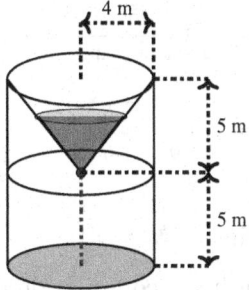

At what rate is the water level in the cylindrical tank rising at that point?

2.5.5 *Miscellaneous relationships*

In this section, quantities are related in miscellaneous ways.

Problem 2.196. A bug is walking on the parabola $y = x^2$. At what point on the parabola are the x- and y-coordinates changing at the same rate?

Problem 2.197. A ship is moving on the surface of the ocean in a straight line at 10 km/h. At the same time, an enemy submarine maintains a position directly below the ship while diving at an angle of 30° below the horizontal.

(1) Provide a diagram of this problem situation. Use x as the line of movement of the ship in kilometers and y as the depth of the submarine in kilometers.
(2) How fast is the submarine depth increasing?

2.6 Answers, Hints, and Solutions

Solution 2.1. By definition, $f'(3) = \lim_{h \to 0} \dfrac{\frac{1}{2(3+h)-1} - \frac{1}{2(3)-1}}{h} = \lim_{h \to 0} \dfrac{-2}{5(5+2h)} = -0.08.$

Solution 2.2. $f'(2) = \lim\limits_{x \to 2} \dfrac{x + \frac{1}{x} - \frac{5}{2}}{x-2} = \lim\limits_{x \to 2} \dfrac{(2x-1)(x-2)}{2x(x-2)} = \dfrac{3}{4}$.

Solution 2.3. $f'(1) = \lim\limits_{x \to 1} \dfrac{3x^2 - 4x + 1}{x-1} = 2$.

Solution 2.4. $f'(4) = \lim\limits_{x \to 4} \dfrac{\sqrt{5-x} - 1}{x-4} = -0.5$. Rationalize the numerator.

Solution 2.5. $f'(x) = \lim\limits_{h \to 0} \dfrac{\sqrt{x+h} - \sqrt{x}}{h} = \dfrac{1}{2\sqrt{x}}$. Rationalize the numerator.

Solution 2.6. $f'(x) = \lim\limits_{h \to 0} \dfrac{\frac{x+h}{(x+h)-2} - \frac{x}{x-2}}{h} = -\dfrac{2}{(x-2)^2}$. Follow through on the subtraction by finding a common denominator.

Solution 2.7. $F'(0) = \lim\limits_{h \to 0} \dfrac{\frac{f(h)\sin^2 h}{h}}{h} = \lim\limits_{h \to 0} \dfrac{f(h)\sin^2 h}{h^2} = f(0)$. Recall that $\lim\limits_{h \to 0} \dfrac{\sin h}{h} = 1$ and that, since f is continuous, $\lim\limits_{h \to 0} f(h) = f(0)$.

Solution 2.8. By definition, $\lim\limits_{h \to 0} \dfrac{\sin^7\left(\frac{\pi}{6} + \frac{h}{2}\right) - \frac{1}{2^7}}{h} =$

$\dfrac{d}{dx}\left(\sin^7 \dfrac{x}{2}\right)\Big|_{x = \frac{\pi}{3}} = \dfrac{7}{2} \cdot \sin^6 \dfrac{\pi}{6} \cdot \cos\dfrac{\pi}{6} = \dfrac{7\sqrt{3}}{256}$.

Solution 2.9. $\lim\limits_{x \to 0} \dfrac{f(1+x) - f(1)}{x} = f'(1) = g(1) = 2$.

Solution 2.10. $\dfrac{d}{dx}\left(\sqrt{x} + x^7\right)\Big|_{x=1} = \dfrac{15}{2}$.

Solution 2.11. $f(x) = x^6$, $a = 2$.

Solution 2.12. Suppose that $f : I \to \mathbb{R}$ is a continuous function and that $a \in I$ is such that f is differentiable at each $x \in (a - \varepsilon, a) \cup (a, a + \varepsilon)$, for some $\varepsilon > 0$, but is not differentiable at a. We distinguish two common cases:

If both limits $\lim\limits_{x \to a^+} \dfrac{f(x) - f(a)}{x-a}$ and $\lim\limits_{x \to a^-} \dfrac{f(x) - f(a)}{x-a}$ exist but are not equal, we say that the graph of f has a corner at the point $(a, f(a))$.

If $\lim\limits_{x \to a} |\dfrac{f(x) - f(a)}{x-a}| = \infty$ we say that the line $x = a$ is a vertical tangent to the graph of the function $y = f(x)$. We say that the graph of f has a cusp at the point $(a, f(a))$.

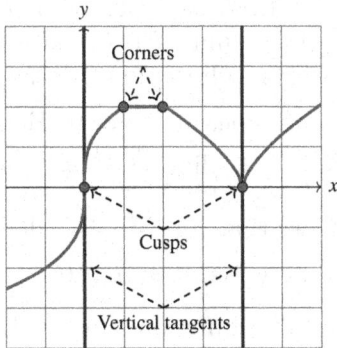

Solution 2.13. We graph the function f to determine the answers.

(1) (a) No; (b)Yes.
(2) (a) Yes; (b) Yes.
(3) (a) No; (b) No.

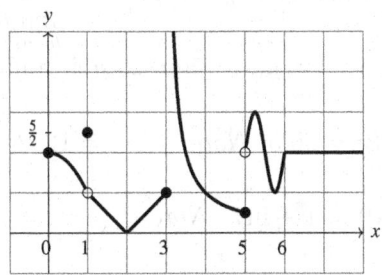

Solution 2.14.

(1) Check that $\lim\limits_{x \to 1^-} f(x) = \lim\limits_{x \to 1^+} f(x) = f(1)$.

(2) Establish that $\lim\limits_{x \to 1^-} \dfrac{\frac{5+x}{2}-3}{x-1} = 0.5$ and $\lim\limits_{x \to 1^+} \dfrac{(2+\sqrt{x})-3}{x-1} = 0.5$ to conclude $f'(1) = 0.5$. See the following figure.

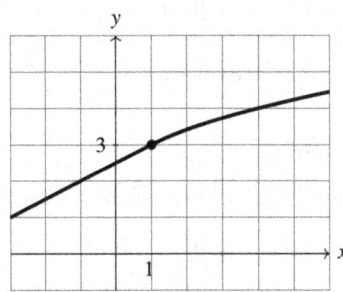

Solution 2.15. Observe that, as a polynomial, f is continuous on the interval $(-\infty, 0)$. As a sum of two continuous functions, f is continuous on $(0, \infty)$. From $\lim\limits_{x \to 0^-} f(x) = \lim\limits_{x \to 0^+} f(x) = 1 = f(0)$, it follows that f is continuous at $x = 0$. Hence, f is continuous on $\mathbb{R}$. From $\lim\limits_{x \to 0^+} \frac{f(x) - f(0)}{x - 0} = \lim\limits_{x \to 0^+} \frac{\sin(x)}{x} = 1$ and $\lim\limits_{x \to 0^-} \frac{f(x) - f(0)}{x - 0} = \lim\limits_{x \to 0^-} (x + 2) = 2$, it follows that $\lim\limits_{x \to 0} \frac{f(x) - f(0)}{x - 0}$ does not exist, i.e., that the function f is not differentiable at $x = 0$.

Solution 2.16. False. The function $f(x) = |x - 1|$ is continuous everywhere, but is not differentiable at $x = 1$. The graph of the function f has a corner at the point $(1, f(1)) = (1, 0)$.

Solution 2.17. True. Recall that, if a function g is differentiable at $x = x_0$, then $\lim\limits_{h \to 0} \frac{g(x_0 + h) - g(x_0)}{h}$ exists. By the Product Rule of limits,

$$\lim_{h \to 0} (g(x_0 + h) - g(x_0)) = \lim_{h \to 0} \frac{g(x_0 + h) - g(x_0)}{h} \cdot \lim_{h \to 0} h = g'(x_0) \cdot 0 = 0.$$

Therefore, $\lim\limits_{h \to 0} g(x_0 + h) = g(x_0)$, and so g is continuous at $x = x_0$.

Solution 2.18. $f'(2) = 96.5$. *Note:* $f'(x) = 12x^3 + \dfrac{1}{x}$.

Solution 2.19. $f'(0) = 1 + \ln 5$. *Note:* $f'(x) = \dfrac{2x+1}{\sqrt{1-(x^2+x)^2}} + 5^x \ln 5$.

Solution 2.20. $y' = -\dfrac{10 \cdot 7^{1-5x} \ln 7}{\sin\left(2 \cdot 7^{1-5x}\right)}$.

Solution 2.21. $y' = -\dfrac{4e^{-4x}}{\sqrt{1-e^{8x}}}, \; x < 0$.

Solution 2.22. $y' = \dfrac{2}{\sqrt{x}} \sinh \sqrt{x} \cdot e^{4 \cosh \sqrt{x}}$.

Solution 2.23. $y' = -2x e^{\cos x^2} \sin x^2$.

Solution 2.24. $y' = x^{19} \left(20 \arctan x + \dfrac{x}{1+x^2} \right)$.

Solution 2.25. $y' = 6(2x + 1)^2$.

Solution 2.26. $y' = \sec(\sinh x) \tan(\sinh x) \cosh x$.

Solution 2.27. $f'(x) = (6x - 3x^2 - 1)e^{-x}$.

Solution 2.28. $g'(z) = \dfrac{z \cos \sqrt{z^2+1}}{\sqrt{z^2+1}}$.

Solution 2.29. $h'(y) = -\dfrac{y \tan y + 1}{2y} \sqrt{\dfrac{\cos y}{y}}$.

Solution 2.30. $f'(x) = \dfrac{1-x^2}{(x^2+1)^2}$.

Solution 2.31. $f'(x) = \dfrac{1}{2\sqrt{x}(1+x)}$.

Solution 2.32. $f'(x) = \dfrac{5(x^{10}-1)}{2x^6}$. *Note*: Use the identity $\cosh(\ln t) = \dfrac{t^2+1}{2t}$.

Solution 2.33. $f'(x) = 3 \cdot 10^{3x} \cdot \ln 10$.

Solution 2.34. $f'(x) = x^9(10 \tanh x + x \operatorname{sech}^2 x)$.

Solution 2.35. $y' = \dfrac{(2x^2-1)\sin x - x \cos x}{x^2 \sin^2 x} \cdot e^{x^2+1}$.

Solution 2.36. $f'(x) = -3 \tan 3x$.

Solution 2.37. $f'(x) = 2^{2x+1} \ln 2 - \dfrac{4x}{3\sqrt[3]{x^2+1}}$.

Solution 2.38. $f'(x) = 4x \tan(x^2) \cdot \sec^2\left(x^2\right)$.

Solution 2.39. $f'(x) = 5 + 5x^4 + 5^x \ln 5 + \dfrac{1}{5\sqrt[5]{x^4}}$.

Solution 2.40. $f'(x) = \coth x$.

Solution 2.41. $f'(x) = (\cos x - x \sin x)e^{x \cos x}$.

Solution 2.42. $f'(x) = \dfrac{1}{1+\cos x}$.

Solution 2.43. $f'(x) = 2x \sin^2(2x^2) + 4x^3 \sin(4x^2)$.

Solution 2.44. $y' = \dfrac{x \sec \sqrt{x^2+1} \tan \sqrt{x^2+1}}{\sqrt{x^2+1}}$.

Solution 2.45. $y' = -(e^{-2x} + 4e^{-8x})$.

Solution 2.46. $y' = \dfrac{1}{x\sqrt{x^2-1}}$.

Solution 2.47. $f'(x) = -3e^{3x-4}\sin(e^{3x-4})$.

Solution 2.48. $y' = 0$.

Solution 2.49. $h'(t) = -\dfrac{1}{3}\sec^2\left(\dfrac{t}{3}\right) \cdot e^{-\tan\left(\frac{t}{3}\right)}$.

Solution 2.50. $f'(y) = \dfrac{\ln 3 \cdot (\arcsin y)^{\frac{\ln 3}{\ln 7}-1}}{\ln 7 \cdot \sqrt{1-y^2}}$. Note that $f(y) = \left(3^{\log_3(\arcsin y)}\right)^{\frac{1}{\log_3 7}} = (\arcsin y)^{\frac{\ln 3}{\ln 7}}$.

Solution 2.51. $y' = \dfrac{(2x-3)\cos(x^2+7x)}{x^2-3x+8} - (2x+7)\sin(x^2+7x)\ln(x^2 - 3x + 8)$. Note that $y = \ln(x^2 - 3x + 8) \cdot \cos(x^2 + 7x)$.

Solution 2.52. $g'(x) = -\dfrac{3x^2+4x+3}{2\sqrt{x+1}(x^2-3)^2}\sinh\left(\dfrac{\sqrt{x+1}}{x^2-3}\right)$.

Solution 2.53. $f'(c) = -3^{\cos x} \cdot (\sin x \cdot \ln 3 + 2) \cdot e^{-2x}$.

Solution 2.54. $f'(x) = -5^{\cos x} \cdot \left(\ln 5 + \cos x \cdot \csc^2 x\right)$.

Solution 2.55. $f'(x) = \dfrac{(1+x)\cos(x^2)+2x^2\sin(x^2)}{\cos^2(x^2)} \cdot e^x$.

Solution 2.56. $f'(x) = \dfrac{(1+\ln x)\sin(2x+3)-2x\ln x\cos(2x+3)}{\sin^2(2x+3)}$.

Solution 2.57. $f'(x) = -\dfrac{(x^2+x)\sin x+2x+1}{(x^2+x)^2} \cdot e^{\cos x}$.

Solution 2.58. $f'(x) = \dfrac{5^{\log_2(\pi)}e^{\cos x}(1-\cos x)\sin x}{\cos^2 x}$.

Solution 2.59. $f'(x) = (x + 6)x^5 e^x + 10e^{2x}$.

Solution 2.60. $f'(x) = -3x^2\sin(\sin(x^3))\cos(x^3)$.

Solution 2.61. $g'(z) = \dfrac{1}{(2z+1)\ln 10\sqrt{\log|2z+1|}}$.

Solution 2.62. $y' = \dfrac{((1+x\tan x)\ln x - 1)\sec x}{10\ln^2 x}$.

Solution 2.63. $y' = \left(2\cdot 7^{2x}\ln 7 - \dfrac{1}{2\sqrt{x}}\right)\cosh(7^{2x} - \sqrt{x})$.

Solution 2.64. $f^{(4)}(2) = 0$. *Note:* $f(x) = (x+2)(x^2+4)$.

Solution 2.65. $g''(t) = 12(4t^2+3)^{-3/2}$.

Solution 2.66. $g^{(11)}(\theta) = \dfrac{1}{2^{11}}\cdot\sin\left(\dfrac{\theta}{2}\right)$.

Solution 2.67. $\dfrac{d^5y}{dx^5} = -\dfrac{5!}{x^6} - 2^5\sin 2x$.

Solution 2.68. $\dfrac{d^2y}{dx^2} = \dfrac{2(1-3x^4)}{(1+x^4)^2}$.

Solution 2.69. $y'' = e^{e^x+x}\cdot(e^x+1)$.

Solution 2.70. $f'''(x) = 8\cosh(2x)$.

Solution 2.71. $y'' = -8(x+2)^{-3}$.

Solution 2.72. $f'(x) = (-\sin x + g'(x)\cdot\cos x)\cdot e^{g(x)}$.

Solution 2.73. $f'(x) = (\cos x)\ln g(x) + \dfrac{g'(x)\cdot\sin x}{g(x)}$.

Solution 2.74. $f'(x) = -\dfrac{6}{x^7}$.

Solution 2.75. Observe that $g(x) = f^{-1}(x)$ and use the formula for derivative of an inverse function.

Solution 2.76. Rationalize the denominator to obtain $y = x^2 - \sqrt{x^4-1}$.

Solution 2.77. Differentiate $F(x) = (f(x))^2 - (g(x))^2$ and conclude that, since $F'(x) = 0$, $F = f^2 - g^2$ must be a constant function.

Solution 2.78. Differentiate the function $f\cdot g$ twice.

Solution 2.79. From $h(1) = f(1)g(1)$ and $h'(1) = f'(1)g(1) + f(1)g'(1)$, it follows that $g'(1) = 8$.

Solution 2.80. $2f(g(1)) \cdot f'(g(1)) \cdot g'(1) = 120$.

Solution 2.81. (1) $S'(3) = \dfrac{F'(3)G(3) - F(3)G'(3)}{[G(3)]^2} = -\dfrac{1}{4}$. (2) $T'(0) = F'(G(0)) \cdot G'(0) = 0$. (3) $U'(3) = \dfrac{F'(3)}{F(3)} = -\dfrac{1}{2}$.

Solution 2.82. (1) $h'(2) = -4$. (2) $k'(2) = -\dfrac{1}{4}$. (3) $p'(2) = -2$. (4) $r'(0) = -3$.

Solution 2.83. $f'(0) = \lim\limits_{x \to 0} \dfrac{xg(x)}{x} = g(0) = 8$. Observe that, since g is not differentiable, we cannot use the Product Rule.

Solution 2.84. $a = e^2$.

Solution 2.85. $a = 2$.

Solution 2.86. $a = 2.5$.

Solution 2.87. $m = e$, $b = 0$. Solve $\lim\limits_{x \to 1^-} e^x = \lim\limits_{x \to 1^+} (mx + b)$ and $\lim\limits_{x \to 1^-} \dfrac{e^x - e}{x - 1} = \lim\limits_{x \to 1^+} \dfrac{mx + b - (m+b)}{x-1}$, for m and b.

Solution 2.88. $A = 0$, $B = 1$. Solve $\lim\limits_{x \to 0^-} (A \sin x + B \cos x) = \lim\limits_{x \to 0^+} (x^2 + 1)$ and $\lim\limits_{x \to 0^-} \dfrac{(A \sin x + B \cos x) - B}{x - 0} = \lim\limits_{x \to 0^+} \dfrac{(x^2 + 1) - 1}{x - 0}$ for A and B.

Solution 2.89. (1) $f'(x) = \sec^2 x$. This follows from $\tan x = \dfrac{\sin x}{\cos x}$ by using the Quotient Rule. (2) From $g(x) = \arctan x = f^{-1}(x)$, it follows that $f(g(x)) = x$, $x \in \mathbb{R}$. Hence, $f'(g(x)) \cdot g'(x) = 1$. We conclude that $g'(x) = \cos^2(g(x)) = \dfrac{1}{1 + (\tan(g(x)))^2} = \dfrac{1}{1 + x^2}$. (3) From $h'(x) = 2x \sec(x^2) + \dfrac{2x}{1 + x^4}$, it follows that $h'\left(\dfrac{\sqrt{\pi}}{2}\right) = 2\sqrt{\pi} + \dfrac{16\sqrt{\pi}}{16 + \pi^2}$.

Solution 2.90. $f'(0) = 0$. Recall that any extension of the function $x \mapsto \sin\left(\dfrac{1}{x}\right)$ is not continuous, and therefore not differentiable, at $x = 0$, so we cannot use the Product Rule. Observe that, for $h \neq 0$, $\left| \dfrac{f(h) - f(0)}{h} \right| = \left| \dfrac{h^2 \sin \frac{1}{h}}{h} \right| = \left| h \sin \dfrac{1}{h} \right| \leq |h|$. Use the Squeeze Theorem to conclude that f is differentiable at $x = 0$.

Solution 2.91. $f'(0) = 0$. Observe that we do not know if the function I is differentiable at $x = 0$, so we cannot use the Product Rule. Since the function I is bounded, there is $M > 0$ such that $|I(x)| \leq M$, for all $x \in \mathbb{R}$. Then, for any $h \neq 0$, $\left| \frac{f(h)-f(0)}{h} \right| = \left| \frac{h^2 I(h)}{h} \right| = |hI(h)| \leq M|h|$. Use the Squeeze Theorem to conclude that f is differentiable at $x = 0$.

Solution 2.92. From $f'(x) = 2 - \sin x > 0$ for all $x \in \mathbb{R}$, it follows that f is a monotone increasing function and, thus, one-to-one. Let $g(0) = \alpha$. From $f(\alpha) = f(g(0)) = 0$, it follows that $2\alpha + \cos \alpha = 0$. Use technology to obtain $\alpha \approx -0.45$ radians. Then $g'(0) = \dfrac{1}{f'(g(0))} = \dfrac{1}{2-\sin \alpha} \approx 0.41$.

Solution 2.93. From $y \ln x = x \ln y$ it follows that $\dfrac{dy}{dx} \ln x + \dfrac{y}{x} = \ln y + \dfrac{x}{y} \dfrac{dy}{dx}$. Hence, $\dfrac{dy}{dx} = \dfrac{y(x \ln y - y)}{x(y \ln x - x)}$.

Solution 2.94. $y' = \dfrac{2x+2y^2}{3-4xy}$.

Solution 2.95. $y' = \dfrac{xy+y^2-1}{3y^3+3xy^2-x^2-xy+1}$.

Solution 2.96. $y' = \dfrac{ex^{e-1}-e^x}{e^y-ey^{e-1}}$.

Solution 2.97. $\dfrac{dy}{dx} = \dfrac{y-x^4}{y^4-x}$.

Solution 2.98. $y' = \dfrac{y^2-3x^2y-2x}{x^3-2xy}$.

Solution 2.99. $\dfrac{dy}{dx} = \dfrac{3^x \ln 3 + \sinh y}{e^y - x \cosh y}$.

Solution 2.100. $\dfrac{dy}{dx} = \dfrac{\cosh x - 2xy}{x^2 - \sin y}$.

Solution 2.101. $y' = \dfrac{y(2x+\cos y - 1)}{xy \sin y + 1}$.

Solution 2.102. $y' = \dfrac{y \sin x - \sin y}{x \cos y + \cos x}$.

Solution 2.103. $y' = \dfrac{2x+\cos y - 1}{x \sin y + \cos y}$.

Solution 2.104. $y' = \dfrac{(y^2+1)(2x+\sin x)}{(1+y^2)e^y-1}$.

Solution 2.105. $y' = \dfrac{2xy\sin(x^2)+\sin(y^2)}{\cos(x^2)-2xy\cos(y^2)}$.

Solution 2.106. $y' = -\dfrac{3\sqrt[4]{2\ln x}}{8x\ln^2 x}$. Observe that $y = (2\ln x)^{-3/4}$, $x > 1$, so implicit differentiation is not needed.

Solution 2.107. $h'(0) = 2$.

Solution 2.108. Write $y = \ln x$ and conclude that $e^y = x$, with $y \in \mathbb{R}$. From $e^y \cdot \dfrac{dy}{dx} = 1$, conclude that $\dfrac{dy}{dx} = e^{-y} = \dfrac{1}{x}$.

Solution 2.109. *Method 1*: Write $y = \sin^{-1}(x)$ and conclude that $\sin y = x$ with $y \in \left(-\dfrac{\pi}{2}, \dfrac{\pi}{2}\right)$. Then differentiate implicitly and solve for $\dfrac{dy}{dx}$. *Method 2*: Let $f(x) = \sin x$, $x \in \left(-\dfrac{\pi}{2}, \dfrac{\pi}{2}\right)$. Then, for $x \in (-1, 1)$, we have $(f^{-1})'(x) = \dfrac{1}{\cos(f^{-1}(x))}$. Suppose that $x \in (0, 1)$. Let $\alpha = f^{-1}(x)$. Consider the right triangle with hypotenuse of length 1 and an angle measured α radians. The length of the leg opposite to the angle α equals $\sin \alpha = x$, which implies $\dfrac{d}{dx}\left(\sin^{-1} x\right) = \dfrac{1}{\sqrt{1-x^2}}$.

Solution 2.110.
$$y'(u) = \frac{1}{3}\left(\frac{1}{u+1} + \frac{1}{u+2} - \frac{2u}{u^2+1} - \frac{2u}{u^2+2}\right)\left(\frac{(u+1)(u+2)}{(u^2+1)(u^2+2)}\right)^{1/3}.$$

Solution 2.111. $y' = \left(\dfrac{3\ln(x+2)}{x} + \dfrac{3\ln x}{x+2} - \dfrac{x}{x^2+1}\right)\dfrac{(x+2)^{3\ln x}}{(x^2+1)^{1/2}}$.

Solution 2.112. $y' = \left(2 - \dfrac{6x}{x^2+1} - \dfrac{5\cos x}{1+\sin x}\right)\dfrac{e^{2x}}{(x^2+1)^3(1+\sin x)^5}$.

Solution 2.113. $f'(x) = \left(\dfrac{2}{x-1} - \dfrac{3}{x+1}\right)\dfrac{(x-1)^2}{(x+1)^3} = \dfrac{(5-x)(x-1)}{(x+1)^4}$.

Solution 2.114. $y' = \left(\dfrac{5}{x} + 3x^2 + \dfrac{2x}{3(x^2+1)} - \dfrac{4}{x+1}\right)\dfrac{x^5 e^{x^3}\sqrt[3]{x^2+1}}{(x+1)^4}$.

Solution 2.115. $g'(x) = \left(\pi - \dfrac{6x\sin(3x^2)}{2+\cos(3x^2)} - \dfrac{1}{2x}\right)\dfrac{(2+\cos(3x^2))e^{\pi x}}{3\sqrt{x}}$.

Solution 2.116. $g'(x) = \dfrac{x}{x^2+1} + 4\cot x.$

Solution 2.117. $y'(0) = 0$. Note that $\ln y = \frac{1}{2}\ln(1+2x) - \frac{1}{3}\ln(1+3x) + \cdots - \frac{1}{101}\ln(1+101x).$

Solution 2.118. $f'(x) = (\ln x + 1)\cdot x^x.$

Solution 2.119. $y' = 3(\ln x + 1)\cdot x^{3x}.$

Solution 2.120. $f'(x) = \dfrac{\ln x + 2}{2\sqrt{x}}\cdot x^{\sqrt{x}}.$

Solution 2.121. $y' = 2\ln x \cdot x^{\ln x - 1}.$

Solution 2.122. $y' = \left(\cosh x \cdot \ln x + \dfrac{\sinh x}{x}\right)\cdot x^{\sinh x}.$

Solution 2.123. $f'(x) = (2\ln x + 1)\cdot x^{x^2+1}.$

Solution 2.124. $f'(x) = \left(\dfrac{\ln x}{1+x^2} + \dfrac{\arctan x}{x}\right)\cdot x^{\arctan x}.$

Solution 2.125. $y' = \left(\dfrac{\cos x}{x\ln x} - \sin x \ln\ln x\right)(\ln x)^{\cos x}.$

Solution 2.126. $f'(x) = \left(\ln(x+2) + \dfrac{x}{x+2}\right)\cdot (x+2)^x.$

Solution 2.127. $y' = \left(e^x \ln x + \dfrac{e^x}{x}\right)\cdot x^{e^x}.$

Solution 2.128. $y' = \left(\dfrac{\ln x}{\sqrt{1-x^2}} + \dfrac{\arcsin x}{x}\right)\cdot x^{\arcsin x}.$

Solution 2.129. $f'(x) = (\ln(\cos x) - x\tan x)\cos^x(x).$

Solution 2.130. $f'(x) = \left((x^{-1}+2x)\ln\tan x + \dfrac{\ln x + x^2}{\sin x \cos x}\right)(\tan x)^{\ln x + x^2}.$

Solution 2.131. (1) $f'(1) = -7$. (2) The slope of the tangent line is -7. (3) $y = -7x + 3.$

Solution 2.132. $y = -3x + 3$. Note that $F(0) = f(g(0)) = 3$ and $F'(0) = f'(g(0))\cdot g'(0) = -3.$

Solution 2.133. $y = 3x + 5$.

Solution 2.134. Solve $y' = \cosh x = 1$. The point is $(0,0)$.

Solution 2.135. Solutions of the equation $-a^3 = 3a^2(4 - a)$ are $a = 0$ and $a = 6$. The points are $(0,0)$ and $(6, 216)$.

Solution 2.136. $y = -4x - 5$.

Solution 2.137. $y = \pm 2\sqrt{3}x - 3$.

Solution 2.138. $y = \sqrt{2}x + 1 - \dfrac{\pi}{4}$.

Solution 2.139. $y = x - 1$.

Solution 2.140. $y = e^4(4x - 7)$.

Solution 2.141. $\left. \dfrac{dy}{dx} \right|_{\left(\frac{1}{2}, 1\right)} = \dfrac{4}{3}$.

Solution 2.142. $y'(3) = \dfrac{10}{21}$.

Solution 2.143. From $e^y \left(\dfrac{dy}{dx} \ln(x + y) + \dfrac{1 + \frac{dy}{dx}}{x+y} \right) = -\left(y + \dfrac{dy}{dx} \right) \cdot$
$\sin(xy)$, it follows that $\left. \dfrac{dy}{dx} \right|_{x=1} = -1$.

Solution 2.144. (1) $\left. \dfrac{dy}{dx} \right|_{\left(\frac{\pi}{2}, \frac{\pi}{2}\right)} = -1$. (2) $x + y = \pi$.

Solution 2.145. (1) $(0,0)$, $(0, \pm 2)$. If $x = 0$, then we need to solve the equation $y^2(y^2 - 4) = 0$. (2) $y' = \dfrac{x(2x^2 - 5)}{2y(y^2 - 2)}$. (3) Observe that the point $(\sqrt{5}, 0)$ lies on the curve. By (2), $y' \to \infty$ when $(x, y(x) < 0) \to (\sqrt{5}, 0)$, and $y' \to -\infty$ when $(x, y(x) > 0) \to (\sqrt{5}, 0)$. Hence, the line $x = \sqrt{5}$ is a vertical tangent line to devil's curve at the point $(\sqrt{5}, 0)$.

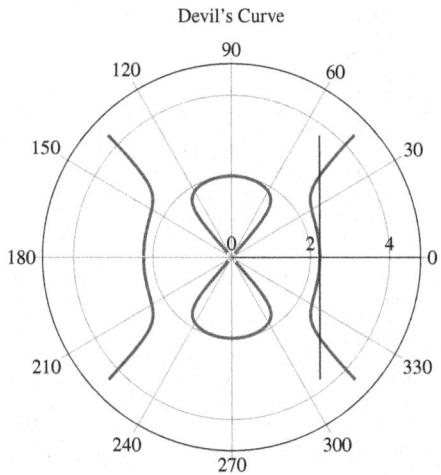

Devil's Curve

Solution 2.146. $y = \dfrac{3x}{4} - 10$.

Solution 2.147. $x + y = 0$.

Solution 2.148. $y = 0$.

Solution 2.149. $y = \dfrac{x}{3} + \dfrac{2\pi}{3}$.

Solution 2.150. The graph crosses the x-axis at the points $(\pm\sqrt{3}, 0)$. The claim follows from the fact that $2x - y - xy' + 2yy' = 0$ implies that if $x = \pm\sqrt{3}$ and $y = 0$, then $y' = 2$.

Solution 2.151. Observe that the point $(3, 1)$ does not lie on the graph of the function $y = 17 - 2x^2$.

(1) There are two lines passing through $(3, 1)$ tangent to the graph, as shown in the following figure.
(2) All lines through $(3, 1)$ are given by $y = 1 + k(x - 3)$. Determine all k so that the equation $17 - 2x^2 = 1 + k(x - 3)$ has a unique solution. This yields $k = -16$ or $k = -8$.
(3) $y = -16x + 49$ and $y = -8x + 25$.

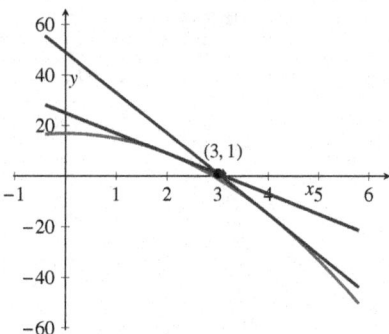

Solution 2.152. The centre is at $\left(0, \dfrac{5}{4}\right)$. The question is to find $a \in \mathbb{R}^+$ such that the circle $x^2 + (y-a)^2 = 1$ and the parabola $y = x^2$ have the same tangent lines at their two intersection points. In other words, we are looking for $a \in \mathbb{R}$ such that, for some $x \in \mathbb{R}\backslash\{0\}$, each of the following conditions is met: $x^2 + (x^2 - a)^2 = 1$ and $x + 2x(x^2 - a) = 0$.

Solution 2.153. $c = \pm 6$.

Solution 2.154. $a = 2 \pm \sqrt{3}$.

Solution 2.155. Yes. Solve the equation $y' = 20$.

Solution 2.156. Note that $y' = 3(x - 1)^2$.

(1) Two lines, none of them horizontal, are perpendicular to each other if the product of their slopes is -1. Thus, to determine all points on the graph C where the tangent line is perpendicular to the line L, solve the equation $-\dfrac{1}{3} \cdot 3(x - 1)^2 = -1$ to obtain $x = 0$ or $x = 2$. The lines are $y = 3x - 1$ and $y = 3x - 5$.

(2)

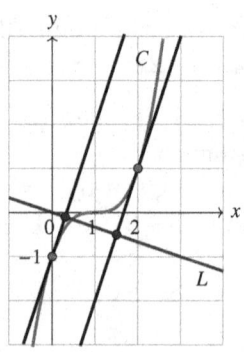

Solution 2.157. $x = 0$, $x = 2\sqrt[3]{4}$. Note that for a tangent line to be vertical it is necessary that $\frac{dy}{dx} = \frac{6y-3x^2}{3y^2-6x}$ is not defined at the point of interest. We solve $3y^2 - 6x = 0$ for x and substitute into the original expression. This yields $y^6 = 16y^3$ and gives two points on the curve $(0,0)$ and $(2\sqrt[3]{4}, 2\sqrt[3]{2})$ with possible vertical tangent lines. Since $2 \cdot 2\sqrt[3]{2} - (2\sqrt[3]{4})^2 = 4\sqrt[3]{2} - 8\sqrt[3]{2} \neq 0$, we conclude that $\lim\limits_{(x,y)\to(2\sqrt[3]{4}, 2\sqrt[3]{2})} \left| \frac{dy}{dx} \right| = \infty$ and that there is a vertical tangent line at $(2\sqrt[3]{4}, 2\sqrt[3]{2})$.

For each real number $m > 1$, let A_m be the intersection point of the curve and the ray $y = mx$, $x > 0$. Then $A_m = \left(\frac{6m}{1+m^3}, \frac{6m^2}{1+m^3} \right)$ and $A_m \to (0,0)$ when $m \to \infty$. From $\frac{dy}{dx}\Big|_{A_m} = \frac{m(m^3-2)}{2m^3-1} \to \infty$ when $m \to \infty$, we conclude that there is a vertical tangent line at $(0,0)$.

Solution 2.158. (1) $\frac{dy}{dx} = -\frac{2xy}{x^2+2ay}$. (2) Solve the system of equations $1 + a = b$, $-\frac{2}{1+2a} = -\frac{4}{3}$ to obtain $a = \frac{1}{4}$ and $b = \frac{5}{4}$.

Solution 2.159. From $\frac{dy}{dx} = 1 - \sqrt{\frac{k}{x}} = -\sqrt{\frac{y}{x}}$ when $x > 0$ and $y > 0$, it follows that the tangent line L to the curve at any of its points (a, b) with $a > 0$ is given by $y - b = -\sqrt{\frac{b}{a}} \cdot (x - a)$. The sum of the x-intercept and the y-intercept of L is given by $(a + \sqrt{ab}) + (b + \sqrt{ab}) = (\sqrt{a} + \sqrt{b})^2 = k$.

Solution 2.160. From $\frac{2}{3\sqrt[3]{x}} + \frac{2y'}{3\sqrt[3]{y}} = 0$, we conclude that $y' = -\sqrt[3]{\frac{y}{x}}$, $x, y \neq 0$. Thus, the tangent line to the curve at the point (a, b) with $a, b \neq 0$ on the curve is given by $y - b = -\sqrt[3]{\frac{b}{a}}(x - a)$. Its x- and y-intercepts are $\left(a + \sqrt[3]{ab^2}, 0 \right)$ and $\left(0, b + \sqrt[3]{a^2b} \right)$. The square of the length of the segment of the tangent line cut off by the coordinate axes is $\left(a + \sqrt[3]{ab^2} \right)^2 + \left(b + \sqrt[3]{a^2b} \right)^2 = a^2 + 2a\sqrt[3]{ab^2} + b\sqrt[3]{a^2b} + b^2 + 2b\sqrt[3]{a^2b} + a\sqrt[3]{ab^2} = \left(\sqrt[3]{a^2} + \sqrt[3]{b^2} \right)^3 = k^2$. The length of each segment is $|k|$.

Solution 2.161. Let $x = x(t)$ be the distance between the foot of the ladder and the wall and $y = y(t)$ be the distance between the top of the ladder and the ground at time t in minutes. It is given that, at any time t, $\frac{dx}{dt} = \frac{1}{2}$ m/min. From $x^2 + y^2 = 144$, it follows that $x\frac{dx}{dt} + y\frac{dy}{dt} = 0$. Thus, when $x(t) = 4$ m, we have that $y(t) = 8\sqrt{2}$ m and $4 \cdot \frac{1}{2} + 8\sqrt{2} \cdot \frac{dy}{dt} = 0$. Therefore, the top of the ladder is sliding at the rate of $\frac{dy}{dt} = -\frac{\sqrt{2}}{8}$ m/min. Alternatively, we say that the ladder is falling at the rate of $\frac{\sqrt{2}}{8}$ m/min.

Solution 2.162. Let $x = x(t)$ be the height of the rocket and $y = y(t)$ be the distance between the rocket and the radar station at time t in hours. It is given that $x^2 = y^2 - 16$ at any time t. Thus, $x\frac{dx}{dt} = y\frac{dy}{dt}$ at any time t. At the instant when $y = 5$ miles and $\left.\frac{dy}{dt}\right|_{y=5} = 3600$ miles/h, we have that $x = 3$ miles. We conclude that, when R is $y = 5$ miles away from S, the vertical speed of the rocket is $v = \left.\frac{dx}{dt}\right|_{y=5} = 6000$ miles/h.

Solution 2.163. Let $x = x(t)$ be the distance between the dock and the bow of the boat and $y = y(t)$ be the length of the rope between the pulley and the bow at time t in seconds. Both x and y are measured in meters. It is given that $\frac{dy}{dt} = -1$ m/s. From $x^2 + 1 = y^2$, it follows that $\frac{dx}{dt} = \frac{y}{x}\frac{dy}{dt}$ m/s. Since $y = 10$ implies $x = \sqrt{99}$, we conclude that $\left.\frac{dx}{dt}\right|_{y=10} = -\frac{10}{\sqrt{99}}$ m/s. Hence, when 10 m of rope is out, the boat is approaching the dock at the rate of $\frac{10}{\sqrt{99}}$ m/s.

Solution 2.164. After time t in hours, the plane is $480t$ km away from the point directly above the observer. Thus, at time t, the distance between the observer and the plane is $D = \sqrt{3^2 + (480t)^2}$ km. We differentiate $D^2 = 9 + 230,400t^2$ with respect to t to obtain $D\frac{dD}{dt} = 230,400t$. Since 30 s is the same amount of time as $\frac{1}{120}$ h, it follows that the distance between the observer and the plane after 30 s is $D = 5$ km. At this instant, the distance is increasing at the rate of $\left.\frac{dD}{dt}\right|_{t=\frac{1}{120}} = 384$ km/h.

Solution 2.165. (1) From $z^2 = 64 + 4t^2$, it follows that $zz' = 4t$. If $z = 10$, then $t = 3$ and $z'(3) = 1.2$ m/s. (2) Since the height of the kite after t seconds is $2t$ meters, it follows that $\tan x = \dfrac{2t}{8}$. Thus, $\dfrac{x'}{\cos^2 x} = \dfrac{1}{4}$. If $y = 6$, then $t = 3$ and $\tan x = \dfrac{3}{4}$. It follows that $\cos x = \dfrac{4}{5}$ and, at that instant, the rate of change of x is given by $x' = x'(3) = \dfrac{4}{25}$ m/s.

Solution 2.166. Let $D = D(t)$ be the distance between the girl's hand and the kite in feet at time t in seconds. It is given that $\dfrac{dD}{dt} = 2$ ft/s. From the relationship $D^2 = x^2 + 100^2$, it follows that $D\dfrac{dD}{dt} = x\dfrac{dx}{dt}$. If $D = 125$ ft, then $x = 75$ ft. Therefore, $\dfrac{dx}{dt}\Big|_{D=125} = \dfrac{10}{3} \approx 3.33$ ft/s.

Solution 2.167. Let $x = x(t)$ be the distance in meters between the boy and the balloon at time t in seconds. Then $(x(t))^2 = (8t)^2 + (36 + 4t)^2$. From $2x(t)x'(t) = 128t + 8(36 + 4t)$ and $x(3) = 24\sqrt{5}$ m, it follows that $x'(3) = \dfrac{16}{\sqrt{5}}$ m/s.

Solution 2.168. Let $D = D(t)$ be the distance in meters between the girl and the balloon at time t in seconds. Let $x = x(t)$ and $y = y(t)$ be the horizontal and vertical distances between the girl and the balloon, respectively, at that time. Note that $x(0) = 10$ and assume that $y(0) = 0$. Observe that $\dfrac{dx}{dt} = -3$ m/s and $\dfrac{dy}{dt} = 2$ m/s for $t > 0$. From the relationship $D^2 = x^2 + y^2$, it follows that $D\dfrac{dD}{dt} = x\dfrac{dx}{dt} + y\dfrac{dy}{dt}$. At time $t = 2$, we have $x = 4$, $y = 4$, and $D = 4\sqrt{2}$. Therefore, $\dfrac{dD}{dt}\Big|_{t=2} = -\dfrac{1}{\sqrt{2}}$ m/s.

Solution 2.169. If $x = x(t)$ is the edge length of the cube in centimeters at time t in seconds, then the length of the diagonal is given by $z = \sqrt{3}\, x$. Observe that it is given that $\dfrac{dx}{dt} = -3$ cm/s. It follows that $\dfrac{dz}{dt} = \sqrt{3}\, \dfrac{dx}{dt} = -3\sqrt{3}$ cm/s.

Solution 2.170. The distance between the boy and the girl is $z = \sqrt{x^2 + y^2}$ m, where $x = x(t)$ and $y = y(t)$ are the distances in meters covered by the boy and the girl respectively at time t in seconds. The question is to evaluate $z'(6)$. We differentiate $z^2 = x^2 + y^2$ with respect to t to obtain $zz' = xx' + yy'$. From $x(6) = 9$, $y(6) = 12$, $z(6) = 15$, $x'(t) = 1.5$, and $y'(t) = 2$, it follows that $z'(6) = 2.5$ m/s.

Solution 2.171. The distance between the two ships is $z = \sqrt{x^2 + (60 - y)^2}$ miles, where $x = x(t)$ and $y = y(t)$ are the distances in miles covered by ship A and ship B respectively at time t in hours. The question is to evaluate $z'(4)$. We differentiate $z^2 = x^2 + (60-y)^2$ with respect to t to obtain $zz' = xx' - (60 - y)y'$. From $x(4) = 60$, $y(4) = 49$, $z(4) = 61$, $x'(t) = 15$, and $y'(t) = 12.25$, it follows that $z'(4) = \dfrac{765.25}{61} \approx 12.55$ miles/h.

Solution 2.172. The square of the distance between the two puppies is $z^2 = (12 - x)^2 + (15 - y)^2$ m, where $x = x(t)$ and $y = y(t)$ are the distances in meters covered by Tilley and Koosen respectively at time t in seconds. The question is to evaluate $\left.\dfrac{dz}{dt}\right|_{z=5}$. We differentiate $z^2 = (12 - x)^2 + (15 - y)^2$ with respect to t to obtain $z\dfrac{dz}{dt} = -(12 - x)\dfrac{dx}{dt} - (15 - y)\dfrac{dy}{dt} = -2(12 - x) - 3(15 - y)$. Let T be the instant when the puppies are 5 m apart, i.e., let T be such that $z(T) = 5$. From $25 = (12 - 2T)^2 + (15 - 3T)^2 = 4(6 - T)^2 + 9(5 - T)^2$, we conclude that $T = 4$. From $x(4) = 8$ and $y(4) = 12$, it follows that $\left.\dfrac{dz}{dt}\right|_{z=5} = -\dfrac{17}{5}$ m/s.

Solution 2.173. Let $x = x(t)$ be the distance between the police car and the intersection in kilometers and $y = y(t)$ be the distance between the SUV and the intersection in kilometers at time t in hours. The distance between the two cars is $z = \sqrt{x^2 + y^2}$ km. The question is to evaluate $\left.\dfrac{dy}{dt}\right|_{t=T}$ at the instant $t = T$ when $x = 0.6$ km and $y = 0.8$ km. Recall that $\dfrac{dz}{dt} = 20$ km/h and $\dfrac{dx}{dt} = -60$ km/h. We differentiate $z^2 = x^2 + y^2$ with respect to t to obtain $z\dfrac{dz}{dt} = x\dfrac{dx}{dt} + y\dfrac{dy}{dt}$. Therefore, $\left.\dfrac{dy}{dt}\right|_{t=T} = 70$ km/h.

Solution 2.174. Let $z = z(t)$ be the distance between the runner and cyclist and $x = x(t)$ be the horizontal distance between the two individuals in meters at time t in seconds. Then $z(t)$ can be expressed as $z^2 = x^2 + 900$, which leads to $z\dfrac{dz}{dt} = x\dfrac{dx}{dt}$. The reduction in horizontal distance between the two individuals is given by $\dfrac{dx}{dt} = -5 - 2 = -7$ m/s. Observe that when $t = 1$ minute, then

$x = 500 + (-7) \cdot 60 = 80$ m and $z = 10\sqrt{73}$ m. Therefore, $\dfrac{dz}{dt}\Big|_{t=1\text{min}} = -\dfrac{56}{\sqrt{73}}$ m/s.

Solution 2.175. Let $\alpha(t)$ be the angle between the x-axis and the line l in radians at time t in seconds. Then $\tan \alpha = x - 4 + \dfrac{8}{x}$ and $\sec^2 \alpha \cdot \dfrac{d\alpha}{dt} = \left(1 - \dfrac{8}{x^2}\right) \dfrac{dx}{dt}$. Therefore, $\dfrac{d\alpha}{dt}\Big|_{x=3} = -\dfrac{3}{17}$ rad/s.

Solution 2.176. Let $x = x(t)$ be the distance between the bottom of the ladder and the wall in feet at time t in seconds. It is given that $\dfrac{dx}{dt} = 2$ ft/s at any time t. Let $\theta = \theta(t)$ be the angle between the top of the ladder and the wall in radians, at time t. Then $\sin \theta = \dfrac{x(t)}{15}$. It follows that $\cos \theta \cdot \dfrac{d\theta}{dt} = \dfrac{1}{15} \dfrac{dx}{dt}$. Thus, when $\theta = \dfrac{\pi}{3}$, the rate of change of θ is $\dfrac{d\theta}{dt}\Big|_{\theta=\frac{\pi}{3}} = \dfrac{4}{15}$ rad/s.

Solution 2.177. Let $y = y(t)$ be the distance covered by the boat in meters at time t in seconds. From $y = 5 \tan \theta$, we obtain that $\dfrac{dy}{dt} = 5 \sec^2 \theta \cdot \dfrac{d\theta}{dt}$ at any time t. At the instant when $\theta = 60° = \dfrac{\pi}{3}$ radians, we have that $v = \dfrac{dy}{dt} = 5 \cdot \sec^2 \dfrac{\pi}{3} \cdot 2 = 40$ m/s.

Solution 2.178. Let $x = x(t)$ be the horizontal distance in kilometers of the train with respect to the camera and $D = D(t)$ be the distance in kilometers between the camera and the train at time t in minutes. From the relationship $\tan \theta = \dfrac{x}{0.5}$, it follows that $\dfrac{d\theta}{dt} = 2 \cos^2 \theta \cdot \dfrac{dx}{dt}$. When $D = 1$, then $\cos \theta = 0.5$. Therefore, $\dfrac{d\theta}{dt}\Big|_{D=1} = 1.5$ rad/min.

Solution 2.179. Let $x = x(t)$ be the horizontal distance between the airplane and the radar station at time t in seconds. Then x is the leg in a right-angled triangle opposite to the complement of the elevation angle θ. Since the other leg is of length 1000 m, it follows that $x = 1000 \tan \left(\dfrac{\pi}{2} - \theta\right) = 1000 \cot \theta$. From $\dfrac{d\theta}{dt} = -0.1$ rad/s, we obtain $\dfrac{dx}{dt} = -\dfrac{1000}{\sin^2 \theta} \dfrac{d\theta}{dt} = \dfrac{100}{\sin^2 \theta}$ m/s. Hence, if $\theta = \dfrac{\pi}{4}$, the speed of the plane is $\dfrac{dx}{dt}\Big|_{\theta=\frac{\pi}{4}} = 200$ m/s.

Solution 2.180. Let $\theta = \theta(t)$ be the elevation angle in radians and $H = H(t)$ be the distance between the helicopter and the ground in meters at time t in seconds. From $\tan\theta = \dfrac{H(t)}{80}$ and $\dfrac{dH}{dt} = 2$, it follows that $\dfrac{d\theta}{dt} = \dfrac{\cos^2\theta}{40}$. When $H = 60\,\mathrm{m}$, then $\tan\theta = \dfrac{3}{4}$ and $\cos\theta = \dfrac{4}{5}$. Thus, when the helicopter is $60\,\mathrm{m}$ above the ground, the elevation angle of the observer's line of sight to the helicopter is changing at the rate of $\dfrac{2}{125}$ rad/s.

Solution 2.181. Let the point L represent the lighthouse, let the light beam shine on the point $A = A(t)$ on the shoreline, and let $x = x(t)$ be the distance in kilometers between A and P at time t in minutes. Let $\theta = \theta(t)$ be the measure in radians of $\angle PLA$. It is given that $x = 3\tan\theta$ and $\dfrac{d\theta}{dt} = 8\pi$ rad/min. The question is to evaluate $\dfrac{dx}{dt}$ at the instant when $x = 1$. First, we note that $\dfrac{dx}{dt} = 3\sec^2\theta\,\dfrac{d\theta}{dt}$. Secondly, at the instant when $x = 1$, we have that $\tan\theta = \dfrac{1}{3}$, which implies $\cos\theta = \dfrac{3}{\sqrt{10}}$. Hence, when shining on a point $1\,\mathrm{km}$ away from P, the light beam moves along the shoreline at the rate of $\dfrac{80\pi}{3}$ km/min.

Solution 2.182. Let ℓ be the horizontal ray with the initial point at the centre of the wheel and on the same side of wheel's vertical axis as the point P. Let $\theta = \theta(t) \in [0, 2\pi)$ be the angle between the ray ℓ and the ray with the initial point at the centre of the wheel and passing through the rider's position at time t in minutes. Let $x = x(t)$ be the horizontal distance and $y = y(t)$ be the vertical distance in meters of the rider from the centre of the Ferris wheel at time t. By our choice of the ray ℓ, for all points close to P, $y = 10\sin\theta$. It follows that, for all points close to P, $\dfrac{dy}{dt} = 10\cos\theta \cdot \dfrac{d\theta}{dt}$. When $x = 6$ meters, then $\cos\theta = \dfrac{3}{5}$. The fact that the wheel is rotating at 1 revolution/min means that $\dfrac{d\theta}{dt} = 2\pi$ rad/min. Therefore, $\dfrac{dy}{dt}\Big|_{x=6} = 12\pi$ m/min ≈ 2.26 km/h.

Solution 2.183. (1) Observe that $\dfrac{d\theta}{dt} = 12\pi$ rad/s. Using the Law of Sines, we find that $3\sin\alpha = \sin\theta$. Hence, $3\cos\alpha \cdot \dfrac{d\alpha}{dt} = \cos\theta \cdot \dfrac{d\theta}{dt}$. When $\theta = \dfrac{\pi}{3}$, then $\sin\alpha = \dfrac{\sqrt{3}}{6}$ and $\dfrac{d\alpha}{dt}\Big|_{\theta=\frac{\pi}{3}} = \dfrac{12\pi}{\sqrt{33}}$ rad/s. (2) By the

Law of Cosines, $120^2 = x^2 + 40^2 - 80x \cos\theta$. It follows that $x = 40 \left(\cos\theta + \sqrt{8 + \cos^2\theta}\right)$. (3) $\frac{dx}{dt} = -480\pi \left(1 + \frac{\cos\theta}{\sqrt{8+\cos^2\theta}}\right)$ $\sin\theta$ rad/s.

Solution 2.184. Let $x = x(t)$ be the distance between the lamppost and the child and $s = s(t)$ be the length of the child's shadow in meters at time t in seconds. Using similar triangles, we find that $\frac{10}{1.5} = \frac{x+s}{s}$, which is the same as $s = \frac{3}{17}x$. It follows that $\frac{ds}{dt} = \frac{3}{17}\frac{dx}{dt}$. Since $\frac{dx}{dt} = -0.25$ m/s, we obtain $\frac{ds}{dt}\Big|_{x=4} = -\frac{3}{68}$ m/s.

Solution 2.185. Let $x = x(t)$ be the distance between the light and the wall and $p = p(t)$ be the height of the man's shadow in meters at time t in seconds. Using similar triangles, we find that $\frac{p-1}{1} = \frac{x}{x-4}$. It follows that $\frac{dp}{dt} = -\frac{4}{(x-4)^2}\frac{dx}{dt}$, with $\frac{dx}{dt} = -2$ m/s. Therefore, $\frac{dp}{dt}\Big|_{x=8} = 0.5$ m/s.

Solution 2.186. (1) The volume of the water in the coffee filter is $V = \frac{1}{3}\pi r^2 h$, where $h = h(t)$ is the depth of water and $r = r(t)$ is the radius of the water surface in centimeters at time t in minutes. The radius r can be represented as $r = \frac{Rh}{H}$ using similar triangles. Therefore, $V = \frac{\pi}{3}\frac{R^2 h^3}{H^2}$. (2) From $\frac{dV}{dt} = \pi h^2 \left(\frac{R}{H}\right)^2 \frac{dh}{dt}$, $\frac{dV}{dt} = -10$ cm^3/min, and $\frac{dh}{dt} = -2$ cm/min, it follows that $\frac{R}{H} = \frac{1}{8}\sqrt{\frac{5}{\pi}}$.

Solution 2.187. Let r denote the radius of the circular containment area in meters at time t in minutes. It is given that $\frac{dr}{dt} = -5$ m/min. From the fact that the area at time t is given by $A = r^2\pi$, where $r = r(t)$, it follows that $\frac{dA}{dt} = 2r\pi\frac{dr}{dt} = -10r\pi$ m^2/min. Hence, when $r = 50$ m, then the area shrinks at the rate of 500π m^2/min.

Solution 2.188. The area $A = A(t)$ of the inscribed rectangle is $A = 4xy$ square centimeters, where $x = x(t)$ and $y = y(t)$ are the coordinates of the point P in centimeters at time t in seconds. Recall that $x = \cos\theta$ and $y = \sin\theta$. It follows that $A(t) = 2\sin 2\theta$, so $\frac{dA}{dt} = 4\cos 2\theta \cdot \frac{d\theta}{dt}$. Therefore, $\frac{dA}{dt}\Big|_{\theta=\frac{\pi}{3}} = -4$ cm^2/s. At the instant when $\theta = \frac{\pi}{3}$, since the corresponding rate is negative, the area is decreasing.

Solution 2.189. Let $x = x(t)$ be the edge length in centimeters at time t in minutes. Then the volume is $V = V(t) = x^3$ cm^3 and the surface area is $S = S(t) = 6x^2$ cm^2. It is given that $\frac{dV}{dt} = 3x^2 \frac{dx}{dt} = 10$ cm^3/min at any time t. Therefore, at the instant when $x = 8$ cm, the edge is increasing at the rate of $\frac{5}{96}$ cm/min. This fact together with $\frac{dS}{dt} = 12x \frac{dx}{dt}$ implies that the surface area is increasing at the rate $\frac{dS}{dt}\big|_{x=8} = 5$ cm^2/min.

Solution 2.190. Let $x = x(t)$ be the edge length of the cube in centimeters at time t in minutes. The volume $V = V(t)$ is $V = x^3$ cm^3, which implies $\frac{dV}{dt} = 3x^2 \frac{dx}{dt}$. The surface area $S = S(t)$ is $S = 6x^2$ cm^2, which implies $\frac{dx}{dt} = \frac{1}{12x} \frac{dS}{dt}$. Since $\frac{dS}{dt} = 6$ cm^2/min, we have $\frac{dx}{dt}\big|_{x=5} = \frac{1}{10}$ cm/min. Therefore, $\frac{dV}{dt}\big|_{x=5} = 7.5$ cm^3/min.

Solution 2.191. Let $x = x(t)$ be the edge length of the ice cube in meters at time t in seconds. The volume $V = V(t)$ is $V = x^3$ cube meters, so, from $\frac{dV}{dt} = 3x^2 \frac{dx}{dt}$, it follows that $\frac{dx}{dt} = \frac{1}{3x^2} \frac{dV}{dt}$ m/s. Therefore, $\frac{dx}{dt}\big|_{x=2} = -\frac{5}{12}$ m/s.

Solution 2.192. Let $H = H(t)$ be the height of the box and $x = x(t)$ be the length of a side of the base in meters at time t in seconds. It is given that $\frac{dH}{dt} = 2$ m/s. From $V = V(t) = Hx^2$, it follows that $\frac{dV}{dt} = 2x \frac{dx}{dt} H + x^2 \frac{dH}{dt} = -5$ m^3/s. Therefore, $\frac{dx}{dt}\big|_{x=H=8} = -\frac{133}{128}$ m/s.

Solution 2.193. Let $H = H(t)$ be the height of the pile in meters, $r = r(t)$ be the radius of the base in meters, and $V = V(t)$ be the volume of the cone in cubic meters at time t in seconds. It is given that $H = r$, which implies $V = \frac{H^3 \pi}{3}$. Hence, $\frac{dV}{dt} = H^2 \pi \frac{dH}{dt} = 1$ m^3/s. It follows that $\frac{dH}{dt}\big|_{H=2} = \frac{1}{4\pi}$ m/s.

Solution 2.194. Let $H = H(t)$ be the height of water in meters, $r = r(t)$ be the radius of the water surface in meters, and $V = V(t)$ be the volume of water in the cone in cubic meters at time t in seconds. Since $r = \frac{3H}{5}$, it follows that $V = \frac{3H^3 \pi}{25}$. Therefore, $\frac{dH}{dt}\big|_{H=3} = -\frac{50}{81\pi}$ m/s.

Solution 2.195. Let $V = V(t)$ be the volume of water in the cone in cubic meters, $h = h(t)$ be the height of water in the cone in meters, $w = w(t)$ be the volume of water in the cylinder in cubic meters, and $H = H(t)$ be the height of water in the cylinder in meters at time t in minutes. From $V = \frac{16}{75}\pi h^3$, it follows that $\frac{dV}{dt} = -\frac{dw}{dt} = \frac{16}{25}\pi h^2 \frac{dh}{dt}$.

Hence, $\left.\frac{dw}{dt}\right|_{h=3} = \frac{72\pi}{25}\,\text{m}^3/\text{min}$. From $w = \pi r^2 H = 16H\pi$, it follows that $\left.\frac{dH}{dt}\right|_{h=3} = \frac{1}{16\pi}\left.\frac{dw}{dt}\right|_{h=3} = \frac{9}{50}\,\text{m}/\text{min}$.

Solution 2.196. Solve for x in the system of equations $\frac{dy}{dt} = 2x\frac{dx}{dt}$ and $\frac{dy}{dt} = \frac{dx}{dt}$ to obtain $x = \frac{1}{2}$, $y = \frac{1}{4}$.

Solution 2.197. (1) Let the positive direction of the x-axis be the line of the movement of the ship. If $(x(t), 0)$ is the position of the ship at time t, then the position of the submarine is given by $(x(t), y(t))$ with $y(t) = -\frac{\sqrt{3}}{3}x(t) + c$ for some negative constant c.

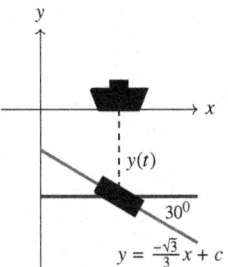

(2) $\frac{dy}{dt} = -\frac{10\sqrt{3}}{3}\,\text{km/h}$.

Chapter 3

Functions and Their Graphs

3.1 Introduction

Use the following definitions, theorems, and properties to solve the problems contained in this chapter.

Absolute Maximum and Minimum. A function $f : D \to \mathbb{R}$ has an *absolute maximum* at $c \in D$ if $f(c) \geq f(x)$ for all $x \in D$. The number $f(c)$ is called the *absolute maximum value* of f on D. A function $f : D \to \mathbb{R}$ has an *absolute minimum* at $c \in D$ if $f(c) \leq f(x)$ for all $x \in D$. The number $f(c)$ is called the *absolute minimum value* of f on D. If $f(c)$ is an absolute maximum (minimum) value of f, then we say that $(c, f(c))$ is an *absolute maximum (minimum) point* on the graph of f. We use the term an *absolute extremum* to refer to both an absolute maximum and an absolute minimum.

Local Maximum and Minimum. A function f has a *local maximum* at c if $f(c) \geq f(x)$, for all x in an open interval containing c. The number $f(c)$ is called a local maximum value of f. Observe that c is necessarily an interior point of the domain. A function f has a *local minimum* at c if $f(c) \leq f(x)$ for all x in an open interval containing c. The number $f(c)$ is called a local minimum value of f. If $f(c)$ is a local maximum (minimum) value of f, then we say that $(c, f(c))$ is a *local maximum (minimum) point* on the graph of f. We use the term a *local extremum* to refer to both a local maximum and a local minimum.

Fermat's Theorem. If a function f has a local maximum or minimum at c and $f'(c)$ exists, then $f'(c) = 0$.

Critical Number. A *critical number* of a function $f : D \to \mathbb{R}$ is a number c in the interior of D such that either $f'(c) = 0$ or $f'(c)$ does not exist. We say that a critical number c of f is a local maximum (minimum) if f has a local maximum (minimum) at c.

Increasing/Decreasing Test. Suppose that f is a differentiable function on an open interval I.

(1) If $f'(x) > 0$ for all $x \in I$, then the function f is increasing on I.

(2) If $f'(x) < 0$ for all $x \in I$, then the function f is decreasing on I.

The First Derivative Test. Suppose that c is a critical number of a continuous function f and that f is differentiable on $(c - \varepsilon, c) \cup (c, c + \varepsilon)$, for some $\varepsilon > 0$.

(1) If the derivative f' changes from positive to negative at c, then the function f has a local maximum at c.

(2) If the derivative f' changes from negative to positive at c, then the function f has a local minimum at c.

(3) If the derivative f' does not change sign at c, then the function f has no local minimum or maximum at c.

Concavity. If the graph of a differentiable function f lies above all of its tangent lines on an interval I, then it is called *concave upward* on I. If the graph of f lies below all of its tangents on I, then it is called *concave downward* on I.

Concavity Test. Suppose that f is a twice differentiable function on an open interval I.

(1) If $f''(x) > 0$ for all $x \in I$, then the graph of f is concave upward on I.

(2) If $f''(x) < 0$ for all $x \in I$, then the graph of f is concave downward on I.

Inflection Point. A point P on the graph of a function $y = f(x)$ is called an *inflection point* if the graph changes from concave upward to concave downward or from concave downward to concave upward at P.

The Second Derivative Test. Suppose f'' is a continuous function near c.

(1) If $f'(c) = 0$ and $f''(c) > 0$, then the function f has a local minimum at c.

(2) If $f'(c) = 0$ and $f''(c) < 0$, then the function f has a local maximum at c.

Asymptotes. An *asymptote* of a curve is a line that, for any positive number d, contains a ray with the property that the distance between any point on the ray and the curve is less than d.

The graph of a function f has a *horizontal asymptote* $y = b$ if $\lim_{x \to \infty} f(x) = b$ or $\lim_{x \to -\infty} f(x) = b$. We say the graph of f approaches the asymptote from above if this limit is b^+ and approaches the asymptote from below if this limit is b^-. It is possible that the graph of f approaches the asymptote neither from above nor below, e.g., through oscillation.

The graph of a function $y = f(x)$ has a *vertical asymptote* $x = a$ if $\lim_{x \to a^-} f(x) = +\infty$ or $-\infty$ and/or $\lim_{x \to a^+} f(x) = +\infty$ or $-\infty$.

A line $L(x) = mx + b$ is a *slant asymptote* of the graph of a function $y = f(x)$ if $\lim_{x \to \infty} [f(x) - L(x)] = 0$ or $\lim_{x \to -\infty} [f(x) - L(x)] = 0$. We say the graph of f approaches the line $y = mx + b$ from above if this limit is 0^+, and approaches $y = mx + b$ from below if this limit is 0^-. It is possible that the graph of f approaches its slant asymptote neither from above nor below.

3.2 Properties of Functions and Their Graphs

3.2.1 *Routine questions*

Use the appropriate definitions and theorems as well as your knowledge of derivatives and their properties to solve the following problems:

Problem 3.1. The function $f(x) = ax^3 + bx$ has a local extreme value of 2 at $x = 1$. Determine whether this extremum is a local maximum or a local minimum.

Problem 3.2. For what values of the constants a and b does the function $f(x) = \ln a + bx^2 - \ln x$ have an extremum value $f(2) = 1$?

Problem 3.3. Prove that the polynomial $f(x) = x^{151} + x^{37} + x + 3$ has neither a local maximum nor a local minimum.

Problem 3.4. For what values of the constants a and b is $(1,6)$ a point of inflection of the graph of the polynomial $y = x^3 + ax^2 + b + 1$?

Problem 3.5. Give an example of a function with one critical number that is the abscissa of an inflection point.

Problem 3.6. Give an example of a function that satisfies $f(-1) = f(1) = 0$ and $f'(x) > 0$, for all x in the domain of f'.

Problem 3.7. Determine the value of a so that $f(x) = \frac{x^2+ax+5}{x+1}$ has a slant asymptote $y = x + 3$.

Problem 3.8. Suppose that f is a function satisfying the following conditions: $f(0) = -3$, $\lim_{x \to 3^+} f(x) = +\infty$, $\lim_{x \to 3^-} f(x) = -\infty$, $\lim_{x \to \infty} f(x) = 2$, $\lim_{x \to -\infty} f(x) = -1$, $f'(x) < 0$ for all $x \neq 3$, $f''(x) > 0$ for all $x > 3$, and $f''(x) < 0$ for all $x < 3$. Sketch a graph of the function f with all asymptotes and intercepts clearly labelled.

Problem 3.9. The graphs of four functions, $f, g, h,$ and k, are shown in the following figure:

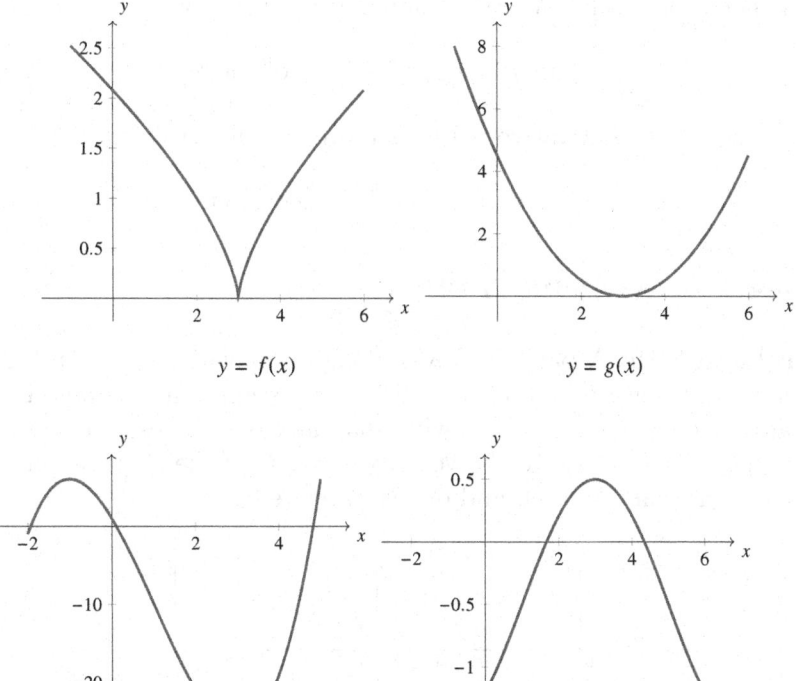

$y = f(x)$

$y = g(x)$

$y = h(x)$

$y = k(x)$

For each question, tick the box if the corresponding function has the stated property. Note that you can tick more than one box.

(1) The derivative of the function is zero at $x = -1$.

$f \; \square \; g \; \square \; h \; \square \; k \; \square$ None of them $\square$

(2) There is a number in the domain of the function where the second derivative does not exist.

$f \; \square \; g \; \square \; h \; \square \; k \; \square$ None of them $\square$

(3) The derivative of the function is negative on the interval $(-2, 0)$.

$f \; \square \; g \; \square \; h \; \square \; k \; \square$ None of them $\square$

(4) The function has a critical number when $x = 3$.

$$f \,\square\, g \,\square\, h \,\square\, k \,\square\, \text{None of them} \,\square$$

(5) The second derivative is positive over the interval $(2, 5)$.

$$f \,\square\, g \,\square\, h \,\square\, k \,\square\, \text{None of them} \,\square$$

3.2.2 *Not-so-routine questions*

Problem 3.10. A particle moves along a line with a position function $s = s(t)$ where s is measured in meters and t in seconds. Four graphs are shown in the following diagram: one corresponds to the function $s = s(t)$, one to the velocity $v = v(t)$ of the particle, one to its acceleration $a = a(t)$, and one is unrelated.

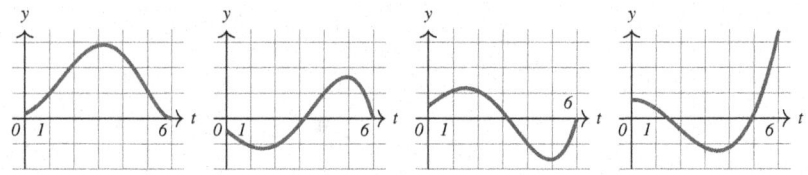

(1) Identify the graphs of s, v, and a.
(2) Determine all time intervals when the particle is slowing down and when it is speeding up.
(3) Estimate the total distance traveled by the particle over the interval $[1, 6]$.

Problem 3.11. Plot the graph of a function f that has only one point of discontinuity on its domain $[-4, \infty)$ and that satisfies the following conditions:

$$\lim_{x \to \infty} f(x) = -2, \quad f'(x) < 0 \text{ for } x \in (-4, -1), \quad f''(x) < 0 \text{ for } x \in (-4, -1),$$
$$\lim_{x \to 0^-} f(x) = \infty, \quad f'(x) > 0 \text{ for } x \in (-1, 0), \quad f''(x) > 0 \text{ for } x \in (-1, 0),$$
$$f(0) = 2, \qquad\qquad f'(x) > 0 \text{ for } x \in (0, 2), \quad f''(x) < 0 \text{ for } x \in (0, 4),$$
$$\qquad\qquad\qquad f'(x) < 0 \text{ for } x \in (2, \infty), \quad f''(x) > 0 \text{ for } x \in (4, \infty).$$

(1) Interpret the above conditions of f by identifying asymptotes, intervals of monotonicity, and intervals of concavity.

(2) Determine all inflection points. For each inflection point $(a, f(a))$, determine if it is possible that $f''(a) = 0$.

(3) Determine all critical numbers. Lastly, draw the graph of f.

Problem 3.12. A function $f : \mathbb{R} \to \mathbb{R}$ is differentiable everywhere and its graph passes through the origin. The graph of the function f' is depicted in the following figure. Assume that the graph of the function f' is below the x-axis and concave downward at all points not shown in this graph.

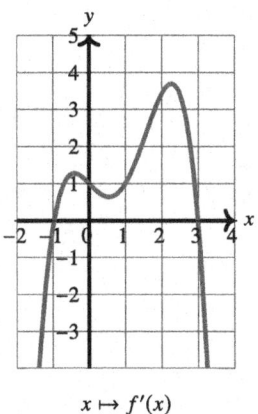

$x \mapsto f'(x)$

(1) Does the function f have a local maximum? If so, determine the approximate coordinates of the local maximum point(s).

(2) Does the function f have a local minimum? If so, determine the approximate coordinates of the local minimum point(s).

(3) Does the function f have any inflection points? If so, determine the approximate coordinates of the inflection point(s).

(4) Determine the interval(s) on which the function f is decreasing.

(5) Determine the interval(s) on which f'' is decreasing.

(6) If f is a polynomial function, what is the least possible degree of f?

Problem 3.13. The graphs of four functions, p', q', r', and s', are shown in the following figure as labelled.

For each question, tick the box if the corresponding function, p, q, r, or s, has the stated property. Note that you can tick more than one box.

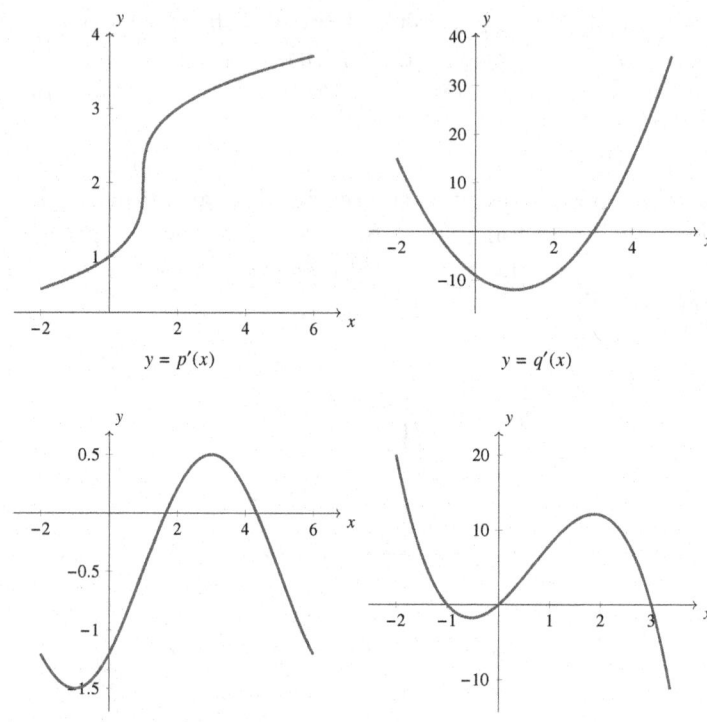

y = p'(x)   y = q'(x)

y = r'(x)   y = s'(x)

(1) The function is increasing over the interval $(1, 3)$.

$p \square \, q \square \, r \square \, s \square$ None of them $\square$

(2) The function has a critical number when $x = 3$.

$p \square \, q \square \, r \square \, s \square$ None of them $\square$

(3) The graph of the function has an inflection point when $x = 1$.

$p \square \, q \square \, r \square \, s \square$ None of them $\square$

(4) The graph of the function is concave upward over the interval $(0, 2)$.

$p \square \, q \square \, r \square \, s \square$ None of them $\square$

(5) There is a number in the domain of the function where the second derivative does not exist.

$p \square \, q \square \, r \square \, s \square$ None of them $\square$

Problem 3.14. A function and its first and second derivatives are given by

$$f(x) = \frac{x^2}{x-1}, \quad f'(x) = \frac{x(x-2)}{(x-1)^2}, \quad f''(x) = \frac{2}{(x-1)^3}.$$

(1) Determine the critical numbers of f.
(2) Determine the intervals on which the function is increasing or decreasing. Classify each critical number as either a local maximum, a local minimum, or neither.
(3) Determine where the graph of f is concave upward and where it is concave downward. Identify any inflection points.
(4) What is the end behavior of f (i.e., what is happening for large positive and negative x-values)? Identify any horizontal or slant asymptotes of f, if any.
(5) Indicate which of the following graphs is the graph of the function $y = f(x)$. Also, identify the critical numbers and inflection points, if any, on the graph you have chosen and write in the x-coordinates of these points.

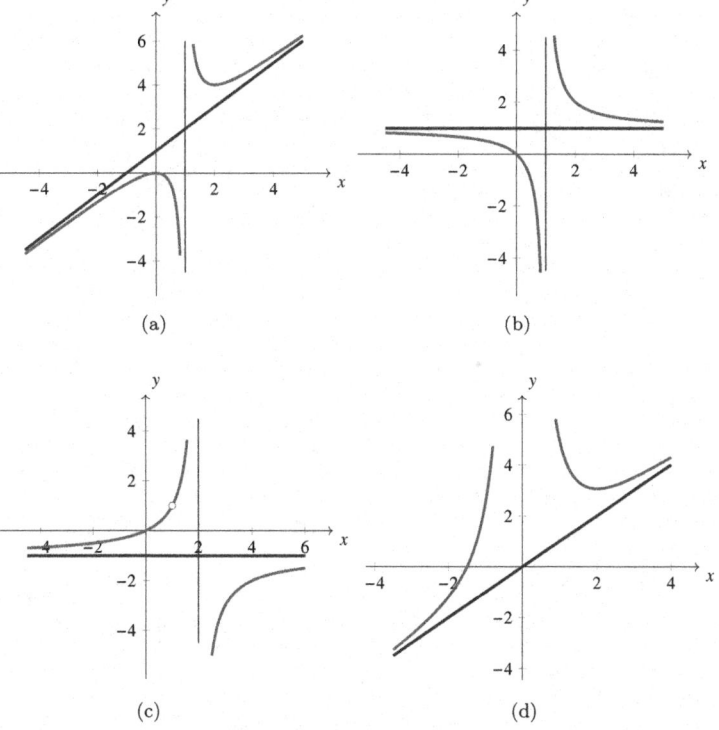

(a)

(b)

(c)

(d)

Problem 3.15. Consider the function $f(x) = x^3 e^{-x+5}$ and its first and second derivatives: $f'(x) = x^2(3-x)e^{-x+5}$ and $f''(x) = x(x^2 - 6x + 6)e^{-x+5}$.

(1) Determine the critical numbers of f.
(2) Determine the intervals on which the function is increasing and those on which the function is decreasing. Classify each critical number as either a local maximum, a local minimum, or neither.
(3) Determine where the graph of f is concave upward and where it is concave downward. Identify any inflection points.
(4) What is the end behavior of f (i.e., what is happening as $x \to \infty$ and $x \to -\infty$)?
(5) Indicate which of the graphs in the following figure is the graph of $y = f(x)$. Also, identify the critical numbers and inflection points, if any, on the graph you have chosen and write in the x-coordinates of these points.

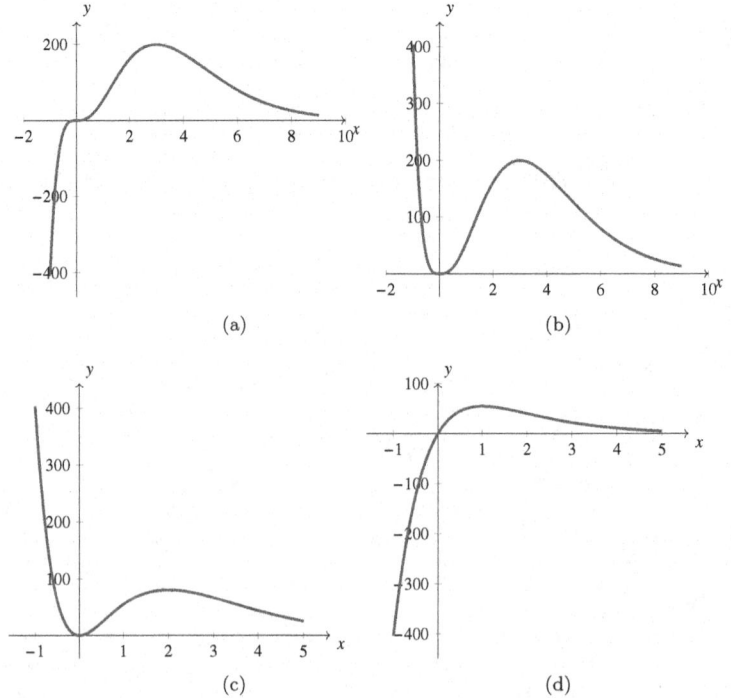

(a)   (b)

(c)   (d)

3.3 Sketching Graphs of Functions

For each of the functions given below, complete these steps to draw a sketch of the graph of the function:

(i) State the domain of the function.
(ii) Is the function even or odd?
(iii) Determine the x-intercepts and the y-intercepts, if any.
(iv) Determine intervals where the function is increasing and intervals where the function is decreasing.
(v) Determine all critical numbers and classify each of them as a local maximum, a local minimum, or neither.
(vi) Determine intervals where the graph of the function is concave upward and intervals where the graph is concave downward.
(vii) Determine any inflection points.
(viii) Determine all asymptotes of the graph of the function (vertical, horizontal, and slant).
(ix) Draw a sketch of the graph of the function based on your results above.
(x) Identify any absolute extrema, if any.

Problem 3.16. $f(x) = 3x^4 - 8x^3 + 10$.

Problem 3.17. $f(x) = x^3 - 2x^2 - x + 1$.

Problem 3.18. $f(x) = \dfrac{x^2}{x^2 - 1}$.

Problem 3.19. $f(x) = \dfrac{x^2 - 1}{x}$.

Problem 3.20. $f(x) = \dfrac{x^2}{x^2 + 9}$.

Problem 3.21. $f(x) = \dfrac{x^3 - 2x}{3x^2 - 9}$.

Problem 3.22. $f(x) = \dfrac{x}{x^2 + 4}$.

Problem 3.23. $f(x) = \dfrac{x^2 - 2}{x^4}$.

Problem 3.24. $f(x) = \dfrac{x^2 - 4x}{(x+4)^2}.$

Problem 3.25. $f(x) = \dfrac{4 - 4x}{x^2}.$

Problem 3.26. $f(x) = \dfrac{x^2 + 2}{x^2 - 4}.$

Problem 3.27. $f(x) = \dfrac{18(x-1)}{x^2}.$

Problem 3.28. $f(x) = \dfrac{x}{x^2 - 1}.$

Problem 3.29. $f(x) = \dfrac{x+3}{\sqrt{x^2+1}}.$

Problem 3.30. $f(x) = \dfrac{x-3}{\sqrt{x^2-9}}.$

Problem 3.31. $f(x) = (5 - 2x)x^{2/3}.$

Problem 3.32. $f(x) = 4x^{1/3} + x^{4/3}.$

Problem 3.33. $f(x) = \dfrac{1}{(1+e^x)^2}.$

Problem 3.34. $f(x) = e^{-2x^2}.$

Problem 3.35. $f(x) = x^2 e^{-x}.$

Problem 3.36. $f(x) = e^{1/x}.$

Problem 3.37. $f(x) = x^x.$

3.4   Answers, Hints, and Solutions

Solution 3.1. From $f'(x) = 3ax^2 + b$, it follows that $f'(1) = 0$ and $f(1) = 2$ provide two equations for the two unknowns: $3a+b = 0$ and $a+b = 2$. Hence, $a = -1$ and $b = 3$. From $f''(x) = 6ax$, we determine that $f''(1) < 0$. By the Second Derivative Test, we conclude that the function f has a local maximum at $x = 1$.

Solution 3.2. Observe that the function f is differentiable on its domain, the interval $(0, \infty)$. We need to determine a and b so that $x = 2$ is a critical number of the function f, i.e., f' changes sign at $x = 2$ and $f(2) = 1$. From $f'(x) = 2bx - \frac{1}{x}$, it follows that $f'(2) = 0$ if $b = \frac{1}{8}$. From $f'(x) = \frac{x^2 - 4}{4x}$, we conclude that, in the case $b = \frac{1}{8}$, f' changes sign at $x = 2$ from negative to positive. Hence, f has a local minimum at $x = 2$. From $1 = f(2) = \ln a + \frac{1}{2} - \ln 2$, it follows that $a = 2\sqrt{e}$.

Solution 3.3. Observe that $f'(x) = 151x^{150} + 37x^{36} + 1 > 0$ for all $x \in \mathbb{R}$. Hence, the polynomial f is always increasing and cannot have a local extremum.

Solution 3.4. From $y'' = 6x + 2a$ it follows that $y''(1) = 0$ if $a = -3$. Observe that, in the case $a = -3$, the second derivative changes sign at $x = 1$. From $y(1) = 6$ it follows that $b = 7$.

Solution 3.5. For example, the function $f(x) = x^3$ has a critical number $x = 0$. The point $(0, 0)$ is an inflection point on the graph of the function f.

Solution 3.6. For example, $f(x) = x - \text{sign}(x)$, $x \in \mathbb{R}$. Observe that the function f is not continuous at $x = 0$ and hence not differentiable, so the domain of the function f' is the set $\mathbb{R} \backslash \{0\}$. Clearly, $f(1) = 1 - 1 = 0$, $f(-1) = -1 - (-1) = 0$, and $f'(x) = 1 > 0$ if $x \neq 0$.

Solution 3.7. For $x \neq -1$,

$$f(x) - (x + 3) = \frac{x^2 + ax + 5}{x + 1} - \frac{(x + 3)(x + 1)}{x + 1} = \frac{(a - 4)x + 2}{x + 1}.$$

It follows that only if $a = 4$, $(f(x) - (x + 3)) \to 0$ as $|x| \to \infty$. Furthermore, for $a = 4$, we see that $f(x) = (x + 3) + \frac{2}{x+1}$ and so we conclude that $f(x)$ approaches $y = x + 3$ from above when $x \to \infty$ and from below when $x \to -\infty$.

Solution 3.8. It is given that the y-intercept is the point $(0, -3)$. Note that the given function has a vertical asymptote $x = 3$, a horizontal asymptotes $y = -1$ when $x \to -\infty$, and a horizontal asymptotes $y = 2$ when $x \to \infty$.

Also, the function f is decreasing on $(-\infty, 3)$ and $(3, \infty)$. Finally, the graph of f is concave upward on $(3, \infty)$ and concave downward on $(-\infty, 3)$. See the following figure for a possible graph.

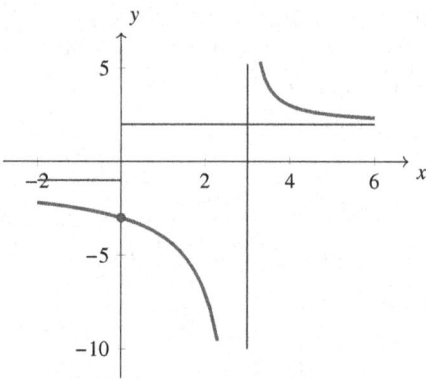

Solution 3.9. (1) h and k; (2) f; (3) f and g; (4) f, g, h, and k; (5) g and h.

Solution 3.10. (1) Observe that the function depicted by the far-left graph has a local maximum and that it flattens at $t = 6$. Hence, the middle-right graph depicts its derivative, as it has one zero, changes from positive to negative, and approaches 0 at $t = 6$. Observe that this derivative function has a local maximum followed by a local minimum. Hence, the graph to the far-right depicts its derivative, as it has two zeros and changes its sign from positive to negative to positive. Therefore, the position function s is depicted by the far-left graph, the velocity function v is depicted by the middle-right graph, and the acceleration function a is depicted by the far-right graph.

(2) Recall that a particle is speeding up if $v(t) \cdot a(t) > 0$ and slowing down if $v(t) \cdot a(t) < 0$. We read from the given graphs that the particle is speeding up on $(0, 1.5)$ and $(3.1, 5)$ and slowing down on $(1.5, 3.1)$ and $(5, 6)$. (3) The total distance is approximated as $|s(3.1) - s(0)| + |s(6) - s(3.1)| \approx |3 - 0.2| + |0 - 3| = 5.8$ m. Another way to approximate the total distance traveled by the particle is to approximate the total area of the regions between the graph of v and the t-axis over the interval $[1, 6]$. We can do that by counting the squares covered by the two regions.

Solution 3.11. (1) The graph has a horizontal asymptote $x = -2$ and a vertical asymptote $y = 0$. The following table summarizes the rest of the given information.

Interval	$(-4,-1)$	$(-1,0)$	$(0,2)$	$(2,4)$	$(4,\infty)$
Monotonity	↘	↗	↗	↘	↘
Concavity	⌢	⌣	⌢	⌢	⌣

(2) There are two inflection points, $(-1, f(-1))$ and $(4, f(4))$. We know that f is differentiable at $x = 4$ but we do not know if $f''(4)$ is defined. Hence, if f is twice differentiable at $x = 4$, then $f''(4) = 0$.
(3) We note that $x = -1$ is a critical number of the function f as well. By the First Derivative Test, the function f has a local minimum at $x = -1$. If $f''(-1) = 0$, then $f'(-1)$ exists and $f'(-1) = 0$.

This would imply that the graph of f is **above** the tangent line in a small neighborhood of $x = -1$, which contradicts the fact that the curve **crosses** its tangent line at each inflection point. It follows that $f'(-1)$ does not exist and, therefore, $f''(-1)$ does not exist. For a possible graph see the following figure.

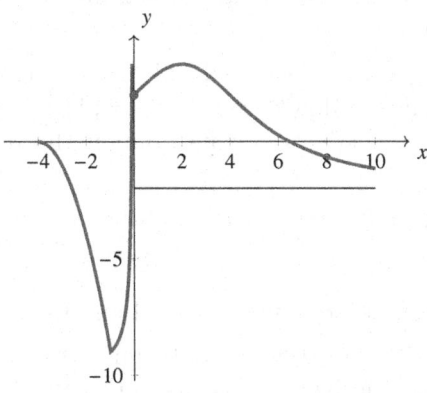

Solution 3.12. Recall that the given graph depicts the function $x \mapsto f'(x)$.

(1) The point $(3, f(3))$ is a local maximum because, from the graph, $f'(3) = 0$ and, at $x = 3$, the first derivative changes sign from positive to negative. Since $f(0) = 0$, the value of $f(3)$ is equal to the area between the graph of the function f' and the line segment determined

by the origin and the point $(3,0)$.[a] By using the grid lines, a possible estimate is $f(3) \approx 4.55$.

(2) $(-1, f(-1))$ and $f(-1) \approx 1$.

(3) Observe from the graph that $f''(x) = 0$ at $x \approx -0.5$, $x \approx 0.5$, and $x \approx 2.3$, and f'' changes sign at these numbers. So, inflection points are (close to) $(-0.5, f(-0.5))$, $(0.5, f(0.5))$, and $(2.3, f(2.3))$. From the graph we obtain possible estimates, $f(-0.5) \approx 0.6$, $f(0.5) \approx 0.35$, and $f(2.3) \approx 2.8$.

(4) $(-\infty, -1)$ and $(3, \infty)$.

(5) This is where $f^{(3)}(x) < 0$, i.e., where the graph of $f'(x)$ is concave downward: $(-\infty, 0)$ and $(1.5, \infty)$.

(6) Observe that the second derivative has three distinct zeros, so if f is a polynomial, then its degree is at least 5.

Solution 3.13. (1) p and s; (2) q and s; (3) q; (4) p, r and s; (5) p.

Solution 3.14. (1) There are two critical numbers, $x = 0$ and $x = 2$.

(2) The function is increasing where $f'(x) > 0$, i.e., on $(-\infty, 0)$ and $(2, \infty)$. The function is decreasing on $(0, 1)$ and $(1, 2)$. By the First Derivative Test, the function f has a local maximum at $x = 0$ and a local minimum at $x = 2$.

(3) Observe that $f''(x) > 0$ if and only if $x > 1$. So, the graph of the function is concave downward on $(-\infty, 1)$ and concave upward on $(1, \infty)$. Since $x = 1$ is not in its domain the function f has no inflection points.

(4) The line $y = x + 1$ is a slant asymptote.

(5) (a).

Solution 3.15. (1) There are two critical numbers, $x = 0$ and $x = 3$.

(2) Observe that $f'(x) > 0$ on $(-\infty, 0)$ and $(0, 3)$ and $f'(x) < 0$ on $(3, \infty)$. By the First Derivative Test, the critical number $x = 0$ is not a local extremum and at the critical number $x = 3$ the function has a local maximum value $f(3) = 27e^2$. The function is increasing on $(-\infty, 3)$ and decreasing otherwise.

---

[a]Recall that when velocity is positive on an interval, we can find the total distance traveled by finding the area under the velocity curve and above the x-axis on the given time interval.

(3) Observe that $f''(x) = 0$ if $x = 0$ and $x = 3 \pm \sqrt{3}$. It follows that $f''(x) > 0$ on $(0, 3 - \sqrt{3})$ and $(3 + \sqrt{3}, \infty)$, so the graph is concave upward there and concave downward otherwise. There are three inflection points $(0, 0)$, $(3 - \sqrt{3}, f(3 - \sqrt{3}))$, and $(3 + \sqrt{3}, f(3 + \sqrt{3}))$.

(4) The line $y = 0$ is a horizontal asymptote when $x \to \infty$.

(5) (a).

Solution 3.16. Observe: $f(x) = 3x^4 - 8x^3 + 10$, $f'(x) = 12x^2(x - 2)$, and $f''(x) = 12x(3x - 4)$.

(i) Domain: $(-\infty, \infty)$.

(ii) Neither even nor odd.

(iii) The y-intercept is $f(0) = 10$. To determine the x-intercepts, we solve the equation $f(x) = 3x^4 - 8x^3 + 10 = 0$. This is a quartic equation and we may use technology to determine that this equation has two solutions: $x \approx 1.3698$ and $x \approx 2.4361$.

(iv) $f'(x) > 0$ for $x \in (2, \infty)$ and $f'(x) < 0$ for $x \in (-\infty, 0) \cup (0, 2)$. So, f is increasing on $(2, \infty)$ and decreasing on $(-\infty, 0)$ and $(0, 2)$.

(v) Critical numbers are $x = 0$ and $x = 2$. Since $f'(x)$ does not change sign at $x = 0$, there is no local maximum or minimum there. Note that $f''(0) = 0$, so the Second Derivative Test is inconclusive. Also note that this implies that f is decreasing on the interval $(-\infty, 2)$. Since $f'(x)$ changes sign from negative to positive at $x = 2$, by the First Derivative Test, the function f has a local minimum at $x = 2$. Note that $f''(2) > 0$, so the Second Derivative Test also implies there is a local minimum at $x = 2$.

(vi) $f''(x) > 0$ for $x < 0$ or $x > \frac{4}{3}$ and $f''(x) < 0$ for $0 < x < \frac{4}{3}$. Thus, the graph of $f(x)$ is concave upward on $(-\infty, 0)$ and $\left(\frac{4}{3}, \infty\right)$, and concave downward on $\left(0, \frac{4}{3}\right)$.

(vii) Since $f''(x) = 0$ for $x = 0$ and $x = \frac{4}{3}$ and $f''(x)$ changes sign at these x-values, the inflection points are $(0, 10)$ and $\left(\frac{4}{3}, \frac{14}{27}\right)$.

(viii) No asymptotes.

(ix) See the following figure.

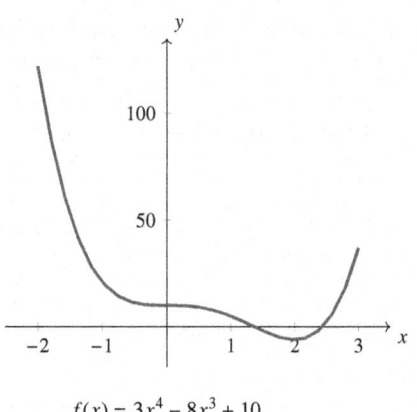

$$f(x) = 3x^4 - 8x^3 + 10$$

(x) The absolute minimum value is $f(2) = -6$.

Solution 3.17. Observe: $f(x) = x^3 - 2x^2 - x + 1$, $f'(x) = 3x^2 - 4x - 1$, and $f''(x) = 6x - 4$.

(i) Domain: $(-\infty, \infty)$.

(ii) Neither even nor odd.

(iii) The y-intercept is $f(0) = 1$. To determine the x-intercepts, we solve the equation $f(x) = x^3 - 2x^2 - x + 1 = 0$. This is a cubic equation and we may use technology to determine that this equation has three solutions: $x \approx -0.80194$, $x \approx 0.55496$, and $x \approx 2.2470$.

(iv) $f'(x) = 3\left(x - \frac{2+\sqrt{7}}{3}\right)\left(x - \frac{2-\sqrt{7}}{3}\right)$. Therefore, $f'(x) > 0$ on $\left(-\infty, \frac{2-\sqrt{7}}{3}\right) \cup \left(\frac{2+\sqrt{7}}{3}, \infty\right)$ and so f is increasing on these intervals, while $f'(x) < 0$ on $\left(\frac{2-\sqrt{7}}{3}, \frac{2+\sqrt{7}}{3}\right)$ and so f is decreasing on this interval.

(v) Critical numbers are $x = \frac{2\pm\sqrt{7}}{3}$. Since $f'(x)$ changes sign from positive to negative at $x = \frac{2-\sqrt{7}}{3}$, the function f has a local maximum at $x = \frac{2-\sqrt{7}}{3}$. Note that $f''\left(\frac{2-\sqrt{7}}{3}\right) < 0$, so the Second Derivative Test also implies that there is a local maximum there. Since $f'(x)$ changes from negative to positive at $x = \frac{2+\sqrt{7}}{3}$, the function f has a local minimum at $x = \frac{2+\sqrt{7}}{3}$. Note that $f''\left(\frac{2+\sqrt{7}}{3}\right) > 0$, so the Second Derivative Test also implies that there is a local minimum.

(vi) $f''(x) < 0$ for $x < \frac{2}{3}$ and $f''(x) > 0$ on $x > \frac{2}{3}$. Thus, the graph of $f(x)$ is concave downward on $\left(-\infty, \frac{2}{3}\right)$ and concave upward on $\left(\frac{2}{3}, \infty\right)$.

(vii) The point of inflection is at $\left(\frac{2}{3}, -\frac{7}{27}\right)$.

(viii) No asymptotes.

(ix) See the following figure.

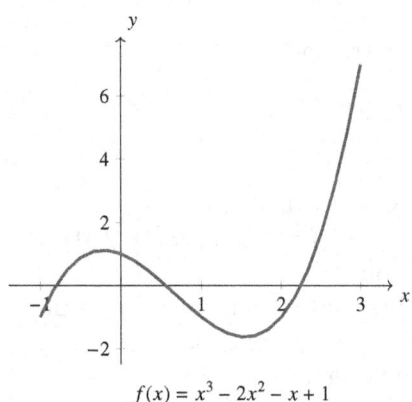

$$f(x) = x^3 - 2x^2 - x + 1$$

(x) No absolute extrema.

Solution 3.18. Observe: $f(x) = \frac{x^2}{x^2-1}$, $f'(x) = -\frac{2x}{(x^2-1)^2}$, and $f''(x) = \frac{2(3x^2+1)}{(x^2-1)^3}$.

(i) Domain: $(-\infty, -1) \cup (-1, 1) \cup (1, \infty)$.

(ii) Observe that $f(-x) = f(x)$ for $x \neq \pm 1$ and conclude that f is an even function.

(iii) The x- and y-intercepts are both 0.

(iv) Increasing on $(-\infty, -1)$ and $(-1, 0)$. Decreasing on $(0, 1)$ and $(1, \infty)$.

(v) Critical number is $x = 0$. By the First Derivative Test, $f(0) = 0$ is a local maximum value of f.

(vi) Concave upward on $(-\infty, -1)$ and $(1, \infty)$. Concave downward on $(-1, 1)$.

(vii) No inflection points.

(viii) See the following figure.

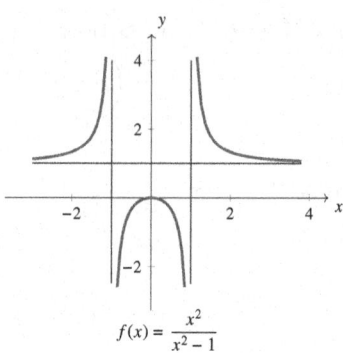

$$f(x) = \frac{x^2}{x^2 - 1}$$

(ix) No absolute extrema.

Solution 3.19. Observe: $f(x) = \frac{x^2-1}{x} = x - \frac{1}{x}$, $f'(x) = 1 + \frac{1}{x^2}$, and $f''(x) = -\frac{2}{x^3}$.

   (i) Domain: $(-\infty, 0) \cup (0, \infty)$.
  (ii) Observe that $f(-x) = -f(x)$ for $x \neq 0$ and conclude that f is an odd function.
 (iii) No y-intercepts, but $x = \pm 1$ are the x-intercepts.
  (iv) Increasing on $(-\infty, 0)$ and $(0, \infty)$.
   (v) No critical numbers.
  (vi) Concave upward on $(-\infty, 0)$ and concave downward on $(0, \infty)$.
 (vii) No inflection points.
(viii) Vertical asymptote: $x = 0$. Slant asymptote: $y = x$, since $f(x) - x = -\frac{1}{x}$; furthermore, $f(x)$ approaches $y = x$ from above as $x \to -\infty$ and approaches $y = x$ from below as $x \to +\infty$.
  (ix) See the following figure.

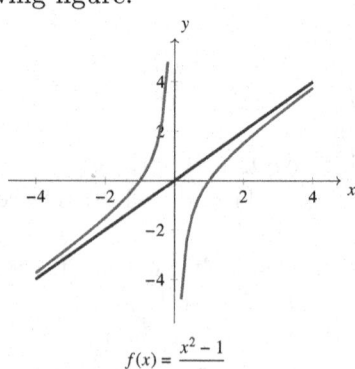

$$f(x) = \frac{x^2 - 1}{x}$$

  (x) No absolute extrema.

Solution 3.20. Observe: $f(x) = \frac{x^2}{x^2+9} = 1 - \frac{9}{x^2+9}$, $f'(x) = \frac{18x}{(x^2+9)^2}$, and $f''(x) = \frac{54(3-x^2)}{(x^2+9)^3}$.

(i) Domain: $(-\infty, \infty)$.
(ii) The function f is even.
(iii) Both x- and y-intercepts are 0.
(iv) Increasing on $(0, \infty)$ and decreasing on $(-\infty, 0)$.
(v) Critical number is $x = 0$. By the First Derivative Test, $f(0) = 0$ is a local minimum of the function f.
(vi) Concave upward on $(-\sqrt{3}, \sqrt{3})$. Concave downward on $(-\infty, -\sqrt{3})$ and $(\sqrt{3}, \infty)$.
(vii) Inflection points are $\left(-\sqrt{3}, \frac{1}{4}\right)$ and $\left(\sqrt{3}, \frac{1}{4}\right)$.
(viii) Horizontal asymptote: $y = 1$, approaching from below as $x \to \pm\infty$.
(ix) See the following figure.

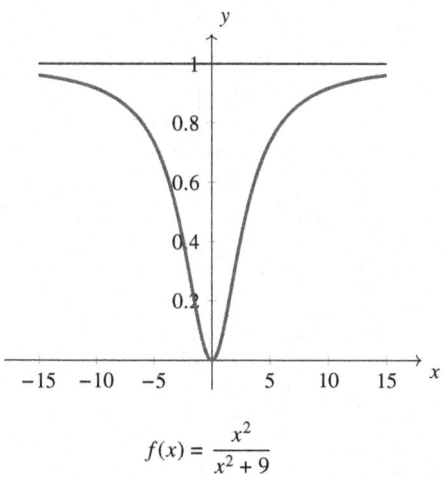

$$f(x) = \frac{x^2}{x^2 + 9}$$

(x) The absolute minimum value is $f(0) = 0$.

Solution 3.21. Observe: $f(x) = \frac{x^3-2x}{3x^2-9}$, $f'(x) = \frac{x^4-7x^2+6}{3(x^2-3)^2} = \frac{(x^2-6)(x^2-1)}{3(x^2-3)^2}$, and $f''(x) = \frac{2x(x^2+9)}{3(x^2-3)^3}$.

(i) Domain: $(-\infty, -\sqrt{3}) \cup (-\sqrt{3}, \sqrt{3}) \cup (\sqrt{3}, \infty)$.
(ii) The function f is odd.
(iii) The y-intercept is $f(0) = 0$. The x-intercepts are 0 and $\pm\sqrt{2}$.

(iv) Increasing on $(-\infty, -\sqrt{6})$, $(-1,1)$, and $(\sqrt{6}, \infty)$. Decreasing on $(-\sqrt{6}, -\sqrt{3})$, $(-\sqrt{3}, -1)$, $(1, \sqrt{3})$, and $(\sqrt{3}, \sqrt{6})$.

(v) Critical numbers: ± 1 and $\pm\sqrt{6}$. By the First Derivative Test, the local maximum values are $f(-\sqrt{6}) \approx -3.265$ and $f(1) = \frac{1}{6}$ and the local minimum values are $f(1) = -\frac{1}{6}$ and $f(\sqrt{6}) \approx 3.265$.

(vi) Concave upward on $(-\sqrt{3}, 0)$ and $(\sqrt{3}, \infty)$. Concave downward on $(-\infty, -\sqrt{3})$ and $(0, \sqrt{3})$.

(vii) Point of inflection: $(0,0)$.

(viii) Vertical asymptotes: $x = \pm\sqrt{3}$. Slant asymptote: $y = \frac{1}{3}x$. Since $f(x) - \frac{1}{3}x = \frac{x}{3x^2+9}$, the graph of f approaches the slant asymptote from below as $x \to -\infty$ and from above as $x \to \infty$.

(ix) See the following figure.

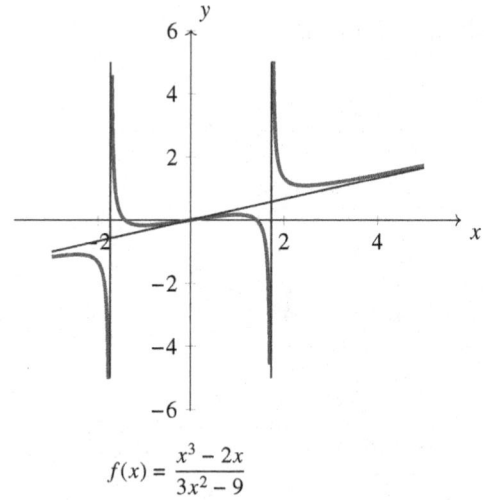

$$f(x) = \frac{x^3 - 2x}{3x^2 - 9}$$

(x) No absolute extrema.

Solution 3.22. Observe: $f(x) = \frac{x}{x^2+4}$, $f'(x) = \frac{4-x^2}{(x^2+4)^2}$, and $f''(x) = \frac{2x(x^2-12)}{(x^2+4)^3}$.

(i) Domain: $(-\infty, \infty)$.

(ii) The function f is odd.

(iii) Both x- and y-intercepts are 0.

(iv) Increasing on $(-2, 2)$. Decreasing on $(-\infty, -2)$ and $(2, \infty)$.

(v) Critical numbers: -2 and 2. By the First Derivative Test, $f(-2) = -0.25$ is a local minimum value and $f(2) = 0.25$ is a local maximum value.

(vi) Concave upward on $(-2\sqrt{3}, 0)$ and $(2\sqrt{3}, \infty)$. Concave downward on $(-\infty, -2\sqrt{3})$ and $(0, 2\sqrt{3})$.

(vii) Inflection points are $\left(-2\sqrt{3}, -\frac{\sqrt{3}}{8}\right)$, $(0, 0)$, and $\left(2\sqrt{3}, \frac{\sqrt{3}}{8}\right)$.

(viii) Horizontal asymptote: $y = 0$, approaching from below as $x \to -\infty$ and from above as $x \to \infty$.

(ix) See the following figure.

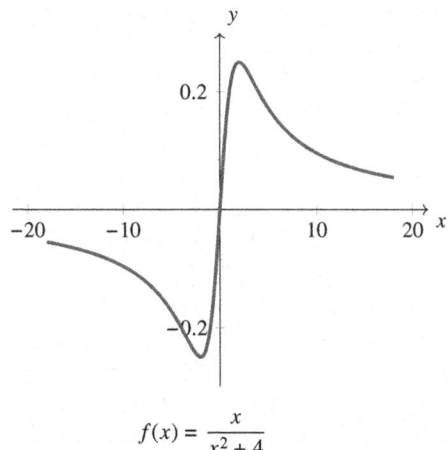

$$f(x) = \frac{x}{x^2 + 4}$$

(x) The absolute maximum value is $f(2) = 0.25$ and the absolute minimum value is $f(-2) = -0.25$.

Solution 3.23. Observe: $f(x) = \frac{x^2 - 2}{x^4}$, $f'(x) = \frac{2(4 - x^2)}{x^5}$, and $f''(x) = \frac{2(3x^2 - 20)}{x^6}$.

(i) Domain: $(-\infty, 0) \cup (0, \infty)$.

(ii) The function f is even.

(iii) The x-intercepts are $-\sqrt{2}$ and $\sqrt{2}$.

(iv) Increasing on $(-\infty, -2)$ and $(0, 2)$. Decreasing on $(-2, 0)$ and $(2, \infty)$.

(v) Critical numbers: -2 and 2. By the First Derivative Test, the local maximum values are $f(-2) = f(2) = 0.125$.

(vi) Concave upward on $\left(-\infty, -\frac{2\sqrt{5}}{\sqrt{3}}\right)$ and $\left(\frac{2\sqrt{5}}{\sqrt{3}}, \infty\right)$. Concave downward on $\left(-\frac{2\sqrt{5}}{\sqrt{3}}, 0\right)$ and $\left(0, \frac{2\sqrt{5}}{\sqrt{3}}\right)$.

(vii) Inflection points are $\left(-\frac{2\sqrt{5}}{\sqrt{3}}, \frac{21}{200}\right)$ and $\left(\frac{2\sqrt{5}}{\sqrt{3}}, \frac{21}{200}\right)$.

(viii) Vertical asymptote: $x = 0$. Horizontal asymptote: $y = 0$, approaching from above as $x \to \pm\infty$.

(ix) See the following figure.

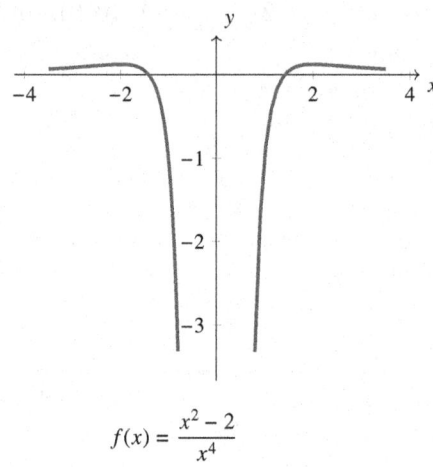

$$f(x) = \frac{x^2 - 2}{x^4}$$

(x) The absolute maximum value is $f(-2) = f(2) = 0.125$.

Solution 3.24. Observe: $f(x) = \frac{x^2-4x}{(x+4)^2}$, $f'(x) = \frac{4(3x-4)}{(x+4)^3}$, and $f''(x) = \frac{24(4-x)}{(x+4)^4}$.

(i) Domain: $(-\infty, -4) \cup (-4, \infty)$.

(ii) The function f is neither even nor odd.

(iii) The x-intercepts are 0 and 4. The y-intercept is 0.

(iv) Increasing on $(-\infty, -4)$ and $\left(\frac{4}{3}, \infty\right)$. Decreasing on $\left(-4, \frac{4}{3}\right)$.

(v) Critical number: $\frac{4}{3}$. By the First Derivative Test, $f\left(\frac{4}{3}\right) = -0.125$ is a local minimum value.

(vi) Concave upward on $(-\infty, -4)$ and $(-4, 4)$. Concave downward on $(4, \infty)$.

(vii) Point of inflection: $(4, 0)$.

(viii) Vertical asymptote: $x = -4$. Horizontal asymptote: $y = 1$, approaching from above as $x \to -\infty$ and from below as $x \to \infty$.

(ix) See the following figure.

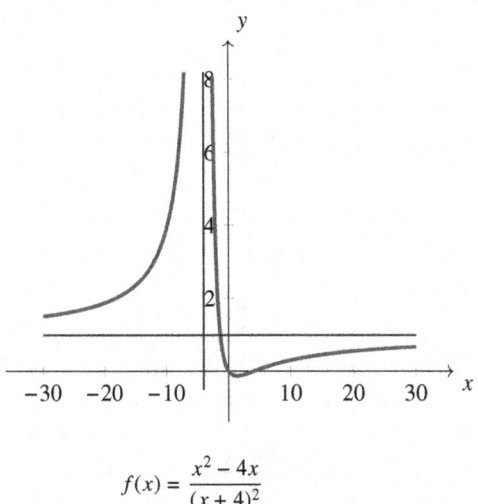

$$f(x) = \frac{x^2 - 4x}{(x+4)^2}$$

(x) The absolute minimum value is $f\left(\frac{4}{3}\right) = -0.125$.

Solution 3.25. Observe: $f(x) = \frac{4-4x}{x^2}$, $f'(x) = \frac{4(x-2)}{x^3}$, and $f''(x)\frac{8(3-x)}{x^4}$.

   (i) Domain: $(-\infty, 0) \cup (0, \infty)$.
  (ii) The function f is neither even nor odd.
 (iii) The x-intercept is 1.
  (iv) Increasing on $(-\infty, 0)$ and $(2, \infty)$. Decreasing on $(0, 2)$.
   (v) Critical number: 2. By the First Derivative Test, $f(2) = -1$ is a local minimum value.
  (vi) Concave upward on $(-\infty, 0)$ and $(0, 3)$. Concave downward on $(3, \infty)$.
 (vii) Point of inflection: $\left(3, -\frac{8}{9}\right)$.
(viii) Vertical asymptote: $x = 0$. Horizontal asymptote: $y = 0$, approaching from above as $x \to -\infty$ and from below as $x \to \infty$.

(ix) See the following figure.

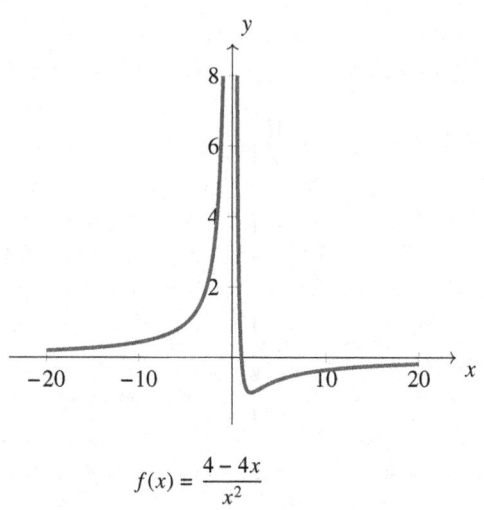

$$f(x) = \frac{4 - 4x}{x^2}$$

(x) The absolute minimum value is $f(2) = -1$.

Solution 3.26. Observe: $f(x) = \frac{x^2+2}{x^2-4}$, $f'(x) = -\frac{12x}{(x^2-4)^2}$, and $f''(x) = \frac{12(3x^2+4)}{(x^2-4)^3}$.

  (i) Domain: $(-\infty, -2) \cup (-2, 2) \cup (2, \infty)$.
 (ii) The function is even.
(iii) The y-intercept is $-\frac{1}{2}$. Observe that the equation $f(x) = 0$ has no solution and conclude that there are no x-intercepts.
 (iv) Increasing on $(-\infty, -2)$ and $(-2, 0)$. Decreasing on $(0, 2)$ and $(2, \infty)$.
  (v) Critical number: 0. By the First Derivative Test, $f(0) = -0.5$ is a local maximum value.
 (vi) Concave upward on $(-\infty, -2)$ and $(2, \infty)$. Concave downward on $(-2, 2)$.
(vii) No inflection points.
(viii) Vertical asymptotes: $x = \pm 2$. Horizontal asymptote: $y = 1$, approaching from above as $x \to \pm\infty$.

(ix) See the following figure.

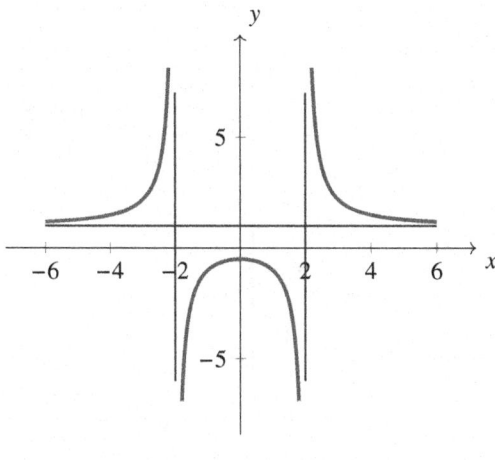

$$f(x) = \frac{x^2 + 2}{x^2 - 4}$$

(x) No absolute extrema.

Solution 3.27. Observe: $f(x) = \frac{18(x-1)}{x^2}$, $f'(x) = \frac{18(2-x)}{x^3}$, and $f''(x) = \frac{36(x-3)}{x^4}$.

  (i) Domain: $(-\infty, 0) \cup (0, \infty)$.

  (ii) The function f is neither even nor odd.

 (iii) The x-intercept is 1.

 (iv) Increasing on $(0, 2)$. Decreasing on $(-\infty, 0)$ and $(2, \infty)$.

  (v) Critical number: 2. By the First Derivative Test, $f(2) = 4.5$ is a local maximum value.

 (vi) Concave upward on $(3, \infty)$. Concave downward on $(-\infty, 0)$ and $(0, 3)$.

(vii) Point of inflection: $(3, 4)$.

(viii) Vertical asymptote: $x = 0$. Horizontal asymptote: $y = 0$, approaching from below as $x \to -\infty$ and from above as $x \to \infty$.

(ix) See the following figure.

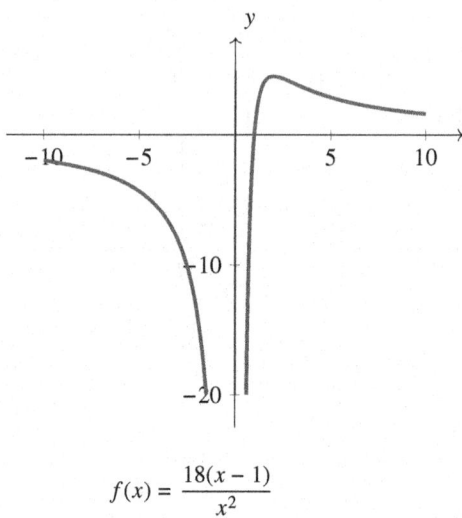

$$f(x) = \frac{18(x-1)}{x^2}$$

(x) The absolute maximum value is $f(2) = 4.5$.

Solution 3.28. Observe: $f(x) = \frac{x}{x^2-1}$, $f'(x) = -\frac{x^2+1}{(x^2-1)^2}$, and $f''(x) = \frac{2x(x^2+3)}{(x^2-1)^3}$.

(i) Domain: $(-\infty, -1) \cup (-1, 1) \cup (1, \infty)$.
(ii) The function f is odd.
(iii) Both x- and y-intercepts are 0.
(iv) Decreasing on each interval in the domain.
(v) No critical numbers.
(vi) Concave upward on $(-1, 0)$ and $(1, \infty)$. Concave downward on $(-\infty, -1)$ and $(0, 1)$.
(vii) Point of inflection: $(0, 0)$.
(viii) Vertical asymptotes: $x = \pm 1$. Horizontal asymptote: $y = 0$, approaching from below as $x \to -\infty$ and from above as $x \to \infty$.

(ix) See the following figure.

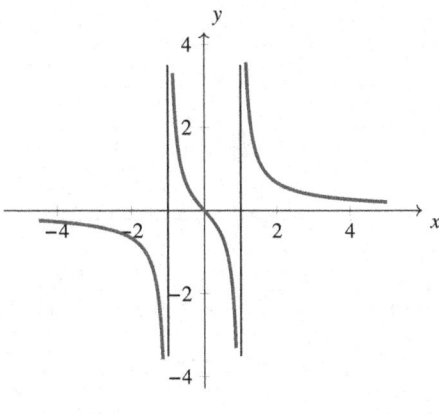

$$f(x) = \frac{x}{x^2 - 1}$$

(x) No absolute extrema.

Solution 3.29. Observe: $f(x) = \frac{x+3}{\sqrt{x^2+1}}$, $f'(x) = \frac{1-3x}{(x^2+1)^{3/2}}$, and $f''(x) = \frac{3(2x^2-x-1)}{(x^2+1)^{5/2}}$.

(i) Domain: $(-\infty, \infty)$.

(ii) The function f is neither even nor odd.

(iii) The x-intercept is -3. The y-intercept is 3.

(iv) Increasing on $\left(-\infty, \frac{1}{3}\right)$. Decreasing on $\left(\frac{1}{3}, \infty\right)$.

(v) Critical number: $\frac{1}{3}$. By the First Derivative Test, $f\left(\frac{1}{3}\right) = \sqrt{10}$ is a local maximum value.

(vi) Concave upward on $\left(-\infty, \frac{1-\sqrt{5}}{4}\right)$ and $\left(\frac{1+\sqrt{5}}{4}, \infty\right)$. Concave downward on $\left(\frac{1-\sqrt{5}}{4}, \frac{1+\sqrt{5}}{4}\right)$.

(vii) Inflection points are $\left(\frac{1-\sqrt{5}}{4}, \frac{13-\sqrt{5}}{\sqrt{32-2\sqrt{5}}}\right)$ and $\left(\frac{1+\sqrt{5}}{4}, \frac{13+\sqrt{5}}{\sqrt{32+2\sqrt{5}}}\right)$.

(viii) Horizontal asymptotes: $y = -1$ approaching from above as $x \to -\infty$ and $y = 1$ approaching from above as $x \to \infty$.

(ix) See the following figure.

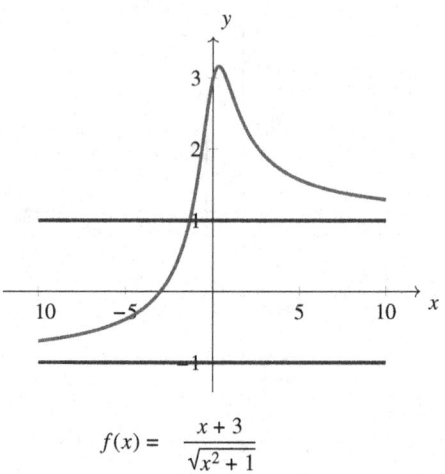

$$f(x) = \frac{x+3}{\sqrt{x^2+1}}$$

(x) The absolute maximum value is $f\left(\frac{1}{3}\right) = \sqrt{10}$.

Solution 3.30. Observe: $f(x) = \dfrac{x-3}{\sqrt{x^2-9}}$, $f'(x) = \dfrac{3(x-3)}{(x^2-9)^{3/2}} = \dfrac{3}{(x+3)\sqrt{x^2-9}}$, and $f''(x) = -\dfrac{3(x-3)(2x-3)}{(x^2-9)^{5/2}}$.

(i) Domain: $(-\infty, -3) \cup (3, \infty)$.

(ii) The function f is neither even nor odd.

(iii) No intercepts. Observe that $x = 3$ is not in the domain of the function.

(iv) Increasing on $(3, \infty)$. Decreasing on $(-\infty, -3)$.

(v) No critical points.

(vi) Since $f''(x) < 0$ for all x in the domain, we conclude that $f(x)$ is concave downward on $(-\infty, -3)$ and $(3, \infty)$.

(vii) No inflection points.

(viii) Note that for $x > 3$, $f(x) = \dfrac{\sqrt{x-3}}{\sqrt{x+3}}$, so $\lim\limits_{x\to 3+} f(x) = 0$. However, for $x < -3$, $f(x) = -\dfrac{\sqrt{3-x}}{\sqrt{-(x+3)}}$, so $\lim\limits_{x\to -3-} f(x) = -\infty$. Thus, the line $x = -3$ is a vertical asymptote. Horizontal asymptotes: $y = -1$ approaching from below as $x \to -\infty$ and $y = 1$ approaching from below as $x \to \infty$.

(ix) See the following figure.

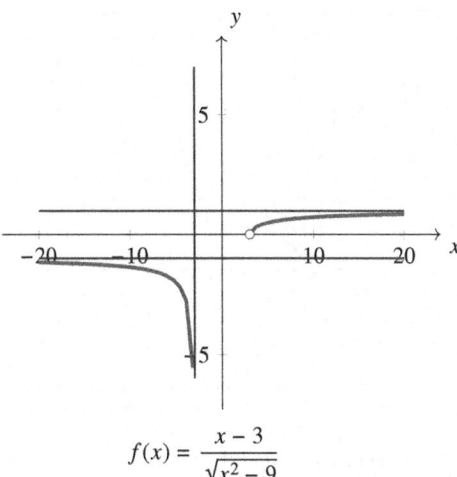

$$f(x) = \frac{x-3}{\sqrt{x^2-9}}$$

(x) No absolute extrema.

Solution 3.31. Observe: $f(x) = (5-2x)x^{2/3}$, $f'(x) = \dfrac{10(1-x)}{3x^{1/3}}$, and $f''(x) = -\dfrac{10(1+2x)}{9x^{4/3}}$.

(i) Domain: $(-\infty, \infty)$.

(ii) The function f is neither even nor odd.

(iii) The x-intercepts are 0 and $\dfrac{5}{2}$. The y-intercept is 0.

(iv) Increasing on $(0,1)$. Decreasing on $(-\infty,0)$ and $(1,\infty)$.

(v) Critical numbers: $0,1$. Observe that f is not differentiable at $x = 0$ and that the point $(0,0)$ is a "corner" on the graph of f. The local maximum value is $f(1) = 3$ and the local minimum value is $f(0) = 0$.

(vi) Concave upward on $\left(-\infty, -\dfrac{1}{2}\right)$. Concave downward on $\left(-\dfrac{1}{2}, 0\right)$ and $(0, \infty)$.

(vii) Point of inflection: $\left(-\dfrac{1}{2}, \dfrac{6}{\sqrt[3]{4}}\right)$.

(viii) No asymptotes.

(ix) See the following figure.

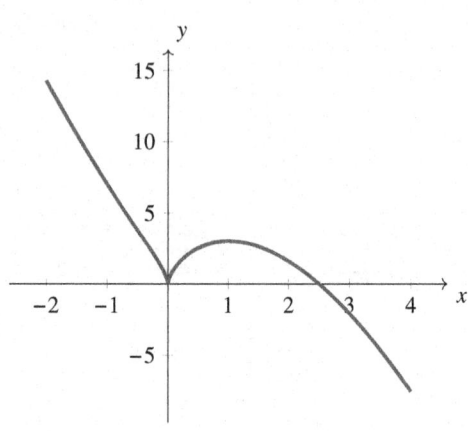

$$f(x) = (5 - 2x)x^{\frac{2}{3}}$$

(x) No absolute extrema.

Solution 3.32. Observe: $f(x) = 4x^{1/3} + x^{4/3}$, $f'(x) = \frac{4}{3}(x+1)x^{-\frac{2}{3}}$, and $f''(x) = \frac{4}{9}(x-2)x^{-\frac{5}{3}}$.

(i) Domain: $(-\infty, \infty)$.

(ii) The function f is neither even nor odd.

(iii) The x-intercepts are -4 an 0. The y-intercept is 0.

(iv) Increasing on $(-1, \infty)$. Decreasing on $(-\infty, -1)$.

(v) Critical numbers are -1 and 0. Observe that f is not differentiable at $x = 0$ and that $x = 0$ is a vertical tangent line to the graph of f at the point $(0, 0)$.
By the First Derivative Test, $f(-1) = -3$ is a local minimum value.

(vi) Concave upward on $(-\infty, 0)$ and $(2, \infty)$. Concave downward on $(0, 2)$.

(vii) Inflection points are $(0, 0)$ and $(2, 6\sqrt[3]{2})$.

(viii) No asymptotes.

(ix) See the following figure.

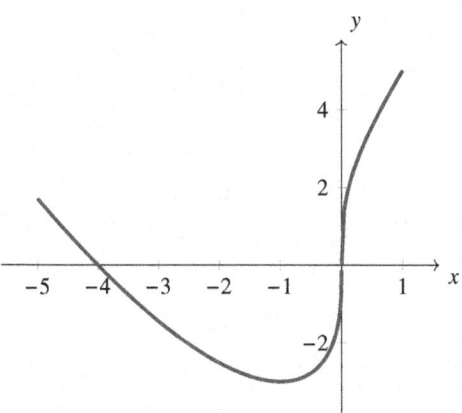

$$y = 4x^{1/3} + x^{4/3}$$

(x) The absolute minimum value is $f(-1) = -3$.

Solution 3.33. Observe: $f(x) = \dfrac{1}{(1+e^x)^2}$, $f'(x) = -\dfrac{2e^x}{(1+e^x)^3}$, and $f''(x) = \dfrac{2e^x(2e^x-1)}{(1+e^x)^4}$.

(i) Domain: $(-\infty, \infty)$.

(ii) The function is neither even nor odd.

(iii) The y-intercept is $f(0) = \dfrac{1}{4}$.

(iv) Decreasing on its domain.

(v) No critical points.

(vi) Concave upward on $(-\ln(2), \infty)$. Concave downward on $(-\infty, -\ln(2))$.

(vii) Point of inflection: $\left(-\ln(2), \dfrac{4}{9}\right)$.

(viii) Horizontal asymptotes: $y = 1$ approaching from below as $x \to -\infty$ and $y = 0$ approaching from above as $x \to \infty$.

(ix) See the following figure.

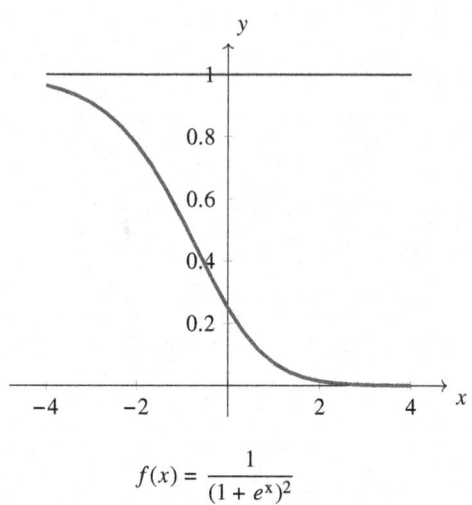

$$f(x) = \frac{1}{(1 + e^x)^2}$$

(x) No absolute extrema.

Solution 3.34. Observe: $f(x) = e^{-2x^2}$, $f'(x) = -4xe^{-2x^2}$, and $f''(x) = 4e^{-2x^2}(4x^2 - 1)$.

   (i) Domain: $(-\infty, \infty)$.
   (ii) The function f is even.
   (iii) The y-intercept is 1.
   (iv) Increasing on $(-\infty, 0)$. Decreasing on $(0, \infty)$.
   (v) Critical number: 0. By the First Derivative Test, $f(0) = 1$ is a local maximum value.
   (vi) Concave upward on $\left(-\infty, -\frac{1}{2}\right)$ and $\left(\frac{1}{2}, \infty\right)$. Concave downward on $\left(-\frac{1}{2}, \frac{1}{2}\right)$.

   (vii) Inflection points are $\left(-\frac{1}{2}, e^{-\frac{1}{2}}\right)$ and $\left(\frac{1}{2}, e^{-\frac{1}{2}}\right)$.

(viii) No asymptotes.

(ix) See the following figure.

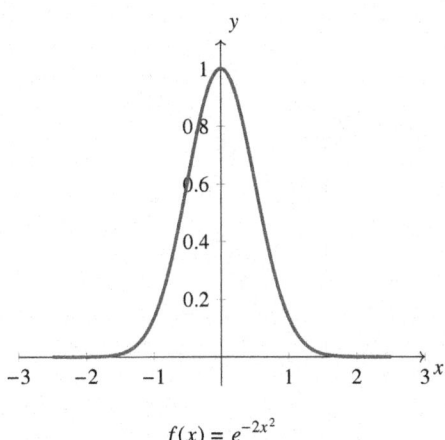

$$f(x) = e^{-2x^2}$$

(x) The absolute maximum value is $f(0) = 1$.

Solution 3.35. Observe: $f(x) = x^2 e^{-x}$, $f'(x) = x(2 - x)e^{-x}$, and $f''(x) = (x^2 - 4x + 2)e^{-x}$.

(i) Domain: $(-\infty, \infty)$.

(ii) The function f is neither even nor odd.

(iii) Both x- and y-intercepts are 0.

(iv) Increasing on $(0, 2)$. Decreasing on $(-\infty, 0)$ and $(2, \infty)$.

(v) Critical numbers: 0 and 2. By the First Derivative Test, $f(0) = 0$ is a local minimum value and $f(2) = 4e^{-2}$ is a local maximum value.

(vi) Concave upward on $(0, 2-\sqrt{2})$ and $(2+\sqrt{2}, \infty)$. Concave downward on $(2 - \sqrt{2}, 2 + \sqrt{2})$.

(vii) Inflection points are $(2-\sqrt{2}, (6-4\sqrt{2})e^{2-\sqrt{2}})$ and $(2+\sqrt{2}, (6+4\sqrt{2})e^{2+\sqrt{2}})$.

(viii) Horizontal asymptote: $y = 0$, approaching from above as $x \to \infty$.

(ix) See the following figure.

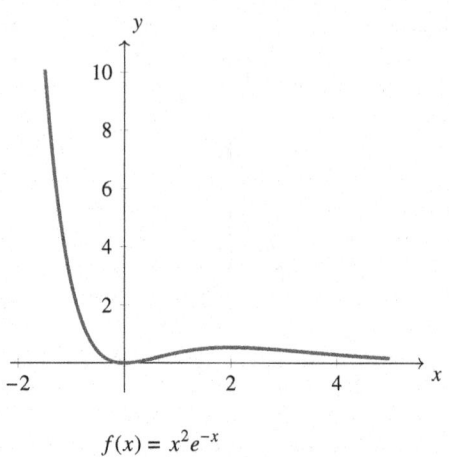

$$f(x) = x^2 e^{-x}$$

(x) The absolute minimum value is $f(0) = 0$.

Solution 3.36. Observe: $f(x) = e^{1/x}$, $f'(x) = -\dfrac{e^{1/x}}{x^2}$, and $f''(x) = \dfrac{e^{1/x}(2x+1)}{x^4}$.

(i) Domain: $(-\infty, 0) \cup (0, \infty)$.
(ii) The function f is neither even nor odd.
(iii) No intercepts.
(iv) Decreasing on $(-\infty, 0)$ and $(0, \infty)$.
(v) No critical numbers.
(vi) Concave upward on $\left(-\dfrac{1}{2}, 0\right)$ and $(0, \infty)$. Concave downward on $\left(-\infty, -\dfrac{1}{2}\right)$.
(vii) Point of inflection: $\left(-\dfrac{1}{2}, e^{-2}\right)$.
(viii) Observe that $\lim\limits_{x \to 0^-} f(x) = 0$ and $\lim\limits_{x \to 0^+} f(x) = \infty$, and conclude that the line $x = 0$ is a vertical asymptote as $x \to 0^+$. Horizontal asymptote: $y = 1$, approaching from below as $x \to -\infty$ and from above as $x \to \infty$.

(ix) See the following figure.

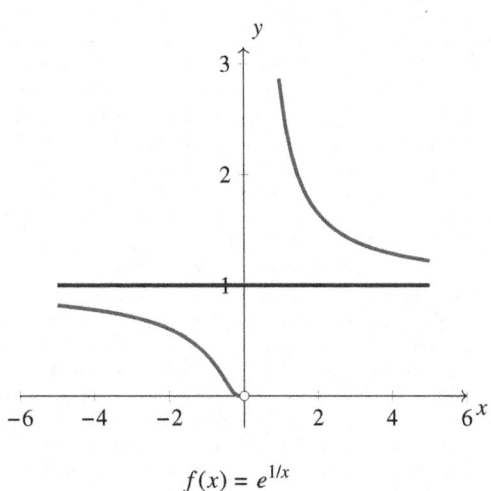

$$f(x) = e^{1/x}$$

(x) No absolute extrema.

Solution 3.37. Recall that, by definition, for $x > 0$, $x^x = e^{x \ln x}$. It follows: $f(x) = x^x$, $f'(x) = x^x(\ln x + 1)$, and $f''(x) = x^x(\ln x + 1)^2 + x^{x-1}$.

(i) Domain: $(0, \infty)$.

(ii) The function f is neither even nor odd.

(iii) No intercepts.

(iv) Increasing on $\left(\frac{1}{e}, \infty\right)$. Decreasing on $\left(0, \frac{1}{e}\right)$.

(v) Critical number: $\frac{1}{e}$. By the First Derivative Test, $f\left(\frac{1}{e}\right) = e^{-\frac{1}{e}} \approx 0.6922$ is a local minimum value.

(vi) Concave upward on its domain.

(vii) No point of inflection.

(viii) Apply L'Hospital's Rule to show that $\lim_{x \to 0^+} x^x = 1$. No asymptotes.

(ix) See the following figure.

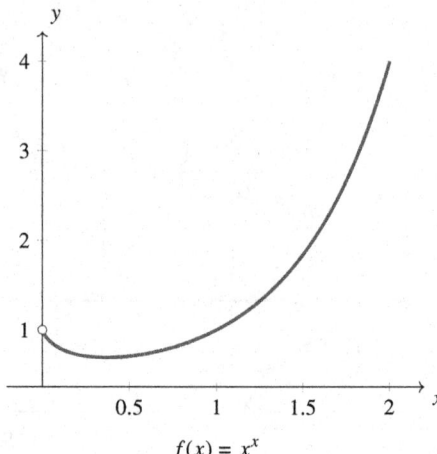

$$f(x) = x^x$$

(x) The absolute minimum value is $f\left(\dfrac{1}{e}\right) = e^{-\frac{1}{e}}$.

Chapter 4

Optimization

4.1 Introduction

Use your knowledge of calculus and the following strategies to solve the problems contained in this chapter.

Optimization Problem. An *optimization problem* is a problem of establishing the objective function $y = f(x)$, $x \in D$ and then finding its maximum value or its minimum value on D.

Closed Interval Method. Follow these steps to determine the absolute maximum and minimum values of a continuous function f on a closed interval $[a, b]$:

(1) Determine the values of f at the critical numbers of f in (a, b).

(2) Determine the values of f at the endpoints of the interval.

(3) The largest of the values from Steps 1 and 2 is the absolute maximum value of f on $[a, b]$, while the smallest of these values is the absolute minimum value.

Strategy for Solving Optimization Problems. To solve an optimization problem, it is common to follow these steps:

(1) Determine the objective function, i.e., a function f of one variable that models the situation described in the given problem.

(2) Determine the critical numbers of f.

(3) Use calculus tools, like the First and the Second Derivative Test and the Closed Interval Method, to determine whether the critical numbers correspond to a local/absolute maximum or minimum of f as required in the problem.

4.2   Routine Questions

Problem 4.1. Determine the absolute maximum and minimum values of $f(x) = 3x^2 - 9x$ on the interval $[-1, 2]$.

Problem 4.2. Determine the absolute and local maximum and minimum values of $f(x) = x^3 - 12x - 5$ on the interval $[-4, 6]$.

Problem 4.3. Let a and b be positive numbers and consider the function $f(x) = x^a (1 - x)^b$ with domain $[0, 1]$. Determine the maximum and minimum values of $f(x)$, if they exist.

Problem 4.4. Determine all critical numbers and all maximum and minimum values of the function $f(x) = |3x - 5|$ on the interval $[-3, 2]$.

Problem 4.5. The sum of two positive numbers is 12. What is the smallest possible value of the sum of their squares?

Problem 4.6. A farmer has 400 ft of fencing with which to build a rectangular pen. The farmer will use part of an existing straight wall 100 ft long as part of one side of the perimeter of the pen. What is the maximum area that can be enclosed?

Problem 4.7. Jessica is 2 km offshore in a boat. She wishes to reach her village that is 6 km down a straight shoreline from the point on the shore nearest to the boat. Jessica can row at 2 km/h and run at 5 km/h. Where should Jessica land her boat to reach the village in the least amount of time?

Problem 4.8. A rectangular box has a square base with edge length of $x > 1$ units. The total surface area of its six sides is 150 square units.

(1) Express the volume V of this box as a function of x.
(2) Determine the domain of the function $V = V(x)$.

(3) Determine the dimensions of the box with the greatest possible volume. What is this greatest possible volume?

Problem 4.9. An open-top box is to have a square base and a volume of $10 \, \text{m}^3$. The cost, per square meter of material, is $5 for the bottom and $2 for the four sides. Let x and y be lengths of the box's width and height respectively. Let C be the total cost of material required to make the box.

(1) Express C as a function of x and determine its domain.
(2) Determine the dimensions of the box so that the cost of materials is minimized. What is this minimum cost?

Problem 4.10. An open-top box is to have a square base and a volume of $13{,}500 \, \text{cm}^3$. Determine the dimensions of the box that minimize the amount of material needed to build such a box.

Problem 4.11. A cylindrical can without a top is made to contain $1000 \, \text{cm}^3$ of liquid. Determine the dimensions that will minimize the cost of the material to make the can.

4.3 Not-So-Routine Questions

Problem 4.12. Determine the point on the curve $x + y^2 = 0$ that is closest to the point $(0, -3)$.

Problem 4.13. A straight piece of wire 40 cm long is cut into two pieces. One piece is bent into a circle and the other is bent into a square. How long should each wire-piece be so that the total area of both the circle and square is minimized?

Problem 4.14. A straight piece of wire 28 cm long is cut into two pieces. One piece is bent into a square and the other piece is bent into a rectangle with aspect ratio three (i.e., the dimensions are a and $3a$ for some $a > 0$). What are the dimensions, in centimeters, of the square and the rectangle so that the sum of their areas is minimized?

Problem 4.15. With a straight piece of wire you are to create an equilateral triangle and a square, or only either one of the shapes. Suppose that the piece of wire is of length L meters and that the triangle has perimeter of length ℓ meters. Determine the value of ℓ that maximizes the total area of both the triangle and the square.

Problem 4.16. A rectangle is inscribed in a semicircle of radius R with one side of the rectangle along the diameter of the semicircle. Show that a perimeter of $2\sqrt{5}R$ is the largest possible perimeter of the rectangle.

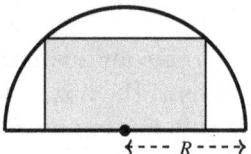

Problem 4.17. A rectangle, with sides parallel to the coordinate axes, is to be inscribed in the region enclosed by the graphs of $y = x^2$ and $y = 4$ so that its perimeter is of maximum length.

(1) Sketch the region under consideration.
(2) Suppose that the x-coordinate of the bottom right vertex of the rectangle is a. Determine a formula that expresses P, the length of the perimeter, in terms of a.
(3) Determine the value of a that gives the maximum value of P, and explain why you know that this value of a gives a maximum.
(4) What is the maximum value of P, i.e., the length of the perimeter of the rectangle?

Problem 4.18. Determine the dimensions of the rectangle of largest area that has its base on the x-axis and its other two vertices above the x-axis and lying on the parabola $y = 12 - x^2$.

Problem 4.19. A $10\sqrt{2}$ ft wall stands 5 ft from a building, see the following figure. Determine the length L of the shortest ladder, supported by the wall, that reaches from the ground to the building.

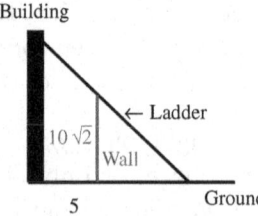

Problem 4.20. An attacking player (Gretzky) is skating with the puck along the boards as shown in the figure following. As Gretzky proceeds, the apparent angle α between the opponent's goal posts first increases, then decreases.

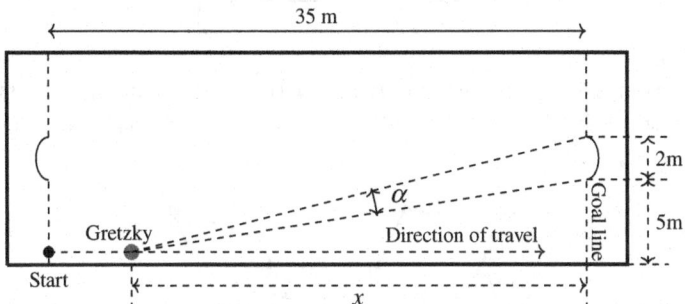

(1) Using the dimensions given in the figure, determine an expression for α in terms of the distance x from Gretzky to the goal line.
(2) Assume that Gretzky's chance of scoring is greatest when α is maximum (this may be the case if the opposing team has "pulled" their goalie). At which distance x, from the goal line, should Gretzky shoot the puck? It is clear that α is very small when $x = 35$ and $x = 0$, so there is no need to check the endpoints of the domain $[0, 35]$.

Problem 4.21. Inside an elliptical sport field, we want to design a rectangular soccer field with a maximum possible area. The sport field is given by the graph of the ellipse $b^2x^2 + a^2y^2 = a^2b^2$. Determine the length $2x$ and width $2y$ of the pitch (in terms of a and b) that maximize the area of the soccer field.

Problem 4.22. The top and bottom margins of a poster are each 6 cm and the side margins are each 4 cm. If the area of the printed material on the poster (that is, the area between the margins) is fixed at $384\,\text{cm}^2$, determine the dimensions of the poster with the smallest total area.

Problem 4.23. A poster is to have an area of $180\,\text{inch}^2$ with 1-inch margins at the bottom and the sides and a 2-inch margin at the top. What dimensions of the poster will give the largest printed area?

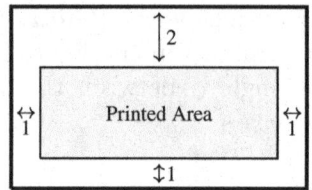

Problem 4.24. A water trough is to be made from a long strip of tin, 6 ft wide, by bending up at the angle θ a 2 ft strip at each side, as shown.

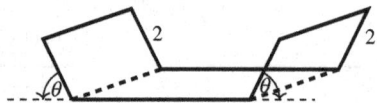

What angle θ would maximize the cross sectional area, and thus the volume, of the trough?

Problem 4.25. Determine the dimension of the right circular cylinder of maximum volume that can be inscribed in a right circular cone of radius R and height H.

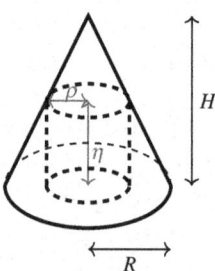

Problem 4.26. A hollow plastic cylinder with a circular base and open top is to be made, and $10\,\mathrm{m}^2$ of plastic is available. Determine the dimensions of the cylinder that give a maximum volume and evaluate the value of the maximum volume.

Problem 4.27. An open-topped cylindrical pot is to have volume $250\,\mathrm{cm}^2$. The material for the bottom of the pot costs 4 cents/cm^2 and that for its curved side costs 2 cents/cm^2. What dimensions will minimize the total cost of this pot?

Problem 4.28. Cylindrical soup cans are to be manufactured to contain a given volume V. No waste is involved in cutting the material for the vertical side of each can, but each top and bottom, which are circles of radius r, are cut from a square that measures $2r$ units on each side. Thus, the material used to manufacture each soup can has an area of $A = 2\pi rh + 8r^2$ square units, where h is the height of the can.

(1) How much material is wasted in making each soup can?
(2) Determine the ratio of the height to diameter for the most economical can (i.e., requires the least amount of material for manufacture).
(3) Use either the First or Second Derivative Test to verify that you have minimized the amount of material used for making each can.

Problem 4.29. Show that the volume of the largest cone that can be inscribed inside a sphere of radius R is $\dfrac{32\pi R^3}{81}$.

Problem 4.30. The sound level y, measured in watts per square meter, varies in direct proportion to the power of the source and inversely as the square of the distance from the source. Therefore, the sound level is given by $y = kPx^{-2}$, where P is the source power in watts, x is the distance from the source in meters, and k is a positive constant. Two beach parties, $100\,\text{m}$ apart, are playing loud music on their portable stereos. The second party's stereo has 64 times as much power as the first. The music approximates the white noise, so the power from the two sources arriving at a point between them adds without any concern about whether the sources are in or out of phase. To what point on the line segment between the two parties should I go if I wish to enjoy as much quiet as possible? Demonstrate that you have found an absolute minimum, not just a local minimum.

4.4 Answers, Hints, and Solutions

Solution 4.1. Note that the function f is continuous on the closed interval $[-1, 2]$. By the Extreme Value Theorem, the function f attains an absolute maximum value and an absolute minimum value on $[-1, 2]$. To find those extrema, we evaluate and compare the values of f at the endpoints of the interval $[-1, 2]$ and at the critical numbers of f that belong to $(-1, 2)$. From $f'(x) = 3(2x - 3)$, it follows that $x = \frac{3}{2}$ is the only such critical number. From $f(-1) = 12$, $f(2) = -6$, and $f\left(\frac{3}{2}\right) = -\frac{27}{4}$, we conclude that the absolute maximum value is $f(-1) = 12$ and the absolute minimum value is $f\left(\frac{3}{2}\right) = -\frac{27}{4}$.

Solution 4.2. Since $f'(x) = 3(x^2 - 4)$, the critical numbers of f in the interval $(-4, 6)$ are $x = \pm 2$. We find that $f(-4) = f(2) = -21$, $f(-2) = 11$, and $f(6) = 139$. Thus, the absolute maximum value is $f(6) = 139$ and the absolute minimum value is $f(-4) = f(2) = -21$. Since $f''(x) = 6x$, by the Second Derivative Test, $f(-2) = 11$ is a local maximum.

Reminder: By our definition, for $f(c)$ to be a local extremum value of a function f on an interval I, it is necessary that c is an interior point of I. This means that there is an open interval $I_1 \subset I$ such that $c \in I_1$.

Solution 4.3. From $f'(x) = ax^{a-1}(1-x)^b - bx^a(1-x)^{b-1} = x^{a-1}(1-x)^{b-1}(a - (a+b)x)$ and the fact that a and b are positive numbers, we conclude that $x = \frac{a}{a+b} \in (0, 1)$ is the only critical number of the function f. Since $f(0) = f(1) = 0$ and $f(x) > 0$ for all $x \in (0, 1)$, it follows that the maximum value of f is $f\left(\frac{a}{a+b}\right) = \left(\frac{a}{a+b}\right)^a \left(\frac{b}{a+b}\right)^b$, while the minimum value is $f(0) = f(1) = 0$.

Another approach is the following. Note that f is continuous on $[0, 1]$ with $f(0) = f(1) = 0$ and that $f(x) > 0$ for $x \in (0, 1)$. By the Extreme Value Theorem, there is $c \in (0, 1)$ such that $f(c) > 0$ is the absolute maximum value of f. Since f is differentiable on $(0, 1)$, c must be a critical number, which we found above to be $x = \frac{a}{a+b}$.

Solution 4.4. From $f(x) = \begin{cases} 3x - 5 & \text{if } x \geq \frac{5}{3} \\ -3x + 5 & \text{if } x < \frac{5}{3} \end{cases}$, it follows that $f'(x) = \begin{cases} 3 & \text{if } x > \frac{5}{3} \\ -3 & \text{if } x < \frac{5}{3} \end{cases}$. Note that the derivative of f is not defined at $x = \frac{5}{3}$ and that $f'(x) \neq 0$ for $x \neq \frac{5}{3}$. The only critical number of the function f on the interval $(-3, 2)$ is $x = \frac{5}{3}$. From $f\left(\frac{5}{3}\right) = 0$, $f(-3) =$

14, and $f(2) = 1$, it follows that the absolute minimum value is $f\left(\frac{5}{3}\right) = 0$ and that the absolute maximum value is $f(-3) = 14$.

Solution 4.5. Let $x \in (0, 12)$ be one of the two numbers. This means that the other number is $12 - x$. The question is to determine the minimum value of the function $f(x) = x^2 + (12 - x)^2$, $x \in (0, 12)$. From $f'(x) = 4(x - 6)$, it follows that $x = 6$ is the only critical number. Since $f''(6) = 4 > 0$, by the Second Derivative Test, it follows that $f(6) = 72$ is the minimum value of the function f.

Solution 4.6. Let x be the length of one side of the fence that is perpendicular to the wall. Note that the length of the side of the fence that is parallel to the wall equals $400 - 2x$ and that this number cannot be larger than 100. The question is to maximize the function $f(x) = x(400 - 2x)$, $x \in [150, 200)$. The only solution of the equation $f'(x) = 4(100 - x) = 0$ is $x = 100$, but this value is not in the domain of the function f. Clearly, $f'(x) < 0$ for $x \in (150, 200)$, which implies that f is decreasing on its domain. The maximum area that can be enclosed is $f(150) = 15{,}000\,\text{ft}^2$.

Solution 4.7. Let P be the point on the shore where Jessica lands her boat, and let x be the distance from P to the point S on the shore that is closest to her initial position. To reach the village, she needs to row the distance $z = \sqrt{4 + x^2}$ and run the distance $y = 6 - x$.

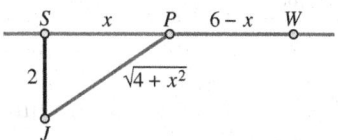

The time needed to row the distance z is $T_1 = \frac{z}{2}$, and the time Jessica needs to row the distance y is $T_2 = \frac{y}{5}$. The question is to minimize the function $T = T(x) = T_1 + T_2 = \frac{\sqrt{4+x^2}}{2} + \frac{6-x}{5}$, $x \in [0, 6]$. From $T'(x) = \frac{x}{2\sqrt{4+x^2}} - \frac{1}{5}$, it follows that the only critical number is $x = \frac{4}{\sqrt{21}}$. From $T(0) = \frac{11}{5} = 2.2$, $T(6) = \sqrt{10}$, and $T\left(\frac{4}{\sqrt{21}}\right) \approx 2.117$, by the Extreme Value Theorem, it follows that the minimum value is $T\left(\frac{4}{\sqrt{21}}\right)$. Thus, Jessica should land $\frac{4}{\sqrt{21}} \approx 0.872$ km from the point initially nearest to the boat.

Solution 4.8. (1) Let y be the height of the box. Then the surface area is given by $S = 2x^2 + 4xy$. From $S = 150$, it follows that $y = \frac{1}{2}\left(\frac{75}{x} - x\right)$. The volume of the box is $V = V(x) = \frac{x}{2}\left(75 - x^2\right)$.

(2) Recall that $x > 1$. Observe that the height $y = \frac{1}{2}\left(\frac{75}{x} - x\right)$ must be a positive number. This implies that the domain of the function V is the interval $(1, 5\sqrt{3})$.

(3) Note that $\frac{dV}{dx} = \frac{3}{2}(25 - x^2)$, which implies that $x = 5$ is the only critical number of the function $V = V(x)$. Since $\frac{d^2V}{dx^2} = -3x < 0$ for all $x \in (1, 5\sqrt{3})$, it follows that the maximum value is $V(5) = 125$ cubic units.

Solution 4.9. (1) Note that the height of the box is given by $y = \frac{\text{Volume}}{\text{Area of the base}} = \frac{10}{x^2}$. The cost function in dollars is given by $C(x) = $ (cost of the base) $ + 4$(cost of a side) $ = 5x^2 + 4\left(2x \cdot \frac{10}{x^2}\right) = 5x^2 + \frac{80}{x}$, $x > 0$.

(2) From $C'(x) = 10x - \frac{80}{x^2}$, it follows that $x = 2$ is the only critical number. By the Second Derivative Test, the box with dimensions $2 \times 2 \times \frac{5}{2}$ will minimize the cost. The minimum cost will be $C(2) = \$60$.

Solution 4.10. Let x be the length and the width of the box. Then its height is given by $y = \frac{13500}{x^2}$. It follows that the surface area is $S = x^2 + 4xy = x^2 + \frac{54000}{x}$ cm^2, $x > 0$. The question is to minimize S. From $\frac{dS}{dx} = 2x - \frac{54000}{x^2}$ and $\frac{d^2S}{dx^2} = 2 + \frac{3 \cdot 54000}{x^3} > 0$ for all $x > 0$, it follows that a local and the absolute minimum is $S(30) = 2700$ cm^2.

Solution 4.11. Let r be the radius of the base of the can and h be its height. Recall that the volume of the can is given by $V = r^2\pi h$ and that its surface area (do not forget that this can has no top) is given by $S = r^2\pi + 2r\pi h$. From $V = 1000$, we conclude that $h = \frac{1000}{r^2\pi}$. Since the cost to make the can is given by $C = (\$/\text{unit area}) \cdot (\text{surface area})$, it follows that we need to find $r > 0$ that minimizes the surface area function $S = S(r) = r^2\pi + \frac{2000}{r}$ for $r > 0$. From $S'(r) = 2r\pi - \frac{2000}{r^2}$, it follows that the only critical number is $r = \frac{10}{\sqrt[3]{\pi}}$. From $S''(r) = 2\pi + \frac{4000}{r^3} > 0$ for $r > 0$, it follows that the dimensions $r = h = \frac{10}{\sqrt[3]{\pi}}$ centimeters minimize the cost of the material needed to make the can.

Solution 4.12. The distance between a point $(x, y) = (-y^2, y)$ on the curve and the point $(0, -3)$ is given by $d = \sqrt{(x-0)^2 + (y-(-3))^2} = \sqrt{y^4 + (y+3)^2}$. Since the function $x \mapsto \sqrt{x}$ is monotone increasing, the question of finding the closest point is equivalent to minimizing the function $f(y) = d^2 = y^4 + (y+3)^2$, $y \in \mathbb{R}$.

From $f'(y) = 2(2y^3 + y + 3) = 2(y+1)(2y^2 - 2y + 3)$, it follows that $y = -1$ is the only critical number of the function f. From $f''(-1) = 14 > 0$, by the Second Derivative Test, we obtain that $f(-1) = 5$ is the local and also the absolute minimum value of f. Thus, the closest point is $(-1, -1)$.

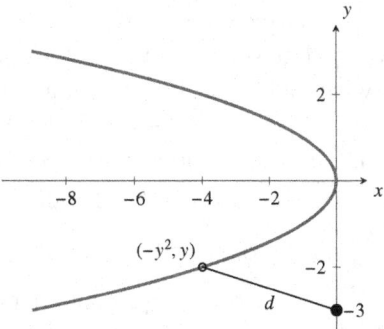

Solution 4.13. Let x be the radius of the circle. The question is to minimize the function $f(x) = \pi x^2 + \left(\frac{40 - 2\pi x}{4}\right)^2$, $x \in \left(0, \frac{20}{\pi}\right)$, based on the information that there are *two* pieces. The only critical number of the function f is $x = \frac{20}{\pi+4}$. To minimize the total area, the two pieces should be of the length $\frac{40\pi}{\pi+4}$ cm and $\frac{160}{\pi+4}$ cm.

Solution 4.14. Let x be the length of a side of the square, and let y be the length of the shorter side of the rectangle. From the constraint $4x + 8y = 28$, we obtain that $y = \frac{28 - 4x}{8}$. The question is to minimize the function $f(x) = x^2 + 3y^2 = x^2 + \frac{3(7-x)^2}{4}$, $x \in (0, 7)$, based on the information that there are *two* pieces. The critical number is $x = 3$, so the wire should be cut into pieces of lengths 12 cm and 16 cm.

Solution 4.15. Observe that the length of a side of the equilateral triangle is $\frac{\ell}{3}$ and that the length of a side of the square is $\frac{L-\ell}{4}$. The question is to *maximize* the function $f(\ell) = \frac{\sqrt{3}}{36}\ell^2 + \frac{(L-\ell)^2}{16}$, $\ell \in [0, L]$.

Note that $f'(\ell) = \frac{\sqrt{3}}{18}\ell - \frac{L-\ell}{8}$ and $f''(\ell) = \frac{\sqrt{3}}{18} + \frac{1}{8}$. Since $f''(\ell) > 0$, for $\ell \in [0, L]$, the maximum value of $f(\ell)$ must occur at $\ell = 0$ and/or at $\ell = L$. From $\frac{\sqrt{3}L^2}{36} = f(L) < f(0) = \frac{L^2}{16}$, we conclude that the maximum total area is obtained when only the square is constructed.

Note: Observe that $\ell = \frac{9L}{4\sqrt{3}+9}$ is a critical number of the function f. By the Second Derivative Test, this value of ℓ will *minimize* the total area of both the triangle and the square.

Solution 4.16. Let the rectangular coordinate system Oxy be such that the origin O is at the centre of the semicircle, that the diameter of the semicircle belongs to the x-axis, and that the positive ray of the y-axis intersects the semicircle. Hence, the semicircle is given by $y = \sqrt{R^2 - x^2}$, $x \in [-R, R]$. The set of vertices of any rectangle inscribed in this semicircle, with one side lying along the diameter of the circle, is given by $\{(x, 0), (x, \sqrt{R^2 - x^2}), (-x, \sqrt{R^2 - x^2}), (-x, 0)\}$, for some $x \in (0, R)$. The question is to maximize the function $P = P(x) = 4x + 2\sqrt{R^2 - x^2}$, $x \in (0, R)$. From $P'(x) = 4 - \frac{2x}{\sqrt{R^2-x^2}}$, it follows that $x = \frac{2R}{\sqrt{5}}$ is the only critical number of the function P. Since $P''(x) = -\frac{2R^2}{(R^2-x^2)^{3/2}} < 0$, by the Second Derivative Test, the absolute maximum is $P\left(\frac{2R}{\sqrt{5}}\right) = 2\sqrt{5}R$.

Solution 4.17. The optimal rectangle must have its top side lying along the line $y = 4$ and the corners of its lower side lying on the parabola, so we will assume this to be true for all rectangles under consideration.

(1) See the following figure.

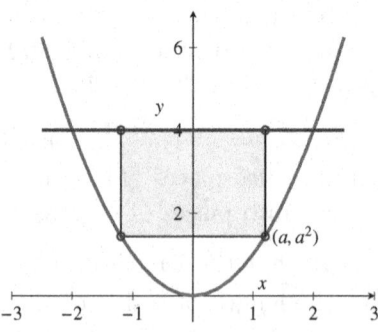

(2) $P = 4a + 2(4 - a^2) = 2(4 + 2a - a^2)$, $a \in (0, 2)$.

(3) From $\frac{dP}{da} = 4(1-a)$, it follows that $a = 1$ is the only critical number of the function P. By the Second Derivative Test, the fact that $\frac{d^2P}{da^2} = -4 < 0$ for all $a \in (0,2)$ implies that $P(1)$ is the maximum value. (4) $P(1) = 10$ units.

Solution 4.18. Let $(x,0)$ be the bottom right vertex of the rectangle. The question is to maximize $f(x) = 2x(12-x^2)$, $x \in (0, 2\sqrt{3})$. The only critical number of the function f is $x = 2$. The length of the base of the rectangle with the largest area is 4 units. The area for such a rectangle is $f(2) = 32$ square units. The height is $12 - 2^2 = 8$ units.

Solution 4.19. $L = 15\sqrt{3}$ ft. Let x be the distance from the wall to the base of the ladder, and let y be the height of the point where the ladder reaches the building.

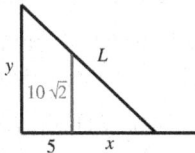

To minimize $L^2 = f(x) = (x+5)^2 + y^2$, use the fact that $\frac{x}{10\sqrt{2}} = \frac{x+5}{y}$ by similar triangles and determine the first derivative of the function f.

Solution 4.20. (1) Note that from Gretzky's point of view, the angle θ from the side along the boards to the left goalpost is the sum of two angles $\theta = \theta_1 + \alpha$, where θ_1 is the angle from the side along the boards to the right goalpost.

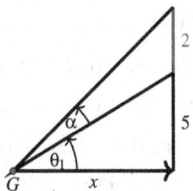

In other words, $\alpha = \theta - \theta_1$. It follows that $\alpha = \alpha(x) = \tan^{-1}\left(\frac{7}{x}\right) - \tan^{-1}\left(\frac{5}{x}\right)$, $x \in (0, 35)$.

(2) From $\frac{d\alpha}{dx} = \frac{5}{25+x^2} - \frac{7}{49+x^2} = \frac{2(35-x^2)}{(25+x^2)(49+x^2)}$ and the fact that $\alpha(x) > 0$, it follows that the only critical number of the function α is

$x = \sqrt{35}$. Gretzky should shoot the pack when he is $\sqrt{35} \approx 5.916\,\mathrm{m}$ from the goal line.

Solution 4.21. Observe that the largest field will have its corners lying on the ellipse. Let $(x, y) = \left(x, \frac{b}{a}\sqrt{a^2 - x^2}\right)$ be the upper right vertex of the rectangle. The question is to maximize the function $A(x) = 2x \cdot 2y = \frac{4b}{a}x\sqrt{a^2 - x^2}$, $x \in (0, a)$. From $A'(x) = \frac{4b}{a}\frac{a^2 - 2x^2}{\sqrt{a^2 - x^2}}$, it follows that the only critical number of the function A is $x = \frac{a}{\sqrt{2}}$.

By the First Derivative Test, $A\left(\frac{a}{\sqrt{2}}\right) = 2ab$ is a local maximum. Since $\lim\limits_{x \to 0^+} A(x) = \lim\limits_{x \to a^-} A(x) = 0$, by the Extreme Value Theorem, it follows that $A\left(\frac{a}{\sqrt{2}}\right) = 2ab$ is the absolute maximum value of the function A. To maximize the area of the soccer field, its length should be $a\sqrt{2}$ units and its width should be $b\sqrt{2}$ units.

Solution 4.22. Let a be the length of the printed area on the poster. Then the width of this area is $b = \frac{384}{a}$. It follows that the length of the poster is $x = a+8$ and the width of the poster is $y = b+12 = \frac{384}{a}+12$.

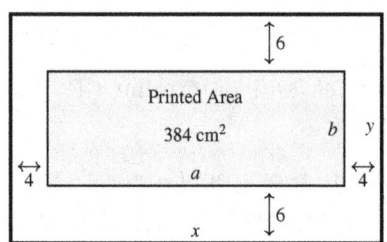

The question is to minimize the function $f(a) = xy = (a + 8)\left(\frac{384}{a} + 12\right) = 12\left(40 + a + \frac{256}{a}\right)$, $a > 0$. Since $f'(a) = 12\left(1 - \frac{16}{a}\right)\left(1 + \frac{16}{a}\right)$, it follows that $f'(a) < 0$ for $a \in (0, 16)$ and $f'(a) > 0$ for $a \in (16, \infty)$. By the First Derivative Test, the function has the absolute minimum at $a = 16$. The dimensions of the poster with the smallest area are $x = 24\,\mathrm{cm}$ and $y = 36\,\mathrm{cm}$.

Solution 4.23. $2\sqrt{30} \times 3\sqrt{30}$ square inches. Maximize the function $A(x) = (x-2)\left(\frac{180}{x} - 3\right)$, where x represents the length of the poster.

Solution 4.24. We need to maximize the area of the trapezoid with the height $h = 2\sin\theta$ and parallel sides of lengths $a = 2$ and $c = 2 + 2 \cdot 2\cos\theta = 2 + 4\cos\theta$.

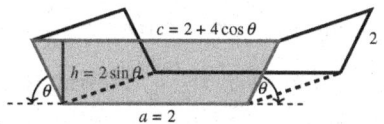

Thus, we maximize the function $A = A(\theta) = \frac{2+(2+4\cos\theta)}{2} \cdot 2\sin\theta =$ $4(\sin\theta + \sin\theta\cos\theta)$, $\theta \in (0, \pi)$. From $\frac{dA}{d\theta} = 4(\cos\theta + \cos^2\theta - \sin^2\theta) =$ $4(2\cos^2\theta + \cos\theta - 1) = 4(2\cos\theta - 1)(\cos\theta + 1)$, it follows that the critical number is $\theta = \frac{\pi}{3}$. By the First Derivative Test and the Extreme Value Theorem, a local and the absolute maximum value of the cross sectional area of the trough is $A\left(\frac{\pi}{3}\right) = 3\sqrt{3} \approx 5.196\,\mathrm{ft}^2$.

Solution 4.25. Let r be the radius of the base of a cylinder inscribed in the cone, and let h be its height. We use similar triangles to establish that $\frac{H}{R} = \frac{h}{R-r}$. It follows that $h = \frac{H(R-r)}{R}$, so the volume of the cylinder is $V = V(r) = \frac{\pi H}{R}r^2(R - r)$, $r \in (0, R)$.

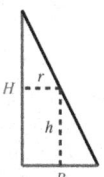

From $\frac{dV}{dr} = \frac{\pi H}{R}r(2R - 3r)$ and $\frac{d^2V}{dr^2} = \frac{2\pi H}{R}(R - 3r)$, it follows that the maximum value of the volume of the cylinder is $V\left(\frac{2R}{3}\right) = \frac{4\pi H R^2}{27}$. The corresponding dimensions are $r = \frac{2R}{3}$ units and $h = \frac{H}{3}$ units.

Solution 4.26. $r = \sqrt{\frac{10}{3\pi}}\,\mathrm{m}$, $h = \frac{5}{\pi}\sqrt{\frac{3\pi}{10}} - \frac{1}{2}\sqrt{\frac{10}{3\pi}} = \sqrt{\frac{10}{3\pi}}\,\mathrm{m}$, $V = \frac{10}{3}\sqrt{\frac{10}{3\pi}}\,\mathrm{m}^3$.

Solution 4.27. Let r be the radius of the base of the pot. Then the height of the pot is $h = \frac{250}{\pi r^2}$. The cost function is $C(r) = 4(\text{area of the bottom}) + 2(\text{area of the side}) = 4\pi r^2 + \frac{1000}{r}$, $r > 0$. The cost function attains its absolute minimum value when $r = \frac{5}{\sqrt[3]{\pi}} \approx 3.414\,\mathrm{cm}$.

Solution 4.28. (1) The surface area of the can is $S = 2\pi rh + 2\pi r^2$. The amount of material wasted is $A - S = 2(4 - \pi)r^2$. (2) From $V = \pi r^2 h$, it follows that the amount of material needed to make a can, of the given volume V, is $A = A(r) = \frac{2V}{r} + 8r^2$. This function attains its minimum value at $r = \frac{\sqrt[3]{V}}{2}$. From $V = \pi r^2 h$, it follows that $\frac{h}{r} = \frac{V}{r^3 \pi}$. The ratio of the height to the diameter for the most economical can is $\frac{h}{r}\Big|_{r=\frac{\sqrt[3]{V}}{2}} = \frac{V}{r^3 \pi}\Big|_{r=\frac{\sqrt[3]{V}}{2}} = \frac{4}{\pi}$. (2) By the Second Derivative Test, from $A''(r) = \frac{4V}{r^3} + 8 > 0$ for $r > 0$, the radius of $r = \frac{\sqrt[3]{V}}{2}$ cm minimizes the waste.

Solution 4.29. From the following figure deduce that $r^2 = R^2 - (h - R)^2 = h(2R - h)$.

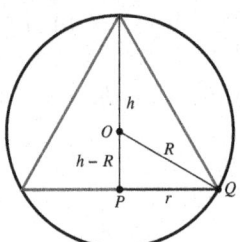

The volume of the cone, as a function of h, is given by $V = \frac{\pi}{3}h^2(2R - h)$, $h \in (0, 2R)$. From $\frac{dV}{dh} = \frac{\pi h}{3}(4R - 3h)$ infer that $h = \frac{4R}{3}$ is the only critical number. Use the fact that $\lim_{h \to 0^+} V(h) = \lim_{h \to 2R^-} V(h) = 0$ to conclude that the function $V = V(h)$ attains its absolute maximum at this critical number.

Solution 4.30. Let P be the source power in watts of the first party's stereo, and let x be the distance between the person and the first party in meters. Since the power of the second party's stereo is $64P$, the sound level is $L(x) = kPx^{-2} + 64kP(100 - x)^{-2}$, $x \in (0, 100)$. From $\frac{dL}{dx} = 2kP\left(\frac{64}{(100-x)^3} - \frac{1}{x^3}\right)$, it follows that $x = 20$ is the only critical number for the function L. Since $\frac{dL}{dx} > 0$ if and only if $x > 20$, it follows that the function L is strictly increasing on the interval $(20, 100)$ and strictly decreasing on the interval $(0, 20)$. By the First Derivative Test, $L(20)$ is a local minimum of L. Since $\lim_{x \to 0^+} L(x) = \lim_{x \to 100^-} L(x) = \infty$, by the Extreme Value Theorem, $L(20) = \frac{kP}{80}$ w/m^2 is the absolute minimum of the sound level. To enjoy as much quiet as possible, I should go to the point that is exactly 20 m away from the first party.

Chapter 5

Other Applications of Differentiation

5.1 Introduction

Use the following definitions, theorems, and properties to solve the problems contained in this chapter.

Rolle's Theorem. Let f be a function that satisfies the following three conditions:

(1) f is continuous on the closed interval $[a, b]$.
(2) f is differentiable on the open interval (a, b).
(3) $f(a) = f(b)$.

Then there is a number c in (a, b) such that $f'(c) = 0$.

The Mean Value Theorem. Let f be a function that satisfies the following conditions:

(1) f is continuous on the closed interval $[a, b]$.
(2) f is differentiable on the open interval (a, b).

Then there is a number c in (a, b) such that $f'(c) = \frac{f(b)-f(a)}{b-a}$
or, equivalently, $f(b) - f(a) = f'(c)(b - a)$.

Method of Linear Approximation. Suppose that a function f is differentiable at $a \in \mathbb{R}$. The linear function $L(x) = f(a) + f'(a)(x - a)$ is called the *linearization of f at a*. By the definition of the derivative, for b close to a, we have that $f(b) \approx L(b) = f(a) + f'(a)(b - a)$. The number $L(b)$ is called the *linear approximation of $f(b)$*.

Differential. Let f be a function differentiable at $x \in \mathbb{R}$. Let $\Delta x = dx$ be a (small) given number. The *differential dy* is defined as $dy = f'(x)\Delta x$.

The following figure depicts the geometrical meaning of the differential: In the right triangle with one leg of length Δx and the adjacent angle α such that $\tan \alpha = f'(x)$, the differential dy is the length of the leg opposite to α. In other words, $dy = L(x + \Delta x) - L(x)$, where the function L is the linearization of f at x. The figure suggests that, for small Δx, $\Delta y = f(x + \Delta x) - f(x) \approx dy$.

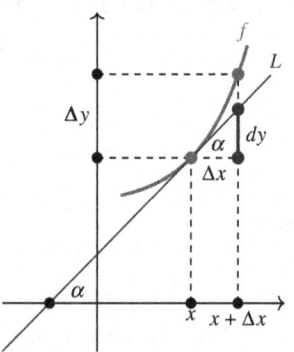

Newton's Method of Approximation. Under certain conditions,[a] it is possible to estimate a solution, say $x = r$, to the equation $f(x) = 0$ by using the following algorithm:

(1) Begin with an initial guess x_0.

(2) Calculate $x_1 = x_0 - \dfrac{f(x_0)}{f'(x_0)}$.

(3) If x_n is known then $x_{n+1} = x_n - \dfrac{f(x_n)}{f'(x_n)}$.

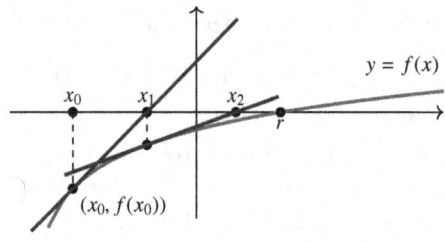

---

[a]See, for example, Burden, R.L., Faires, J.D., and Burden, A.M. (2015). *Numerical Analysis*. Cengage.

(4) If x_n and x_{n+1} agree to k decimal places then x_n approximates the root r up to k decimal places and $f(x_n) \approx 0$.

Antiderivative. A function F is called an *antiderivative* of f on an interval I if $F'(x) = f(x)$ for all $x \in I$. It is common to write $F(x) = \int f(x)dx$.

Initial Value Problem. Let $(x, y) \mapsto F(x, y)$ be a function of two variables and let $a, b \in \mathbb{R}$. The *initial value problem* $y' = F(x, y)$, $y(a) = b$ is a problem of finding all functions $y = f(x)$ such that $f'(x) = F(x, f(x))$, $x \in (a - \varepsilon, a + \varepsilon)$ for some $\varepsilon > 0$ with $f(a) = b$.

Natural Growth/Decay Model. The *natural growth/decay* is modelled by the initial-value problem $\frac{dy}{dt} = ky$, $y(0) = y_0$, $k, y_0 \in \mathbb{R}$.

Newton's Law of Cooling and Heating. *Newton's Law of Cooling and Heating* states that the temperature of a body changes at a rate proportional to the difference in temperature between the body and its surroundings. In other words, $\frac{dT}{dt} = k(T - T_s)$, where k is a constant, $T = T(t)$ is the temperature of the object at time t, and T_s is the temperature of the surroundings.

5.2 Rolle's Theorem and the Mean Value Theorem

Use Rolle's Theorem and the Mean Value Theorem to solve the following problems.

Problem 5.1. Determine all intervals where the function $f(x) = x^3 - 6x^2 + 11x - 6$ satisfies the hypothesis of Rolle's Theorem.

Problem 5.2. The following graph depicts the function $f(x) = 1 - (x - 2)^{\frac{2}{3}}$, $x \in [1, 3]$. Observe that f is a continuous function such that $f(1) = f(3) = 0$, but that the graph of f has no horizontal tangent line. Does this contradict Rolle's Theorem?

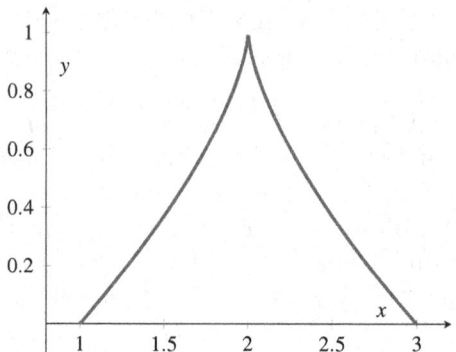

Problem 5.3. Consider a quadratic function $f(x) = ax^2 + bx + c$, $a \neq 0$, $b^2 - 4ac > 0$. Prove that the solution of the equation $f'(x) = 0$ lies between the roots of the equation $f(x) = 0$.

Problem 5.4. Suppose that a polynomial $p(x) = a_n x^n + a_{n-1} x^{n-1} + \cdots + a_1 x$, $n > 1$, $a_n \neq 0$, has a positive root. Prove that the polynomial $q(x) = n a_n x^{n-1} + (n-1) a_{n-1} x^{n-2} + \cdots + a_1$ has a positive root as well.

Problem 5.5. Suppose that a function g is differentiable everywhere. Let h be a function defined by $h(x) = g(x) + g(2 - x)$. Prove that the equation $h'(x) = 0$ has a solution in the interval $(0, 2)$.

Problem 5.6. Prove that the equation $x - \sin x = 1$ has exactly one solution in the interval $\left[\dfrac{\pi}{2}, \pi\right]$.

Problem 5.7. Verify that the function $g(x) = \dfrac{3x}{x+7}$ satisfies the hypothesis of the Mean Value Theorem on the interval $[-1, 2]$. Then find all numbers c that satisfy the conclusion of the Mean Value Theorem.

Problem 5.8. Let a and b be real numbers such that $a < b$. Consider the function $f(x) = |x|$, $x \in [a, b]$. Use the Mean Value Theorem to determine under which condition $|f(b) - f(a)| \neq |a - b|$.

Problem 5.9. Prove the following inequalities:

(1) $na^{n-1}(b - a) \leq b^n - a^n \leq nb^{n-1}(b - a)$ for $0 < a \leq b$ and $n > 1$.
(2) $|\sin b - \sin a| \leq |b - a|$ for all real numbers a and b.

(3) $|\arctan b - \arctan a| \le |b - a|$ for all real numbers a and b.
(4) $1 - \frac{a}{b} < \ln \frac{b}{a} < \frac{b}{a} - 1$ for $0 < a < b$.

Problem 5.10. Two horses start a race at the same time and finish in a tie. Prove that at some time during the race they have the same speed.

Problem 5.11. A car is driving along a rural road where the speed limit is $70\,\text{km/h}$ At 4:30 pm the odometer of the car reads $9075\,\text{km}$. At 4:48 the odometer reads $9098\,\text{km}$. Prove that the driver violated the speed limit at some instant between 4:30 and 4:48 pm.

Problem 5.12. Consider the function

$$f(x) = \begin{cases} \dfrac{3 - x^2}{2} & \text{if } x \in (-\infty, 1] \\ \dfrac{1}{x} & \text{if } x \in (1, \infty). \end{cases}$$

(1) Verify that the function f satisfies the hypothesis of the Mean Value Theorem on any closed interval $[a, b]$, $a < b$.
(2) Find all numbers c that satisfy the conclusion of the Mean Value Theorem on the interval $[0, 2]$.

Problem 5.13. If a function f is such that $-1 \le f'(x) \le 3$ for all $x \in \mathbb{R}$, then what can we say about the number $f(a + 2) - f(a)$, $a \in \mathbb{R}$?

Problem 5.14. Consider the curve $\mathcal{C}$ defined by the equation $y = x^4 - 1$ and observe that the points $A = (0, -1)$ and $B = (1, 0)$ belong to the curve. If M is a point on the arc of the curve $\mathcal{C}$ with the end points A and B, then what is the largest possible area of the triangle $\triangle AMB$?

Problem 5.15. Prove that any function $f : (a, b) \to \mathbb{R}$ such that $f'(x) = 0$ for all $x \in (a, b)$ must be a constant function.

Problem 5.16. Prove that any function $f : (a, b) \to \mathbb{R}$ such that $f'(x) > 0$ for all $x \in (a, b)$ must be an increasing function.

Problem 5.17. Prove that there is a constant c such that $\arctan \frac{1+x}{1-x} = c + \arctan x$ for all $x \in (1, \infty)$. Determine the value of c.

5.3   Approximation

Solve the following problems by using differentials, the linearization of a function at a given number, or Newton's method of approximation.

5.3.1   *Differential*

Problem 5.18. The circumference of a sphere is measured to be 24 cm with a maximum possible error of 0.25 cm. Use the differential dV to estimate the maximum error in the calculated volume V.

Problem 5.19. In a triangle $\triangle ABC$, the sum of two sides is $b+c = 4$ cm and the angle between them is $\alpha = \frac{\pi}{6}$ radians. How much will the area of $\triangle ABC$ increase if the side b increases from 1 cm to 1.1 cm?

Problem 5.20. Consider a circular sector $\mathcal{C}$ of radius $r = 100$ cm and with central angle $\theta = \frac{\pi}{3}$ radians. How will the area of $\mathcal{C}$ change under the following condition?

(1) The radius r increases by 1 cm.
(2) The angle θ decreases by $\frac{\pi}{360}$ radians.

Problem 5.21. The area A of a square of side length s is $A = s^2$. Suppose s increases by an amount $\Delta s = ds$.

(1) Draw a square and then illustrate the quantity dA on your diagram.
(2) If dA is used to approximate ΔA, illustrate the error of approximation on the same diagram.

5.3.2   *Method of linear approximation*

Problem 5.22. Use the linearization of the function $f(x) = \sqrt{(x+4)^3}$ at $a = 0$ to estimate the number $\sqrt{(3.95)^3}$. Is your estimate an overestimate or an underestimate?

Problem 5.23. Use the method of linear approximation to estimate the following numbers. For each estimate, decide if it is an overestimate or an underestimate.

(1) $\sqrt{80}$　　(2) $\sqrt[3]{65}$　　(3) $1001^{1/3}$　(4) $\sqrt[3]{26^2}$

(5) $(63)^{2/3}$　(6) $e^{-0.015}$　(7) $\ln 0.9$.

Problem 5.24. Assume that f is a function such that $f(5) = 2$ and $f'(5) = 4$. Using a linearization of f at $a = 5$, find an approximation of the number $f(4.9)$.

Problem 5.25. Suppose that we know that $g(2) = -4$ and $g'(x) = \sqrt{x^2 + 5}$ for all x. Use the method of linear approximation to estimate the number $g(2.05)$. Is your estimate larger or smaller than the actual value of $g(2.05)$?

Problem 5.26. Use the linearization of the function $f(x) = \sqrt{1 - x}$ at $a = 0$ to estimate the number $\sqrt{0.9}$. Sketch a graph to illustrate the relationship between $f(x) = \sqrt{1 - x}$ and its linearization at $a = 0$.

Problem 5.27. Use the linearization of the function $f(x) = \sqrt{1 + 2x}$ at $a = 0$ to estimate the number $\sqrt{1.1}$. Is your estimate an overestimate or underestimate?

Problem 5.28. Use the linearization of the function $f(x) = \sqrt[3]{x + 8}$ at $a = 0$ to estimate the numbers $\sqrt[3]{7.95}$ and $\sqrt[3]{8.1}$.

Problem 5.29. Use the linearization of the function $f(x) = (1 + x)^{100}$ at $a = 0$ to estimate the number $(1.0003)^{100}$. Is your estimate an overestimate or underestimate?

Problem 5.30. Determine the equation of the tangent line to the graph of the function $f(x) = \sqrt[3]{27 + 3x}$ at $x = 0$. Use your answer to estimate the value of $\sqrt[3]{30}$. Draw a sketch to show how the graph of f and its tangent line behave around the point where $x = 0$.

Problem 5.31. Use the linearization of the function $f(x) = \ln x$ at $a = 1$ to estimate the number $e^{-0.1}$.

Problem 5.32. Let $f(x) = \sqrt{x} + \sqrt[5]{x}$. Use the method of linear approximation to estimate the value of $f(1.001)$. Is your estimate of $f(1.001)$ too high or too low?

Problem 5.33.

(1) Find the linearization of $f(x) = \sin x$ at $a = \dfrac{\pi}{6}$.

(2) Is the differential df larger or smaller than $\Delta f = f\left(\dfrac{2\pi}{9}\right) - f\left(\dfrac{\pi}{6}\right)$?

Problem 5.34. Suppose that the only information we have about a function f is that $f(1) = 5$ and that the graph of its derivative is as shown in the figure.

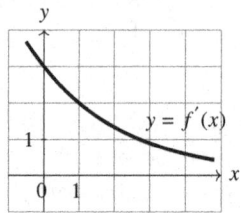

Use the method of linear approximation to estimate $f(0.9)$ and $f(1.1)$. Are your estimates too large or too small?

Problem 5.35. Prove that $\sqrt[n]{a^n + x} \approx a + \dfrac{x}{na^{n-1}}$ for $n \geq 2$, $a > 0$, and x close enough to 0.

Problem 5.36. The curve defined by $e^y + y(x - 3) = x^2 - 15$ passes through the point $A = (4, 0)$.

(1) Determine the values of $\dfrac{dy}{dx}$ and $\dfrac{d^2y}{dx^2}$ at A.
(2) Use linear approximation to estimate the value of y when $x = 3.95$.
(3) Is the true value of y greater or less than the approximation in part (2)? Make a sketch showing how the curve relates to the tangent line near the point A.

5.3.3   *Newton's method of approximation*

5.3.3.1   *Routine questions*

Problem 5.37.

(1) Explain how Newton's method can be used to approximate $\sqrt{5}$.
(2) Explain which of the following choices is the best initial approximation x_0 when using Newton's method as in (1): -1, 0, or 1?
(3) Determine the approximation x_3 of $\sqrt{5}$ by using Newton's method with the initial approximation x_0 you chose in (2).

Problem 5.38. Apply Newton's method to $f(x) = x^{1/3}$ with $x_0 = 1$ and calculate x_1, x_2, x_3, and x_4. Determine a formula for $|x_n|$.

What happens to $|x_n|$ as $n \to \infty$? Draw a picture that shows what is going on.

Problem 5.39. Determine the Newton's method iteration formula to compute an estimate of $\sqrt[3]{68}$. Provide an initial guess. Then explain whether your initial guess will lead to an overestimate or underestimate after the first iteration.

Problem 5.40.

(1) Show that if we apply Newton's method to approximate the value of $\sqrt[5]{k}$, $k \in \mathbb{R}$, we get the following iterative formula $x_{n+1} = \dfrac{x_n}{5}\left(4 + \dfrac{k}{x_n^5}\right)$.

(2) If $x_n = \sqrt[5]{k}$, what is the value of x_{n+1}?

(3) Take $x_0 = 2$ and use the formula in part (a) to find x_1, an estimate of the value of $\sqrt[5]{20}$ that is correct to one decimal place.

Problem 5.41. Use one iteration of Newton's method with $x_0 = \dfrac{\pi}{2}$ to approximate the x-coordinate of the point where the function $g(x) = \sin x$ crosses the line $y = x$.

Problem 5.42. The equation $8x^3 - 12x^2 - 22x + 25 = 0$ has a solution near $x_0 = 1$. Use Newton's method to find a better approximation x_1 to this solution.

Problem 5.43. The tangent line to the graph $y = f(x)$ at the point $A = (2, -1)$ is given by $y = -1 + 4(x - 2)$. It is also known that $f''(2) = 3$.

(1) Assume that Newton's method is used to solve the equation $f(x) = 0$ and that $x_0 = 2$ is the initial guess. Find the next approximation x_1 of the solution.

(2) Assume that Newton's method is used to find a critical number for f and that $x_0 = 2$ is the initial guess. Find the next approximation x_1 of the critical number.

Problem 5.44.

(1) Apply Newton's method to the equation $\dfrac{1}{x} - a = 0$ to derive the following algorithm for finding reciprocals: $x_{n+1} = 2x_n - ax_n^2$.

(2) Use the algorithm from part (1) to calculate $\dfrac{1}{1.128}$ correct to three decimal places starting with the initial approximation $x_0 = 1$.

Problem 5.45.

(1) Apply Newton's Method to the equation $x^2 - a = 0$ to derive the following algorithm for the square roots: $x_{n+1} = \frac{1}{2}\left(x_n + \frac{a}{x_n}\right)$.

(2) Approximate $\sqrt{2}$ by taking $x_0 = 2$ and calculating x_1.

Problem 5.46. Approximate $\sqrt[4]{81.1}$ using the following technique.

(1) The method of linear approximation.
(2) One iteration of Newton's Method.

Problem 5.47. You seek the approximate value of x near to 1.8 for which $\sin x = \frac{x}{2}$. Your first guess is that $x \approx x_0 = \frac{\pi}{2}$. Use one iteration of Newton's Method to find a better approximation to x.

Problem 5.48. For the function $f(x) = x^3 - 3x + 5$ use the Intermediate Value Theorem, and any other tools you need, to determine an interval of length 1 which contains a root of f. Choose the left endpoint of this interval to be x_0. Use this as a starting value to find two new iterations of the root of f by using Newton's Method. Determine from these whether Newton's Method is working.

Problem 5.49. Show that the function $f(x) = x^3 + 3x + 1$ has exactly one root in the interval $\left(-\frac{1}{2}, 0\right)$. Use Newton's Method to approximate the root. Stop when the next iteration agrees with the previous one to two decimal places.

Problem 5.50. Show that the equation $\ln x = -x^2 + 3$ has exactly one solution in the interval $[1, 3]$. Use Newton's Method to approximate the solution of the equation by starting with $x_0 = 1$ and finding x_1.

Problem 5.51. Show that the equation $2x = \cos x$ has exactly one solution. Use Newton's Method to approximate the solution of the equation by starting with $x_0 = 0$ and finding x_1.

Problem 5.52. Show that the equation $e^x = 2 \cos x$ has exactly one positive solution. Use Newton's Method to approximate the solution of the equation by starting with $x_0 = 0$ and finding x_1.

Problem 5.53. Show that there is only one number $a > 1$ such that $a^6 - a - 1 = 0$. Use Newton's Method with the initial approximation

$a_0 = 1$ to find a_1 and a_2. Technology gives $a \approx 1.13472$. Evaluate the number $|a_2 - 1.13472|$ to estimate the error of the approximation obtained after applying Newton's Method twice.

5.3.3.2 *Not-so-routine questions*

Problem 5.54. This question concerns finding zeros of the function

$$f(x) = \begin{cases} \sqrt{x} & \text{if } x \geq 0 \\ -\sqrt{-x} & \text{if } x < 0. \end{cases}$$

(1) If the initial approximation is $x_0 \neq 0$, what iteration formula does Newton's Method give for the next approximation?
(2) The root of the equation $f(x) = 0$ is $x = 0$. Explain why Newton's Method fails to find the root no matter which initial approximation $x_0 \neq 0$ is used. Illustrate your explanation with a sketch.

Problem 5.55. Use Rolle's Theorem to prove that the function $f(x) = \sin\left(\frac{\pi x}{2}\right) - x^2$ has a critical number in the interval $(0, 1)$. Set up the Newton's Method iteration formula to approximate the critical number.

Problem 5.56. A function h is said to have a fixed point at $x = c$ if $h(c) = c$. Suppose that the domain and range of a function f are both the interval $[0, 1]$ with $f(0) \neq 0$ and $f(1) \neq 1$ and that f is continuous on this domain. In addition, suppose that $f'(x) < 1$ for all $x \in (0, 1)$. Prove that f has has exactly one fixed point in $[0, 1]$. Use Newton's Method to determine an iteration formula for the fixed point at $x = c$.

Problem 5.57. Use Newton's Method to determine, correct to 2 decimal places, the x-coordinate of the point on the curve $y = e^x$ which is closest to the origin.

Problem 5.58. Prove that the equation $3^x = 3x$ has exactly two positive solutions. One of those solutions is $r_1 = 1$. Use Newton's Method to determine, correct to two decimal places, the second solution of the equation.

5.4   Antiderivatives and Initial Value Problems

To solve some of the problems in this section you may wish to use a table of antiderivatives from your textbook or another learning resource. Also, recall that the velocity $v = v(t)$ is the derivative of the position function $s = s(t)$ and that the acceleration $a = a(t)$ is the derivative of the velocity.

Problem 5.59. Determine the following antiderivatives.

(1) $\int (1-x)^8 dx.$ (2) $\int \tan^2 x dx.$ (3) $\int \frac{dx}{2+x^2}.$

(4) $\int e^{2x} \cosh x dx.$ (5) $\int \frac{z}{z^2+9} dz.$ (6) $\int \frac{dx}{x+x \ln x}.$

Problem 5.60. Determine a function f such that:

(1) $f'(x) = \dfrac{5}{x} + \dfrac{1}{1+x^2}.$

(2) $f'(x) = \dfrac{1}{2x+1}.$

(3) $f'(x) = \sin x + x^{-2} - e^x$ and $f(\pi) = 1.$

(4) $f'(x) = 2\cos(8x) + e^{3x} - 1$ and $f(0) = 1.$

(5) $f'(x) = 2\cos x + 8x^3 - e^x$ and $f(0) = 7.$

(6) $f''(t) = 2e^t + 3\sin t,$ $f(0) = 0$ and $f'(0) = 0.$

Problem 5.61. Suppose that h is a function such that $h'(x) = x^2 + 2e^x + 3$ and $h(3) = 0.$ What is $h(1)$?

Problem 5.62. Determine a function f such that $f'(x) = kf(x)$ and $f(0) = A$ with $k, A \in \mathbb{R}.$

Problem 5.63. Solve the initial value problems:

(1) $\dfrac{dy}{dx} = 2\sin 3x + x^2 + e^{3x} + 1,$ $y(0) = 0.$

(2) $\dfrac{dy}{dx} = \sqrt{1-y^2},$ $y(0) = 1.$

(3) $\dfrac{dy}{dx} = y^2 + 1,$ $y\left(\dfrac{\pi}{4}\right) = 0.$

(4) $\dfrac{dy}{dx} = 1 + y,$ $y(0) = 3.$

(5) $\dfrac{dx}{dt} = \dfrac{36}{(4t-7)^4},$ $x(2) = 1.$

(6) $\dfrac{dy}{dx} = e^{-y},$ $y(0) = 2.$

(7) $\dfrac{dy}{dt} = 2 - y,$ $y(0) = 1.$

(8) $\dfrac{dy}{dx} = 3\cos 2x + \dfrac{1}{e^{4x}},$ $y(0) = 1.$

Problem 5.64. Check that both $f(t) = 3 + \left(\dfrac{t}{5}\right)^5$ and $g(t) = 3$ are solutions of the initial value problem $y' = (y - 3)^{4/5}$, $y(0) = 3$.

Problem 5.65. Find a function $y = f(x)$ such that $\dfrac{d^2 y}{dx^2} = 6x$ and that the graph of f passes through the point $(0, 1)$ and has a horizontal tangent there.

Problem 5.66. Given the acceleration of a particle is $a(t) = 2e^t + 3 \sin t$, and $s(0) = 0$, $s(\pi) = 0$, find the position function $s(t)$ of the particle.

Problem 5.67. A particle moves in a straight line and has acceleration described by $a(t) = \sin t + 3 \cos t$. The particle's initial velocity is $v(0) = 2 \,\mathrm{cm/s}$ and initial displacement is $s(0) = 0 \,\mathrm{cm}$. Find the particle's position function $s(t)$.

Problem 5.68. A particle starts from rest (that is with initial velocity zero) at the point where $x = 10$ and moves along the x-axis with acceleration $a(t) = 12t$. Find the resulting position function $x(t)$.

Problem 5.69. At time $t = 0$, a car is moving at velocity of $6 \,\mathrm{m/s}$. At that instant, the driver begins to smoothly accelerate so that the acceleration after t seconds is $a(t) = 3t \,\mathrm{m/s^2}$.

(1) Write a formula for the velocity $v(t)$ of the car after t seconds.
(2) How far did the car travel during the time it took to accelerate from 6 m/s to 30 m/s?

Problem 5.70. The skid marks made by a car indicate that its breaks were fully applied for a distance of 160 ft before it came to a stop. Suppose that the car in question had a constant deceleration of $20 \,\mathrm{ft/s^2}$ under the conditions of the skid. How fast was the car traveling when its breaks were applied?

Problem 5.71. A stone is thrown downward off a building 30 m high. Suppose that the stone has an initial velocity of 5 m/s and that it falls with a constant acceleration due to gravity of $10 \,\mathrm{m/s^2}$. Assume that air resistance is negligible.

(1) Find the velocity function v for the falling stone.
(2) Find the position function s for the falling stone.
(3) How many seconds does it take for the stone to hit the ground?
(4) How fast is the stone traveling when it hits the ground?

Problem 5.72. You are standing at ground level and you throw a ball upward into the air with an initial velocity of 64 ft/s. The acceleration due to gravity is 32 ft/s^2 (towards the ground).

(1) How much time is the ball in the air for?
(2) What is the velocity of the ball at the time that it hits the ground?

Problem 5.73. A company is manufacturing and selling key chains. The marginal cost for producing key chains is $0.15 per key chain. A market survey has shown that for every $0.10 increase in the price per key chain, the company will sell 50 key chains less per week. Currently, the company sells 1000 key chains per week against the price that maximizes their profit. What is the price of one key chain?

5.5   Natural Growth and Decay

Recall that the solution of the initial-value problem $y' = ky$, $y(0) = A$, is given by $y = Ae^{kx}$. Use this fact to solve the following problems.

5.5.1   *Routine questions*

Problem 5.74.

(1) An amount of A_0 Canadian dollars (CAD) is invested at a yearly interest of $p\%$. Determine an expression for $A(t)$, the value of the investment in CAD after t years, if the interest is compounded continuously.
(2) Jane invests 10,000 CAD at a yearly interest of $p\%$ compounded continuously. After 4 years, the value of her investment is 15,000 CAD. What is p?

Problem 5.75. Suppose that C, the blood alcohol concentration (BAC), i.e., the amount of alcohol in grams per 100 ml of blood, obeys the decay equation $\frac{dC}{dt} = -0.2C$, where the elimination time is given in hours. If a person has a blood alcohol concentration of 0.1, how long would it take for blood alcohol concentration to drop to 0.05?

Problem 5.76. The mass of a radioactive material decays according to the rule $\dfrac{dm}{dt} = -5m$. Determine the half-life of the radioactive material.

Problem 5.77. Carbon dating is used to estimate the age of an ancient human skull. Let $f(t)$ be the proportion of original ^{14}C atoms remaining in the scull after t years of radioactive decay. Since ^{14}C has a half-life of 5700 years, we have $f(0) = 1$ and $f(5700) = 0.5$.

(1) Write an expression for $f(t)$ in terms of t.
(2) Suppose that only 15% of the original ^{14}C is found to be remaining in the skull. From your previous answer, derive an expression for the estimated age of the skull.

Problem 5.78. On a certain day, a scientist had 1 kg of a radioactive substance X at 1:00 pm. After 6 h, only 27 g of the substance remained. How much of the substance X was there at 3:00 pm that same day?

Problem 5.79. In a certain culture of bacteria, the number of bacteria increased tenfold in 10 h. Assuming natural growth, how long did it take for their number to double?

Problem 5.80. A bacteria population $P(t)$ quadruples every 15 min. The initial bacteria population is $P(0) = 10$.

(1) What is the population after 3 h?
(2) How much time does it take for the population to grow to 1 billion?

Problem 5.81. A freshly brewed cup of coffee has temperature 95°C and is in a 20°C room. When the coffee's temperature is 70°C, it is cooling at the rate of 1°C per minute. When does this occur?

Problem 5.82. A cold drink is taken from a refrigerator and placed outside, where the temperature is 32°C. After 25 min outside, the drink's temperature is 14°C and after 50 min outside its temperature is 20°C. Assuming the temperature of the drink obeys Newton's Law of Cooling, what was the initial temperature of the drink?

Problem 5.83. On Hallowe'en night you go outside to sit on the porch to hand out candy. It is a cold night and the temperature is only 10°C so you have made a cup of hot chocolate to drink. If the hot chocolate is 90°C when you first go out, how long does it take until the hot chocolate is a drinkable (60°C) given that $k = 0.03s^{-1}$?

5.5.2   *Not-so-routine questions*

Problem 5.84. The rate at which a student learns new material is proportional to the difference between a maximum M and the amount $A(t)$ the student already knows at time t. This is called a learning curve.

(1) Write an initial value problem to model the learning curve.
(2) Solve the initial value problem you created in part (1).
(3) It took a student 100 h to learn 50% of the material in a calculus class. If the student would like to know 75% in order to get a B, how much longer should the student study? You may assume that the student began knowing none of the material and that the maximum the student can achieve is 100%.

Problem 5.85. Suppose that C, the blood alcohol concentration (BAC) obeys the decay differential equation $\frac{dC}{dt} = -\frac{1}{k}C$, where the constant $k = 7$ h is called the elimination time. It is estimated that a male weighing 70 kg, who drinks three pints of beer over a period of 1 h, has a BAC of 0.112. The allowed legal concentration for driving is a maximum of 0.08.

(1) If a person has a BAC of 0.112, how long should the person wait before driving in order not to disobey the law.
(2) What is the initial (i.e., at $t = 0$) rate of change in the concentration?

Note: The permissible BAC limit in the Criminal Code of Canada is 0.08 (80 mg of alcohol in 100 ml of blood). Some advocate a lower criminal limit of 0.05 (50 mg of alcohol in 100 ml of blood).

Problem 5.86. Plutonium-239 is part of the highly radioactive waste that nuclear power plans produce. The half-life of Plutonium-239 is 24,110 years. Suppose that 10 kg of Plutonium-239 has leaked into and contaminated a lake. Let $m(t)$ denote the mass of Plutonium-239 that remains in the lake after t years.

(1) Determine an expression for $m(t)$ based on the given information.
(2) How much mass remains in the lake after 1000 years?
(3) Suppose the lake is considered safe for use when only 1 kg of the Plutonium-239 remains. How many years will this take?

Problem 5.87. A bacterial culture starts with 500 bacteria and after 3 h there are 8000. Assume that the culture grows at a rate proportional to its size.

(1) Determine an expression in t for the number of bacteria after t hours.
(2) Determine the number of bacteria after 6 h.
(3) Determine an expression of the form $m\dfrac{\ln a}{\ln b}$ with m, a, and b positive integers for the number of hours it takes the number of bacteria to reach a million.

Problem 5.88. The population of a bacteria culture grows at a rate that is proportional to the size of the population.

(1) Let P denote the population of the culture at time t. Express $\dfrac{dP}{dt}$ in terms of the proportional constant k and P.
(2) If the population is 240 at time $t = 1$ and is 360 at time $t = 2$, determine a formula for the number of bacteria at time t. Take t to be in hours.
(3) How many bacteria were there at time $t = 0$?
(4) What is the value of $\dfrac{dP}{dt}$ when $t = 0$?

Problem 5.89. Assume that a Calculus class in the fall of 2010 had an enrollment of 500 students and in the fall of 2020 had an enrollment of 750 students. Assume also that, if $P(t)$ is the enrollment at time t in years with $t = 0$ corresponding to year 2010, then $P'(t) = kP(t)$ for some constant k. Calculate $P(90)$, the enrollment in the Calculus class in the fall of 2100.

Problem 5.90. A cup of coffee, cooling off in a room at temperature $20°$C, has cooling constant $k = 0.09\,\text{min}^{-1}$.

(1) How fast is the coffee cooling in degrees per minute when its temperature is $T = 80°$C?
(2) Use linear approximation to estimate the change in temperature over the next 6 s when $T = 80°$C.

(3) The coffee is served at a temperature of 90°. How long should you wait before drinking it if the optimal drink-temperature is 65°C?

Problem 5.91. In a murder investigation the temperature of the corpse was 32.5°C at 1:30 pm and 30.3°C an hour later. Normal body temperature is 37°C and the temperature of the surroundings was 20°C. When did the murder take place?

5.6   Answers, Hints, and Solutions

Solution 5.1. As a polynomial, the function f is differentiable and therefore continuous everywhere. From $f(x) = (x-1)(x-2)(x-3)$, we conclude that the function f satisfies the hypothesis of Rolle's theorem on the intervals $[1, 2]$, $[1, 3]$, and $[2, 3]$.

Solution 5.2. From $f'(x) = -\dfrac{2}{3(x-2)^{\frac{1}{3}}}$, we conclude that the function is not differentiable at $x = 2 \in (1, 3)$. Hence, the hypothesis of Rolle's Theorem are not satisfied on the interval $[1, 3]$, which explains why there is no horizontal tangent line.

Solution 5.3. Observe that, as a polynomial, the function f is differentiable everywhere. Since $b^2 - 4ac > 0$, it follows that there are $x_1, x_2 \in \mathbb{R}$, $x_1 < x_2$, such that $f(x_1) = f(x_2) = 0$. By Rolle's Theorem, there is $c \in (x_1, x_2)$ such that $f'(c) = 0$. Since f' is a linear function, this is the only solution of the equation $f'(x) = 0$.

Solution 5.4. Observe that, as a polynomial, the function p is differentiable everywhere. Let $a > 0$ be such that $p(a) = 0$. Since $p(0) = 0$, by Rolle's Theorem, there is $c \in (0, a)$ such that $p'(c) = q(c) = 0$.

Solution 5.5. Note that $h(2) = h(0)$ and apply Rolle's Theorem (or the Mean Value Theorem) for the function h on the closed interval $[0, 2]$.

Solution 5.6. Let $f(x) = x - \sin x - 1$, $x \in \left[\frac{\pi}{2}, \pi\right]$. Observe that the function f is continuous on the interval $\left[\frac{\pi}{2}, \pi\right]$ and differentiable on $\left(\frac{\pi}{2}, \pi\right)$. From $f\left(\frac{\pi}{2}\right) = \frac{\pi}{2} - 2 < 0$ and $f(\pi) = \pi - 1 > 0$, by the Intermediate Value Theorem, there is $a \in \left(\frac{\pi}{2}, \pi\right)$ such that $f(a) = 0$.

Suppose that there is $b \in \left(\frac{\pi}{2}, \pi \right)$, $b \neq a$, such that $f(b) = 0$. Say, $a < b$. By Rolle's Theorem, there is c with $\frac{\pi}{2} < a < c < b < \pi$ such that $f'(c) = 1 - \cos c = 0$, which is not possible. Hence, a is the unique root of the given equation.

Solution 5.7. Since $x + 7 \neq 0$ for all $x \in [-1, 2]$, we conclude that the function g, as a rational function, is continuous on the closed interval $[-1, 2]$ and differentiable on the open interval $(-1, 2)$. Therefore, the function g satisfies the hypothesis of the Mean Value Theorem on the interval $[-1, 2]$. By the Mean Value Theorem, there is $c \in (-1, 2)$ such that $g'(c) = \frac{g(2) - g(-1)}{2 - (-1)}$. We solve $\frac{21}{(c+7)^2} = \frac{7}{18}$ for c to obtain $c = -7 \pm 3\sqrt{6}$. Since $-7 - 3\sqrt{6} < -1$, this value is rejected. Since $\sqrt{6} \in (2, 3)$, it follows that $c = -7 + 3\sqrt{6} \in (-1, 2)$ is the only value that satisfies the conclusion of the Mean Value Theorem.

Solution 5.8. Observe that, if $a \geq 0$ or $b \leq 0$, the function f satisfies the hypothesis of the Mean Value Theorem. This implies that, if $a \geq 0$ or $b \leq 0$, then $\left| \frac{f(b) - f(a)}{b - a} \right| = 1$ by the Mean Value Theorem. If $a < 0 < b$, then $\left| \frac{f(b) - f(a)}{b - a} \right| = \left| \frac{b + a}{b - a} \right|$. Therefore, $|f(b) - f(a)| \neq |a - b|$ if and only if $a < 0 < b$.

Solution 5.9. (1) The inequality is satisfied if $a = b$. Let $n > 1$ and consider the power function $f(x) = x^n$, $x \in \mathbb{R}$. Observe that $f'(x) = nx^{n-1}$, i.e., that f is differentiable everywhere. Let $a, b \in (0, \infty)$, $a < b$. Since f satisfies the hypothesis of the Mean Value Theorem on the interval $[a, b]$, there is $c \in (a, b)$ such that $\frac{b^n - a^n}{b - a} = nc^{n-1}$. Since f is a power function with the exponent greater than 1, from $0 < a < c < b$, it follows that $a^n < c^n < b^n$. Hence, $na^{n-1} < \frac{b^n - a^n}{b - a} < nb^{n-1}$ and the inequality follows.

(2) The inequality is satisfied if $a = b$. Let $a, b \in \mathbb{R}$, $a < b$, and let $f(x) = \sin x$, $x \in [a, b]$. The function f is continuous on the closed interval $[a, b]$ and differentiable on (a, b). By the Mean Value Theorem, there is $c \in (a, b)$ such that $\cos c = \frac{\sin b - \sin a}{b - a}$. Since $|\cos c| \leq 1$ for all real numbers c, it follows that $|\sin b - \sin a| \leq |b - a|$.

(3) Use the fact that the function $f(x) = \arctan x$ satisfies the hypothesis of the Mean Value Theorem on any closed interval.

(4) Let $a, b \in (0, \infty)$, $a < b$, and let $f(x) = \ln x$, $x \in [a, b]$. The function f is continuous on the closed interval $[a, b]$ and differentiable on (a, b). By the Mean Value Theorem, there is $c \in (a, b)$ such that $\frac{1}{c} = \frac{\ln b - \ln a}{b - a} = \frac{\ln \frac{b}{a}}{b - a}$. Since $0 < a < c < b$ implies $\frac{1}{b} < \frac{1}{c} < \frac{1}{a}$, the inequality follows.

Solution 5.10. Let $f(t)$ be the distance that the first horse covers from the start in time t and let $g(t)$ be the distance that the second horse covers from the start in time t. Let T be time in which the two horses finish the race. It is given that $f(0) = g(0)$ and $f(T) = g(T)$. Let $F(t) = f(t) - g(t)$, $t \in [0, T]$. As the difference of two position functions, the function F is continuous on the closed interval $[0, T]$ and differentiable on the open interval $(0, T)$. By the Mean Value Theorem (or by Rolle's Theorem), there is $c \in (0, T)$ such that $F'(c) = \frac{F(T) - F(0)}{T - 0} = 0$. It follows that $f'(c) = g'(c)$, which is the same as to say that at the instant c the two horses have the same speed.

Solution 5.11. Let t, in hours, be the time that has elapsed since 4:30 pm. Observe that $t = \frac{18}{60} = \frac{3}{10}$ h corresponds to 4:48 pm. Let $f(t)$ be the distance, in kilometers, traveled in time t. Hence, $f(0) = 0$ and $f\left(\frac{3}{10}\right) = 9098 - 9075 = 23$. By the Mean Value Theorem, there $c \in \left(0, \frac{3}{10}\right)$ such that $f'(c) = \frac{23}{\frac{3}{10}} \approx 76.6$. This means that he driver violated the speed limit at the instant c.

Solution 5.12. (1) Observe that the function f is differentiable for any $x \neq 1$. From $f(1) = \lim\limits_{x \to 1^-} f(x) = \lim\limits_{x \to 1^-} \frac{3 - x^2}{2} = 1$ and $\lim\limits_{x \to 1^+} f(x) = \lim\limits_{x \to 1^+} \frac{1}{x} = 1$, it follows that the function is continuous at $x = 1$. From, $\lim\limits_{h \to 0^-} \frac{f(1+h) - f(1)}{h} = \lim\limits_{h \to 0^-} \frac{-2h + h^2}{2h} = -1$ and $\lim\limits_{h \to 0^+} \frac{f(1+h) - f(1)}{h} = \lim\limits_{h \to 0^+} \frac{-h}{h(h+1)} = -1$ it follows that $f'(1) = -1$, i.e., the function f is differentiable at $x = 1$. The function f is differentiable everywhere and therefore satisfies the hypothesis of the Mean Value Theorem on any closed interval $[a, b]$, $a < b$.

(2) We solve the equation $f'(c) = \frac{f(2) - f(0)}{2 - 0} = -\frac{1}{2}$, $c \in (0, 2)$. Since $c \neq 1$, this equation is equivalent to the union of two equations,

$f'(c) = -c = -\frac{1}{2}$, $c \in (0,1)$, and $f'(c) = -\frac{1}{c^2} = -\frac{1}{2}$, $c \in (1,2)$. The two numbers c that satisfy the conclusion of the Mean Value Theorem on the interval $[0,2]$ are $c = \frac{1}{2}$ and $c = \sqrt{2}$.

Solution 5.13. Note that all conditions of the Mean Value Theorem are satisfied. For some $c \in (a, a+2)$, $f(a+2)-f(a) = 2f'(c) \in [-2,6]$.

Solution 5.14. The arc of the curve $\mathcal{C}$ with the end points A and B is a graph of the function $f(x) = x^4 - 1$, $x \in [0,1]$.

Since, for $x \in (0,1)$, $f'(x) = 4x^3 > 0$ and $f''(x) = 12x^2 > 0$, the function f is increasing on its domain and its graph is concave downward. This means that all points that lie on the graph of f belong to the region in the plane that is bounded by the chord $\overline{AB}$ and the tangent to the curve that is parallel to this chord.

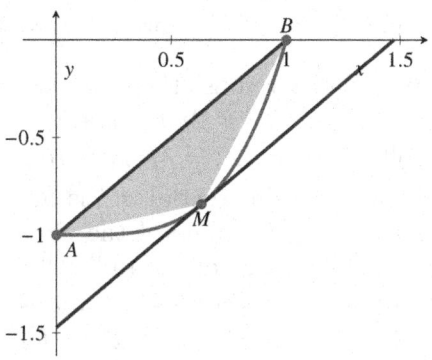

$$f(x) = x^4 - 1, x \in [0,1]$$

By the Mean Value Theorem, the coordinates of the point M at which the tangent line to the curve $\mathcal{C}$ is parallel to the chord $\overline{AB}$ are $\left(4^{-\frac{1}{3}}, 4^{-\frac{4}{3}} - 1\right)$. Clearly, the point M is the farthest away from the chord, so the area of the triangle $\triangle AMB$ is the largest possible.

We use the fact that the distance between the point M and the line $x - y - 1 = 0$, i.e., the line through the points A and B, is given by $d = \frac{|4^{-\frac{1}{3}} - (4^{-\frac{4}{3}}-1)-1|}{\sqrt{2}} = \frac{3}{8\sqrt[6]{2}}$ to obtain that the area of $\triangle AMB$ is $\frac{1}{2} \cdot \sqrt{2} \cdot \frac{3}{8\sqrt[6]{2}} = \frac{3\sqrt[3]{2}}{16}$.

Solution 5.15. Let x_0 be any number in the interval (a,b). By the Mean Value Theorem, for any $x \in (a,b)$, $x \neq x_0$, there is c between

x_0 and x such that $\frac{f(x)-f(x_0)}{x-x_0} = f'(c) = 0$. Hence, for any $x \in (a,b)$, $f(x) = f(x_0)$.

Solution 5.16. Let x_1 and x_2 be any two numbers in the interval (a,b) such that $x_1 < x_2$. By the Mean Value Theorem, there is $c \in (x_1, x_2)$ such that $\frac{f(x_2)-f(x_1)}{x_2-x_1} = f'(c) > 0$. Hence, $f(x_2) > f(x_1)$ for any $x_1, x_2 \in (a,b)$ such that $x_2 > x_1$, which proves that f is an increasing function on the interval (a,b).

Solution 5.17. Let $f(x) = \arctan \frac{1+x}{1-x} - \arctan x$. Then, for $x \in (1, \infty)$, $f'(x) = \frac{1}{1+\left(\frac{1+x}{1-x}\right)^2} \cdot \frac{2}{(1-x)^2} - \frac{1}{1+x^2} = 0$. It follows that there is a constant c such that $\arctan \frac{1+x}{1-x} = c + \arctan x$ for all $x \in (1, \infty)$.

In particular, $c = \lim\limits_{x \to 1+} \left(\arctan \frac{1+x}{1-x} - \arctan x \right) = \frac{\pi}{2} - \frac{\pi}{4} = \frac{\pi}{4}$.

Solution 5.18. If r is the radius, then the circumference of the sphere is $C = 2\pi r$. Hence, $dC = 2\pi dr$. The volume of the sphere is $V = \frac{4}{3}r^3\pi$. It follows that $dV = 4r^2\pi dr$. It is given that $|C - 24| \le 0.25$. Recall that the differential dC is an approximation of the difference between the true value of the circumference and the value obtained by a small change of the radius.[b] In other words, we may take that $|dC| \le 0.25$ cm. From $C = 2\pi r \approx 24$ cm, it follows that $r \approx \frac{12}{\pi}$ cm. From $|dC| = 2\pi|dr| \le 0.25$, it follows that we may take $|dr| \le \frac{1}{8\pi}$ cm. Therefore, the maximum error in the calculated volume V is approximated by $|dV| = 4r^2\pi|dr| \approx 4 \cdot \left(\frac{12}{\pi}\right)^2 \cdot \pi \cdot |dr| \le 4 \cdot \left(\frac{12}{\pi}\right)^2 \cdot \pi \cdot \frac{1}{8\pi} = \frac{72}{\pi^2} \approx 7.29$ cm^3.

Solution 5.19. The area of $\triangle ABC$ is $A = A(b) = \frac{1}{2}bc\sin\alpha = \frac{1}{4}b(4-b)$. Hence, $dA = \frac{1}{2}(2-b)db$. From $A(1.1) - A(1) \approx dA = A'(1) \cdot (1.1 - 1) = 0.05$, we conclude that the area of $\triangle ABC$ will increase by approximately 0.05 cm^2.

Solution 5.20. Recall that the area of a circular sector of radius r and with central angle θ is given by $A = \frac{r^2\theta}{2}$.

---

[b]Think about the thickness of the measuring tape that we may use to measure the circumference of the sphere.

(1) Here, we consider the area A to be a function of r and that $\theta = \dfrac{\pi}{3}$ is fixed. Hence, $A = A(r) = \dfrac{\pi r^2}{6}$. It follows that $A(101) - A(100) \approx dA = A'(100) \cdot (101 - 100) = \dfrac{100\pi}{3}$. Therefore, if the radius r increases by $1\,\mathrm{cm}$, the area will increase by approximately $104.72\,\mathrm{cm}^2$.

(2) Now, we consider the area A to be a function of θ and that $r = 100$ is fixed. Hence, $A = A(\theta) = 5000\,\theta$. It follows that $A\left(\dfrac{\pi}{3} - \dfrac{\pi}{360}\right) - A\left(\dfrac{\pi}{3}\right) \approx dA = A'\left(\dfrac{\pi}{3}\right) \cdot \left(\dfrac{\pi}{3} - \dfrac{\pi}{360} - \dfrac{\pi}{3}\right) = -\dfrac{125\pi}{9}$. Therefore, if the angle θ decreases by $\dfrac{\pi}{360}$ radians, the area will decrease by approximately $43.63\,\mathrm{cm}^2$.

Solution 5.21. Observe that $\Delta A = A(s + \Delta s) - A(s) = (s + \Delta s)^2 - s^2 = 2s\Delta s + (\Delta s)^2$ and $dA = 2s\Delta s$. Therefore, the error is $(\Delta s)^2$.

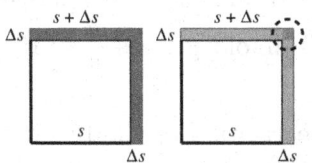

Solution 5.22. Note that $f(0) = 8$. From $f'(x) = \dfrac{3}{2}\sqrt{x + 4}$, it follows that $f'(0) = 3$. The linearization of f at $a = 0$ is $L(x) = 8 + 3x$. For x *close* to 0, we have that $f(x) = \sqrt{(x + 4)^3} \approx L(x)$. Thus, $\sqrt{(3.95)^3} = f(-0.05) \approx L(-0.05) = 8 - 0.15 = 7.85$. Since $f''(x) = \dfrac{3}{4\sqrt{x+4}} > 0$ in the neighborhood of $x = 0$, the graph of the function f is above the tangent line at $x = 0$. Thus, $L(-0.05)$ is an underestimate. *Note:* Technology gives $\sqrt{3.95^3} \approx 7.85046973117$.

Solution 5.23. (1) Let $f(x) = \sqrt{x}$. Then $f'(x) = \dfrac{1}{2\sqrt{x}}$, $f(81) = 9$, and $f'(81) = \dfrac{1}{18}$. The linearization of the function f at $a = 81$ is $L(x) = 9 + \dfrac{1}{18}(x - 81)$. It follows that $\sqrt{80} = f(80) \approx L(80) = 9 - \dfrac{1}{18} = \dfrac{161}{18}$. Since the graph of the function f is concave downward, $L(80)$ is an overestimate. *Note:* Technology gives $\dfrac{161}{18} \approx 8.944444444$ and $\sqrt{80} \approx 8.944271910$.

(2) $\sqrt[3]{65} \approx L(65) = \dfrac{193}{48}$. Overestimate. *Note:* Technology gives $\dfrac{193}{48} \approx 4.020833333333$ and $\sqrt[3]{65} \approx 4.02072575859$.

(3) $L(x) = 10 + \frac{1}{300}(x - 1000)$ implies $1001^{1/3} \approx L(1001) = \frac{3001}{300}$. Overestimate. *Note*: Technology gives $\frac{3001}{300} \approx 10.00333333$ and $\sqrt[3]{1001} \approx 10.00333222$.

(4) Let $f(x) = x^{\frac{2}{3}}$. Then $f'(x) = \frac{2}{3}x^{-\frac{1}{3}}$, $f(27) = 9$, and $f'(27) = \frac{2}{9}$. The linearization of the function f at $a = 27$ is $L(x) = 9 + \frac{2}{9}(x - 27)$. It follows that $\sqrt[3]{26^2} = f(26) \approx L(26) = 9 - \frac{2}{9} = \frac{79}{9}$. From $f''(x) = -\frac{2}{9}x^{-\frac{4}{3}} < 0$, it follows that the graph of the function f is concave downward. Overestimate. *Note*: Technology gives $\frac{79}{9} \approx 8.777777778$ and $\sqrt[3]{26^2} \approx 8.776382955$.

(5) The linearization of the function $f(x) = x^{\frac{2}{3}}$ at $a = 64$ is $L(x) = 16 + \frac{1}{6}(x - 64)$. It follows that $(63)^{2/3} = f(63) \approx L(63) = 16 - \frac{1}{6} = \frac{95}{6}$. Overestimate. *Note*: Technology gives $\frac{95}{6} \approx 15.83333333$ and $\sqrt[3]{63^2} \approx 15.83289626$.

(6) Determine the linearization of the function $f(x) = e^x$ at $a = 0$ to obtain $e^{-0.015} \approx 0.985$. Underestimate. *Note*: Technology gives $e^{-0.015} \approx 0.9851119396$.

(7) The linearization of the function $f(x) = \ln x$ at $a = 1$ is $L(x) = x - 1$. It follows that $\ln 0.9 \approx L(0.9) = -0.1$. Overestimate. *Note*: Technology gives $\ln 0.9 \approx -0.1053605157$.

Solution 5.24. The linearization of the function f at $a = 5$ is $L(x) = 2 + 4(x - 5)$. Thus, $f(4.9) \approx L(4.9) = 2 - 0.4 = 1.6$.

Solution 5.25. The linearization of the function g at $a = 2$ is $L(x) = -4 + 3(x - 2)$. Thus, $g(2.05) \approx L(2.05) = -3.85$. From $g''(2) = \frac{2}{3} > 0$, it follows that the the graph of the function g is concave downward at $a = 2$, i.e., the graph of the function lies below the tangent line. Thus, the estimate is larger than $f(2.05)$.

Solution 5.26. The linearization of the function f at $a = 0$ is $L(x) = 1 - \frac{x}{2}$. It follows that $\sqrt{0.9} \approx L(0.1) = 1 - \frac{1}{20} = \frac{19}{20}$. Recall that the graph of the linearization L is the tangent line to the graph of f at the point $(0, 1)$. See the following figure.

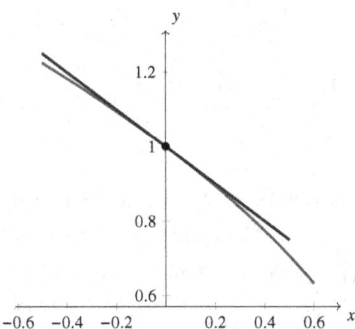

Solution 5.27. The linearization is $L(x) = 1 + x$, which implies $\sqrt{1.1} = f(0.05) \approx L(0.05) = 1.05$. This is an overestimate since the graph of f is concave downward. *Note*: Technology gives $\sqrt{1.1} \approx 1.048808848$.

Solution 5.28. The linearization is $L(x) = 2 + \frac{x}{12}$, which implies $\sqrt[3]{7.95} \approx L(-0.05) = 2 - \frac{1}{240} = \frac{479}{240}$ and $\sqrt[3]{8.1} \approx L(0.1) = 2 + \frac{1}{120} = \frac{241}{120}$. Since the graph of the function f is concave downward, both approximations are overestimates. *Note*: Technology gives $\frac{479}{240} \approx 1.995833333$ and $\sqrt[3]{7.95} \approx 1.995824623$. Also, $\frac{241}{120} \approx 2.00833333333$ and $\sqrt[3]{8.1} \approx 2.008298850$.

Solution 5.29. The linearization is $L(x) = 1 + 100x$, which implies $(1.0003)^{100} \approx L(0.0003) = 1.03$. Since $f''(x) = 100 \cdot 99(1 + x)^{98} > 0$, the graph of the function f is concave upward. Underestimate. *Note*: Technology gives $1.0003^{100} \approx 1.03044989785$.

Solution 5.30. An equation of the tangent line is $y = \frac{x}{9} + 3$, which implies $\sqrt[3]{30} \approx \frac{1}{9} + 3 = \frac{28}{9}$. See the following figure.

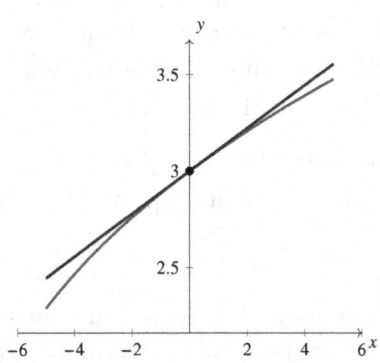

Note: Technology gives $\frac{28}{9} \approx 3.111111111$ and $\sqrt[3]{30} \approx 3.107232506$.

Solution 5.31. The linearization is $L(x) = x - 1$. If $u = e^{-0.1}$ then $\ln u = -0.1 \approx L(u) = u - 1$. Thus, $u \approx 0.9$. *Note*: Technology gives $e^{-0.1} \approx 0.9048374180$.

Solution 5.32. The linearization of the function $f(x) = \sqrt{x} + \sqrt[5]{x}$ at $a = 1$ is $L(x) = 2 + \frac{7}{10}(x - 1)$. It follows that $f(1.001) \approx L(1.001) = 2 + 0.7 \cdot 0.001 = 2.0007$. Note that the domain of the function f is the interval $[0, \infty)$. From $f''(x) = -\frac{1}{4}x^{-\frac{3}{2}} - \frac{4}{25}x^{-\frac{9}{5}}$, it follows that the graph of f is concave downward on the interval $(0, \infty)$. The graph of the function is below the tangent line at $a = 1$, so the estimate $f(1.001) \approx 2.0007$ is too high. *Note*: Technology gives $f(1.001) \approx 2.00069979511$.

Solution 5.33. (1) The linearization is $L(x) = \frac{1}{2} + \frac{\sqrt{3}}{2}\left(x - \frac{\pi}{6}\right)$. (2) Observe that $\frac{2\pi}{9} - \frac{\pi}{6} = \frac{\pi}{18}$, i.e., observe that $\frac{2\pi}{9}$ is "close" to $\frac{\pi}{6}$. From $\Delta f - df = \left(f\left(\frac{2\pi}{9}\right) - f\left(\frac{\pi}{6}\right)\right) - \left(L\left(\frac{2\pi}{9}\right) - f\left(\frac{\pi}{6}\right)\right) = f\left(\frac{2\pi}{9}\right) - L\left(\frac{2\pi}{9}\right)$, conclude that the question is if $L\left(\frac{2\pi}{9}\right)$ is an overestimate or an underestimate of $f\left(\frac{2\pi}{9}\right)$. From $f''\left(\frac{\pi}{6}\right) < 0$, it follows that $L\left(\frac{2\pi}{9}\right)$ is an overestimate, so $\Delta f - df < 0$ and $\Delta f < df$.

Solution 5.34. We read from the graph that $f'(1) = 2$. The linearization of the function f at $a = 1$ is $L(x) = 5 + 2(x - 1)$. It follows that $f(0.9) \approx L(0.9) = 4.8$ and $f(1.1) \approx L(1.1) = 5.2$. Since f' is a decreasing function, $f''(x) < 0$ and the graph of the function f is concave downward. Overestimates.

Solution 5.35. Let $f(x) = \sqrt[n]{a^n + x}$. If n is an even integer, then the domain of the function f is the interval $[-a^n, \infty)$. If n is an odd integer, then f is defined on the set of all real numbers. In both cases, f is differentiable at $x = 0$. From $f'(x) = \frac{1}{n\sqrt[n]{(a^n + x)^{n-1}}}$, it follows that the linearization of the function f at $x = 0$ is $L(x) = a + \frac{x}{na^{n-1}}$. Therefore, for all x sufficiently close to 0, $\sqrt[n]{a^n + x} \approx a + \frac{x}{na^{n-1}}$.

Solution 5.36. (1) $\left.\frac{dy}{dx}\right|_{x=4} = 4$, $\left.\frac{d^2y}{dx^2}\right|_{x=4} = -11$; (2) $y(3.95) \approx -0.2$; (3) Since the curve is concave downward, the tangent line is above the curve and the approximation is an overestimate. *Note*: Observe that, for any fixed $x > 3$, there is a unique $y \in \mathbb{R}$ such that $e^y + y(x-3) = x^2 - 15$.

Solution 5.37. (1) We use Newton's Method to solve the equation $x^2 - 5 = 0$, $x > 0$. Since $f(x) = x^2 - 5$ and $f'(x) = 2x$, Newton's Method yields the iteration formula $x_{n+1} = x_n - \frac{x_n^2 - 5}{2x_n} = \frac{1}{2}\left(x_n + \frac{5}{x_n}\right)$. (2) A rough estimate of $\sqrt{5}$ gives a value that is a bit greater than 2. Thus, we take $x_0 = 1$. (3) $x_1 = 3$, $x_2 = \frac{7}{3}$, $x_3 = \frac{47}{21} \approx 2.23809$. *Note:* Technology gives $\sqrt{5} \approx 2.23606$.

Solution 5.38. Let $f(x) = x^{\frac{1}{3}}$. Then Newton's Method yields the iteration formula $x_{n+1} = x_n - \frac{f(x_n)}{f'(x_n)} = x_n - \frac{x_n^{\frac{1}{3}}}{\frac{1}{3}x_n^{-\frac{2}{3}}} = -2x_n$. So, $|x_{n+1}| = 2|x_n|$. This implies that if $x_0 \neq 0$, $|x_n| = 2^n|x_0| \to \infty$ as $n \to \infty$. Thus, Newton's Method does not work in this case.

The figure depicts what is happening if we apply Newton's Method with the initial guess $x_0 = 1$. Then $x_1 = -2$, $x_2 = 4$, and $x_3 = -8$. The sequence obtained by Newton's Method is going away from the root $r = 0$.

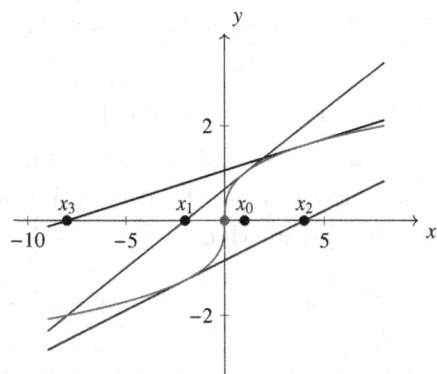

Solution 5.39. The question is to approximate the solution of the equation $x^3 - 68 = 0$. We consider the function $f(x) = x^3 - 68$, $x > 0$, and use Newton's Method to obtain the iteration formula $x_{n+1} = x_n - \frac{x_n^3 - 68}{3x_n^2}$. If we choose $x_0 = 4$, then our initial guess is smaller than the root of the equation $x^3 - 68 = 0$ because $4^3 = 64 < 68$. Since the graph of the function f is concave upward for

$x > 0$, the graph of f is above the tangent line to the graph at the point $(4, f(4))$. This means that the tangent line will cross the x-axis at a point that is greater than $\sqrt[3]{68}$. Hence, $x_1 = 4 - \dfrac{64-68}{48} = \dfrac{49}{12}$ is an overestimate. *Note*: Technology gives $\dfrac{49}{12} \approx 4.08333333333$ and $\sqrt[3]{68} \approx 4.08165510192$.

Solution 5.40. (1) If $f(x) = x^5 - k$, then $f'(x) = 5x^4$ and Newton's Method gives $x_{n+1} = x_n - \dfrac{x_n^5 - k}{5x_n^4} = \dfrac{4x_n^5 + k}{5x_n^4} = \dfrac{x_n}{5}\left(4 + \dfrac{k}{x_n^5}\right)$.
(2) $x_{n+1} = x_n = \sqrt[5]{k}$. (3) From $x_1 = \dfrac{2}{5}\left(4 + \dfrac{20}{2^5}\right) = \dfrac{2}{5} \cdot \dfrac{37}{8} = 1.85$
and $x_2 = 1.821486137$ it follows that the approximation $\sqrt[5]{20} \approx x_1 = 1.85$ is correct to one decimal place. *Note*: Technology gives $\sqrt[5]{20} \approx 1.820564203$.

Solution 5.41. The question is approximate the solution of the equation $\sin x - x = 0$ with $x_0 = \dfrac{\pi}{2}$. Thus, $x_1 = \dfrac{\pi}{2} - \dfrac{\sin\frac{\pi}{2} - \frac{\pi}{2}}{\cos\frac{\pi}{2} - 1} = 1$. Newton's Method with $x_0 = \dfrac{\pi}{2}$ gives $x_7 = 0.08518323251$. Clearly the solution of the given equation is $x = 0$.

Solution 5.42. $x_1 = 1 - \dfrac{-1}{-22} = \dfrac{21}{22}$. *Note*: Technology gives $\dfrac{21}{22} \approx 0.9545454545$ and approximates the solution of the equation as 0.9555894038.

Solution 5.43. (1) Observe that, as the slope of the tangent line, $f'(2) = 4$. Hence, $x_1 = 2 - \dfrac{-1}{4} = \dfrac{9}{4}$. (2) The question is to approximate a solution of the equation $f'(x) = 0$ with the initial guess $x_0 = 2$, $f'(2) = 4$, and $f''(2) = 3$ given. Hence, $x_1 = 2 - \dfrac{4}{3} = \dfrac{2}{3}$.

Solution 5.44. (1) From $f(x) = \dfrac{1}{x} - a$ and $f'(x) = -\dfrac{1}{x^2}$, it follows that $x_{n+1} = x_n - \dfrac{\frac{1}{x_n} - a}{-\frac{1}{x_n^2}} = 2x_n - ax_n^2$. (2) Note that $\dfrac{1}{1.128}$ is the solution of the equation $\dfrac{1}{x} - 1.128 = 0$. Thus, $x_1 = 2 - 1.128 = 0.872$, $x_2 = 2 \cdot 0.872 - 1.128 \cdot 0.872^2 = 0.886286848$, and $x_3 = 0.8865247589$. *Note*: Technology gives $\dfrac{1}{1.128} \approx 0.8865248227$.

Solution 5.45. (2) $x_1 = \dfrac{3}{2}$.

Solution 5.46. (1) The linearization of $f(x) = \sqrt[4]{x}$ at $a = 81$ is $L(x) = 3 + \frac{x-81}{108}$. It follows that $\sqrt[4]{81.1} \approx L(81.1) = 3 + \frac{1}{1080} \approx$ 3.000925. (2) For $g(x) = x^4 - 81.1$ and $x_0 = 3$, Newton's Method gives $x_1 = 3 + \frac{1}{1080} \approx 3.000925$. *Note*: Technology gives $\sqrt[4]{81.1} \approx$ 3.00092549756.

Solution 5.47. $x_1 = \frac{\pi}{2} - \frac{1 - \frac{\pi}{4}}{-\frac{1}{2}} = 2$. *Note*: Technology estimates the positive solution of the equation $\sin x = \frac{x}{2}$ as 1.895494267. Newton's Method with the initial guess $x_0 = \frac{\pi}{2}$ gives $x_2 \approx 1.900995594$.

Solution 5.48. From $f'(x) = 3(x^2 - 1)$, it follows that the critical numbers are $x = \pm 1$. From $f(1) = 3$, $f(-1) = 7$, $\lim\limits_{x \to -\infty} f(x) = -\infty$, and $\lim\limits_{x \to \infty} f(x) = \infty$, it follows that f has only one root and that root belongs to the interval $(-\infty, -1)$. From $f(-2) = 3 > 0$ and $f(-3) = -13 < 0$, we conclude that the root belongs to the interval $(-3, -2)$ by the Intermediate Value Theorem. Let $x_0 = -3$. Then $x_1 = -3 - \frac{-13}{24} = -\frac{59}{24} \approx -2.458333333$ and $x_2 \approx -2.294310576$. It seems that Newton's Method is working, the new iterations are inside the interval $(-3, -2)$, where the root is. *Note*: Technology estimates the solution of the equation $x^3 - 3x + 5 = 0$ as $x \approx -2.279018786$.

Solution 5.49. Observe that $f'(x) = 3(x^2 + 1) > 0$ for all x. Hence, as an always increasing cubic polynomial, f has a unique root. From $f\left(-\frac{1}{2}\right) = -\frac{5}{8} < 0$ and $f(0) = 1 > 0$, the function f has a root in the interval $\left(-\frac{1}{2}, 0\right)$ by the Intermediate Value Theorem. Take $x_0 = -\frac{1}{3}$. Then $x_1 = -\frac{1}{3} - \frac{-\frac{1}{27} + 3 \cdot \left(-\frac{1}{3}\right) + 1}{3\left(\frac{1}{9} + 1\right)} = -\frac{29}{90} \approx -0.3222222222$ and $x_2 \approx -0.3221853550$. *Note*: Technology estimates the solution of the equation $x^3 + 3x + 1 = 0$ as $x \approx -0.3221853546$.

Solution 5.50. Consider the function $f(x) = \ln x + x^2 - 3$, $x \in [1, 3]$. Evaluate $f(1)$ and $f(3)$, and then use the Intermediate Value Theorem to conclude that f has a root in $(1, 3)$. From $f'(x) = \frac{1}{x} + 2x > 0$ for $x \in (1, 3)$, conclude that, as an increasing function, f has a unique root in this interval. From $f(1) = -2$ and $f'(1) = 3$, it follows that $x_1 = \dfrac{5}{3} \approx 1.66$. *Note*: Technology estimates the solution of the equation as $x = 1.592142937$.

Solution 5.51. Consider the function $f(x) = 2x - \cos x$, $x \in \mathbb{R}$. Evaluate $\lim\limits_{x \to -\infty} f(x)$ and $\lim\limits_{x \to \infty} f(x)$, and then use the Intermediate Value Theorem to conclude that f has a root. From $f'(x) = 2 + \sin x > 0$, conclude that, as an increasing function, f has a unique root. From $f(0) = -1$ and $f'(0) = 2$, it follows that $x_1 = 0.5$. *Note*: Technology estimates the solution of the equation as $x = 0.4501836113$.

Solution 5.52. $x_1 = 1$.

Solution 5.53. Consider the function $f(x) = x^6 - x - 1$, $x > 0$ to find $a_1 = 1.2$ and $a_2 \approx 1.143575843$. Also, $|a_2 - 1.1347| \approx 0.008855843$.

Solution 5.54. (1) Observe that $f(x) = \text{sign}(x) \cdot \sqrt{|x|}$ and $f'(x) = \dfrac{1}{2\sqrt{|x|}}$, $x \neq 0$. For $x_0 \neq 0$ Newton's Method gives the iteration formula $x_{n+1} = -x_n$. (2) By induction, $|x_n| = |x_0|$ for all $n \in \mathbb{N}$. See the figure.

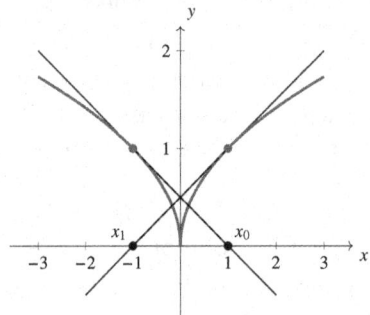

Solution 5.55. From $f(0) = f(1) = 0$, by Rolle's Theorem, it follows that the equation $f'(x) = \dfrac{\pi}{2} \cos\left(\dfrac{\pi x}{2}\right) - 2x = 0$ has a solution in the interval $(0, 1)$. Newton's Method gives the iteration formula $x_{n+1} = x_n + \dfrac{\dfrac{\pi}{2}\cos\left(\dfrac{\pi x_n}{2}\right) - 2x_n}{\dfrac{\pi^2}{4}\sin\left(\dfrac{\pi x_n}{2}\right) + 2}$.

Solution 5.56. Consider the function $g(x) = f(x) - x$, $x \in [0, 1]$. Use the necessary calculus techniques to establish the existence of a unique fixed point, i.e., a unique solution of the equation $g(c) = f(c) - c = 0$, $c \in (0, 1)$. Newton's Method gives the iteration formula $x_{n+1} = x_n - \dfrac{f(x_n) - x_n}{f'(x_n) - 1}$.

Solution 5.57. The square of the distance between the point (x, e^x) and the origin is given by $f(x) = x^2 + e^{2x}$. The question is to minimize this function. To find critical numbers of the function f we need to solve the equation $f'(x) = 2(x + e^{2x}) = 0$, or equivalently, the equation $e^{2x} = -x$. It should be clear that this equation has a unique solution somewhere in the interval $(-1, 0)$. Newton's Method gives the iteration formula $x_{n+1} = x_n - \frac{x_n + e^{2x_n}}{1 + 2e^{2x_n}}$. If $x_0 = -0.5$, then $x_1 \approx -0.42388311523$ and $x_2 \approx -0.4263000539$. The first coordinate of the point on the curve that is closest to the origin is $x \approx -0.4263$. *Note*: Technology approximates the first coordinate as -0.42630275100.

Solution 5.58. Let $f(x) = 3^x - 3x$, $x \in \mathbb{R}$. Since the exponential function $x \mapsto 3^x \ln 3$ is an increasing function on its domain with the range $(0, \infty)$, the equation $f'(x) = 3^x \ln 3 - 3 = 0$ has a unique solution. This means that the function f has a unique critical number $c \in (0, \infty)$. Since $3 > \ln 3 > 1$, from $3^c \ln 3 = 3$, it follows that $1 < 3^c < 3$, which implies $0 < c < 1$. From $f''(c) = 3^c \cdot (\ln 3)^2 > 0$, it follows, by the Second Derivative Test, that the function f has a local minimum at c. Since f is an increasing function on the interval (c, ∞) (keep in mind that c is the only critical number and that the function f has a local minimum there) and since $f(1) = 0$, it follows that $f(c) < 0$. Since $f(0) = 1$, by the Intermediate Value Theorem, there is $r \in (0, c)$ such that $f(r) = 0$. Since f is a decreasing function on the interval $(-\infty, c)$, this root is unique. Therefore, the equation $3^x = 3$ has exactly two solutions, $r_1 = 1$ and $r_2 = r \in (0, 1)$. We use Newton's Method to approximate the number r. From the iteration formula $x_{n+1} = x_n - \frac{3^{x_n} - 3x_n}{3^{x_n} \ln 3 - 3}$ with the initial guess $x_0 = 0.75$, we obtain $x_1 = 0.8095253734$, $x_2 = 0.8247865836$, and $x_3 = 0.8260095086$. Hence, $r \approx 0.826$. *Note*: Technology approximates the number r as 0.8260175600961.

Solution 5.59. (1) $\int (1 - x)^8 dx = \begin{vmatrix} 1 - x = t \\ dx = -dt \end{vmatrix} = -\int t^8 dt = -\frac{t^9}{9} + C = -\frac{1}{9}(1 - x)^9 + C$, $C \in \mathbb{R}$.

(2) $\int \tan^2 x \, dx = \int \frac{1 - \cos^2 x}{\cos^2 x} dx = \tan x - x + C$, $C \in \mathbb{R}$.

(3) $F(x) = \frac{1}{\sqrt{2}} \arctan \frac{x}{\sqrt{2}} + C$, $C \in \mathbb{R}$. Use the substitution $x = t\sqrt{2}$.

(4) $F(x) = \frac{1}{6} e^{3x} + \frac{1}{2} e^x + C$, $C \in \mathbb{R}$. Use the fact that $\cosh x = \frac{e^x + e^{-x}}{2}$.

(5) $F(z) = \frac{1}{2}\ln(z^2 + 9) + C$, $C \in \mathbb{R}$. Use the substitution $z^2 + 9 = t$.

(6) $\int \frac{dx}{x + x\ln x} = \int \frac{dx}{x(1+\ln x)} = \begin{vmatrix} 1 + \ln x = t \\ \frac{dx}{x} = dt \end{vmatrix} = \int \frac{dt}{t} = \ln|1 + \ln x| +$

C, $C \in \mathbb{R}$.

Solution 5.60. (1) $f(x) = \int \left(\frac{5}{x} + \frac{1}{1+x^2} \right) dx = \ln|x| + \arctan x + C$, $C \in \mathbb{R}$.

(2) $f(x) = \frac{1}{2}\ln(2x + 1) + C$, $C \in \mathbb{R}$. Use the substitution $2x + 1 = t$.

(3) From $f(x) = \int (\sin x + x^{-2} - e^x) dx = -\cos x - x^{-1} - e^x + C$ and $f(\pi) = 1$, it follows that $f(x) = -\cos x - x^{-1} - e^x + \pi^{-1} + e^\pi$.

(4) $f(x) = \frac{\sin 8x}{4} + \frac{1}{3}e^{3x} - x + \frac{2}{3}$.

(5) $f(x) = 2\sin x + 2x^4 - e^x + 8$.

(6) $f(t) = 2e^t - 3\sin t + t - 2$. Observe that $f'(t) = 2e^t - 3\cos t + 1$.

Solution 5.61. $h(1) = 2e(1 - e^2) - \frac{44}{3}$. Observe that $h(x) = \frac{x^3}{3} + 2e^x + 3x - 18 - 2e^3$.

Solution 5.62. $f(x) = Ae^{kx}$. The question is to solve the initial value problem $\frac{dy}{dx} = ky$, $y(0) = A$. We separate the variables to obtain $\frac{dy}{y} = kdx$. Finding antiderivatives of both sides gives $\ln|y| = kx + C$, $C \in \mathbb{R}$, which is the same as $|y| = e^C \cdot e^{kx}$. Since $y = f(x)$ is a differentiable function in a neighborhood of $x = 0$ (recall that we are solving an initial value problem), $y = e^C \cdot e^{kx}$ or $y = -e^C \cdot e^{kx}$. Since $f(0) = A$, the answer $y = Ae^{kx}$ follows.

Solution 5.63.

(1) $y = \frac{1}{3}(-2\cos 3x + x^3 + e^{3x} + 1) + x$.

(2) $y = \sin\left(x + \frac{\pi}{2}\right)$.   (3) $y = \tan\left(x - \frac{\pi}{4}\right)$.

(4) $y = 4e^x - 1$.   (5) $x(t) = -3(4t - 7)^{-3} + 4$.

(6) $y = \ln(x + e^2)$.   (7) $y = 2 - e^{-t}$.

(8) $y = \frac{3}{2}\sin 2x - \frac{1}{4e^{4x}} + \frac{5}{4}$.

Solution 5.64. Clearly, $f(0) = 3$. From $f'(t) = 5 \cdot \left(\frac{t}{5}\right)^4 \cdot \frac{1}{5} = \left(\frac{t}{5}\right)^4$ and $(f(t) - 3)^{4/5} = \left(3 + \left(\frac{t}{5}\right)^5 - 3\right)^{4/5} = \left(\left(\frac{t}{5}\right)^5\right)^{4/5} = \left(\frac{t}{5}\right)^4$, it follows that the function f is a solution of the given initial value problem. From $g(0) = 3$ and $(g(t) - 3)^{4/5} = (3 - 3)^{4/5} = 0 = g'(t)$ it follows that the function g is also a solution of the given initial value problem. *Note*: This exercise establishes that a given initial value problem may have more than one solution. Commonly, the conditions under which an initial value problem has a unique solution are discussed in an Introduction to Differential Equations course.

Solution 5.65. $f(x) = x^3 + 1$. It is given that $f(0) = 1$ and $f'(0) = 0$.

Solution 5.66. $s(t) = 2e^t - 3\sin t + \frac{2}{\pi}(1 - e^\pi)t - 2$. The question is to find a function s such that $a(t) = s''(t)$, $s(0) = 0$, and $s(\pi) = 0$.

Solution 5.67. $s(t) = -\sin t - 3\cos t + 3t + 3$.

Solution 5.68. $x(t) = 2t^3 + 10$. It is given that $x(0) = 10$, $x'(0) = v(0) = 0$, and $x''(t) = 12t$.

Solution 5.69. (1) $v(t) = \frac{3}{2}t^2 + 6$. (2) Observe that $v(0) = 6$ m/s and $v(4) = 30$ m/s. From $s(t) = \frac{t^3}{2} + 6t + c$ with $c \in \mathbb{R}$, it follows that $s(4) - s(0) = 8$ m. Recall that, since $v(t) > 0$, the total distance traveled is equal to the area of the region between the graph of the velocity function and the t-axis over the interval $[0, 4]$. By the Fundamental Theorem of Calculus, the total distance is $\int_0^4 v(t)\, dt = s(4) - s(0) = 8$ m.[c]

Solution 5.70. Let $t = 0$ be the instant when the breaks were fully applied and let $t = T > 0$ be the instant when the car stopped. Let $t \mapsto s(t)$ be the position function, i.e., $s(t)$ is the distance in feet the car traveled since the instant $t = 0$. Observe that $s(0) = 0$ and $s(T) = 160$. The fact that the car had a constant deceleration means that $v = v(t) = s'(t)$ is a linear function. Hence, $v(t) = -20t + C$, where $C = v(0)$. This, together with $s(0) = 0$, implies that $s(t) = -10t^2 + Ct$. From $v(T) = -20T + C = 0$ and $s(T) = -10T^2 + CT = 160$, it follows that $C = v(0) = 80\,\text{ft/s}$.

Solution 5.71. Let $s(t)$ be the height of the stone after t seconds. Let $T > 0$ be time at which the stone hits the ground. It is given that

---

[c]Definite integrals and the Fundamental Theorem of Calculus are commonly studied in an Integral Calculus course.

$s(0) = 30$ meters, $s'(0) = v(0) = -5\,\mathrm{m/s}$,[d] and $s''(t) = a(t) = -10$ m/s^2 for $t \in [0, T]$. (1) $v(t) = -10t - 5$. (2) $s(t) = -5t^2 - 5t + 30$. (3) $T = 2$ s. Solve the equation $s(T) = 0$, $T > 0$. (4) $|v(2)| = 25\,\mathrm{m/s}$. Recall that speed is the absolute value of velocity.

Solution 5.72. (1) Let $s(t)$ be the height of the ball after t seconds. It is given that $s(0) = 0$, $s'(0) = v(0) = 64\,\mathrm{ft/s}$, and $s''(0) = v'(0) = a(0) = -32\,\mathrm{f/s}^2$. Thus, $s(t) = -16t^2 + 64t = 16t(4-t)$. From $s(4) = 0$, it follows that the ball is in the air for $4\,\mathrm{s}$. (2) $v(4) = s'(4) = -64\,\mathrm{ft/s}$.

Solution 5.73. Let $x > 0$ be the number of key chains sold per week at the price $p = p(x)$. Let $C = C(x)$ be the cost of manufacturing x key chains. It is given that $\dfrac{dC}{dx} = 0.15$ dollars per key chain and $\dfrac{dp}{dx} = -\dfrac{0.10}{50}$ dollars per key chain. Hence, $C(x) = 0.15x + a$ and $p(x) = -\dfrac{0.10x}{50} + b$ for some constants a and b in dollars. The profit is given by $P = P(x) = \text{Revenue} - \text{Cost} = x \cdot p(x) - C(x) = -\dfrac{0.10x^2}{50} + bx - 0.15x - a$. The quantity that maximizes profit is $x = 1000$ key chains per week and it must be a solution of the equation $\dfrac{dP}{dx} = -\dfrac{0.10x}{25} + b - 0.15 = 0$. Hence, $-\dfrac{0.10 \cdot 1000}{25} + b - 0.15 = 0$ and $b = 4.15$ dollars. The price that maximizes the profit is $p = -\dfrac{0.10 \cdot 1000}{50} + 4.15 = 2.15$ dollars for a key chain. *Note*: Even though key chains are only sold in integer units, we used differentiable functions to construct a mathematical model of the given situation.

Solution 5.74. (1) The function $A = A(t)$ satisfies the differential equation $\dfrac{dA}{dt} = pA$. With the given initial value $A(0) = A_0$, the solution of this equation is $A = A_0 e^{pt}$. (2) Solve $15{,}000 = 10{,}000 \cdot e^{4p}$ to obtain $p = \dfrac{\ln 1.5}{4} \approx 0.101$. Hence, $p \approx 10.1\%$.

Solution 5.75. The blood alcohol concentration at time t can be modelled as $C(t) = c_0 e^{-0.2t}$, the solution of the given equation with the initial value $c_0 = C(0)$. Since it is given that $c_0 = 0.1$, the question is to determine $t > 0$ such that $0.05 = C(t) = 0.1 e^{-0.2t}$. It follows that $t = 5 \ln 2 \approx 3.46\,\mathrm{h}$ is needed for the BAC to drop from 0.1 to 0.05.

---

[d]Observe that $s = s(t)$ is a decreasing function so $s'(t) = v(t) < 0$.

Solution 5.76. The proportion of the radioactive material remaining after time t can be modelled as $m(t) = m_0 e^{-5t}$, the solution of the given equation with the initial value $m_0 = m(0)$. If $m(t) = 0.5$, then $m_0 = 1$, and the model is $m(t) = e^{-5t}$. To determine τ, the half-life of the radioactive material, we need to solve the equation $\frac{1}{2} = e^{-5\tau}$. It follows that $\tau = \frac{\ln 2}{5} \approx 0.138$ units of time.

Solution 5.77. (1) $f(t) = e^{-\frac{t \ln 2}{5700}}$. (2) The question is to solve $0.15 = e^{-\frac{t \ln 2}{5700}}$ for t. Hence, the age of the skull is $t = -\frac{5700 \ln 0.15}{\ln 2} \approx 15{,}600$ years.

Solution 5.78. The model is $A = A(0)e^{-kt}$. It is given that $A(0) = 1$ kg and $A(6) = 0.027$ kg. Hence, $A(t) = e^{\frac{t \ln 0.0027}{6}}$. It follows that, at 3:00, there were $A(2) = e^{\frac{\ln 0.0027}{3}} \approx 0.139$ kg of the substance X.

Solution 5.79. The model is $P = P(t) = P_0 e^{kt}$, where k is a constant, P_0 is the initial population, and t is the elapsed time in hours. It is given that $10P_0 = P_0 e^{10k}$, which implies that $k = \frac{\ln 10}{10}$. The question is to solve $2 = e^{\frac{t \ln 10}{10}}$ for t. Hence, the population will double in $t = \frac{10 \ln 2}{\ln 10} \approx 3.01$ h.

Solution 5.80. (1) Since $P(0) = 10$, the model is $P(t) = 10e^{kt}$, where t is time in hours. Since $t = 15 \min = \frac{1}{4}$ h, from $40 = 10e^{\frac{k}{4}}$, it follows that $k = 4 \ln 4$. Hence, $P(3) = 10e^{12 \ln 4} = 167{,}772{,}160$ bacteria. (2) Solve $10^9 = 10e^{4t \ln 4}$ to obtain $t \approx 3.32$ h.

Solution 5.81. Recall that Newton's Law of Cooling is given by $\frac{dT}{dt} = k(T - T_s)$, where $T = T(t)$ is the temperature of the object in degrees Celsius at time t in minutes, T_s is the temperature of the surroundings in degrees Celsius, and k is the cooling constant. Since it is given that, at some time t_1, $\frac{dT}{dt}\big|_{t=t_1} = -1$ and $T(t_1) = 70$, it follows that $k = -\frac{1}{50}$. The model is $T(t) = 20 + (95 - 20)e^{-t/50}$. Solve the equation $70 = 20 + 75e^{-t_1/50}$ to obtain $t_1 = -50 \ln \frac{2}{3} \approx 20.2$ min.

Solution 5.82. The model is $\frac{dT}{dt} = k(T - 32)$, where $T = T(t)$ is the temperature in degrees Celsius after t minutes and k is a constant.

Hence, $T = 32 + (T_0 - 32)e^{kt}$, where T_0 is the initial temperature of the drink. From $14 = 32 + (T_0 - 32)e^{25k}$ and $20 = 32 + (T_0 - 32)e^{50k}$, it follows that $\frac{3}{2} = e^{-25k}$ and $k = -\frac{1}{25}\ln\frac{3}{2}$. Hence, $T_0 = 32 - 12 \cdot e^{-2\ln\frac{2}{3}} = 5^0\mathrm{C}$.

Solution 5.83. $t \approx 4.5\,\mathrm{min}$.

Solution 5.84. (1) The learning curve is determined by the initial value problem $\frac{dA}{dt} = k(M - A(t))$, $A(0) = 0$, where k is a constant. (2) Take the substitution $T = M - A(t)$ and observe that the initial value problem obtained in (1) becomes $\frac{dT}{dt} = -kT$, $T(0) = M$. Recall that the solutions of this initial value problem is $T = Me^{-kt}$. Hence, $A(t) = M(1 - e^{-kt})$. (3) It is given that $M = 100$, $A(0) = 0$, and $A(100) = 50$. Hence, $A(t) = 100\left(1 - e^{-\frac{t\ln 2}{100}}\right)$. Since $t = 200$ is

the solution of the exponential equation $75 = 100\left(1 - e^{-\frac{t\ln 2}{100}}\right)$, it

follows that the student needs to study for another $100\,\mathrm{h}$.

Solution 5.85. (1) The model is $C(t) = C_0 e^{-\frac{t}{7}}$, where $C_0 = 0.112$. The question is to solve $0.08 = 0.112e^{-\frac{t}{7}}$ for t. Hence, the person should wait for at least $t = -7\ln\frac{5}{7} \approx 2.35$ hours before driving. (2) The question is to evaluate $\left.\frac{dC}{dt}\right|_{t=0} = -\frac{0.112}{7} = -0.016$ BAC units per hour.

Solution 5.86. (1) The model is $m(t) = 10e^{-kt}$, where t is in years, $m(t)$ is in kilograms, and k is a constant that should be determined from the fact that $m(24110) = 5$. Hence, $k = \frac{\ln 2}{24110}$ and $m(t) = 10e^{-\frac{t\ln 2}{24110}}$. (2) $m(1000) = 10e^{-\frac{\ln 2}{24.11}} \approx 9.716\,\mathrm{kg}$. (3) We solve $1 = 10e^{-\frac{t\ln 2}{24110}}$ to obtain that the lake will be safe for use after $t = 24110\frac{\ln 10}{\ln 2} \approx 80{,}091.68$ years.

Solution 5.87. (1) The model is $P = 500e^{kt}$. From $P(6) = 8000 = 500e^{3k}$, it follows that $k = \frac{4\ln 2}{3}$. Thus, $P = 500e^{\frac{4t\ln 2}{3}}$. (2) $128{,}000$ bacteria. (3) Solve the exponential equation $10^6 = 500e^{\frac{4t\ln 2}{3}}$ for the unknown t. It follows that $t = \frac{3(4\ln 2 + 3\ln 5)}{4\ln 2} = 3 \cdot \frac{\ln 2000}{\ln 64} \approx 4.7414\,\mathrm{h}$.

Solution 5.88. (1) $\frac{dP}{dt} = kP$. (2) $P = 160e^{t\ln\frac{3}{2}}$. (3) $P(0) = 160$ bacteria. (4) $\left.\frac{dP}{dt}\right|_{t=0} = 160\ln\frac{3}{2}$ bacteria per hour.

Solution 5.89. The model is $P(t) = 500e^{kt}$, where t is time in years after 2010. From $P(10) = 750$, it follows that $k = \frac{1}{10}\ln\frac{3}{2}$. Thus, the model suggest that the class enrollment in 2100 would be $P(90) = 500e^{\frac{90}{10}\ln\frac{3}{2}} \approx 19{,}000$. Observe that, undoubtedly, this estimate is not realistic. Recall that a mathematical model is just an approximation of a real-world phenomenon. Consequently, you have to critically check any data obtained by a mathematical model before accepting it and using it to make any further conclusions.

Solution 5.90. (1) From Newton's Law of Cooling, $\frac{dT}{dt} = k(T - T_s)$, for $T = 80$, $T_s = 20$, and $k = 0.09$, it follows that $\left.\frac{dT}{dt}\right|_{T=80^0} = -0.09 \cdot (80 - 20) = -5.4°C/\min$. (2) Note that 6 s should be used as $\Delta t = 0.1$ minutes. From $T \approx 80 - 5.4\Delta t = 80 - 5.4 \cdot 0.1$, it follows that the change of temperature will be $T - 80 \approx -0.54°C$. (3) Determine the function $T = T(t)$ that is the solution of the initial value problem $\frac{dT}{dt} = -0.09(T - 20)$, $T(0) = 90$, and then solve the equation $T(t) = 65$ for t. You should wait about $t = -\frac{100}{9}\ln\frac{9}{16} \approx 6.4$ minutes.

Solution 5.91. $t \approx 2.63\,\text{h}$.

Chapter 6

Parametric and Polar Curves

6.1 Introduction

Use the following definitions, techniques, and properties to solve problems contained in this chapter.

Parametric Curves. Let I be an interval of real numbers and let $f : I \to \mathbb{R}$ and $g : I \to \mathbb{R}$ be two, possibly equal, continuous functions on I. A set $C = \{(f(t), g(t)) : t \in I\}$ is called a *parametric curve*.[a]

The variable t is called a *parameter*.

We say that the parametric curve C is defined by *parametric equations* $x = f(t)$, $y = g(t)$, $t \in I$.

We say that $x = f(t)$, $y = g(t)$, $t \in I$ is a *parametrization* of C.

If $I = [a, b]$, then $(f(a), g(a))$ is called the *initial point* of C and $(f(b), g(b))$ is called the *terminal point* of C.

Graph of a Parametric Curve. We observe that the definition of a parametric curve includes two different notions, a pair of continuous functions f and g defined on an interval I and a set of points C that is traced out, in a given rectangular coordinate system, by pairs $(f(t), g(t))$, while t passes through the interval I. We say that the set of points is the *graph of the parametric curve*.

---

[a]Additional conditions are required to guarantee that C matches our intuitive perception of a curve as the trajectory of a moving point. This was commonly studied in a Differential Geometry course.

Derivative of a Parametric Curve. If functions f and g are differentiable, then the *derivative of the parametric curve* defined by $x = f(t)$, $y = g(t)$, $t \in I$ is a function $\frac{dy}{dx} = \frac{\frac{dy}{dt}}{\frac{dx}{dt}} = \frac{g'(t)}{f'(t)}$ defined on the set $\{t \in I : f'(t) \neq 0\}$.

Smooth Curve. A *curve C is smooth* if there is a parametrization $x = f(t)$, $y = g(t)$, $t \in I$ of the curve C such that $\frac{dx}{dt}$ and $\frac{dy}{dt}$ are continuous on I and not simultaneously zero.

Tangent Line. If $x = f(t)$, $y = g(t)$, $t \in I$ is a parametrization of a smooth curve C then, for $a \in I$, the line determined by the equation $f'(a)(y - g(a)) = g'(a)(x - f(a))$ is called the *tangent line to C at the point* $(f(a), g(a))$.

Polar Coordinate System. The *polar coordinate system* is a two-dimensional coordinate system in which each point in the plane, except the fixed point, is associated with a pair of numbers (r, θ) in the following way:

(1) Fix a point O in the plane. We call this fixed point the *pole*.

(2) Fix a ray p starting at O. We call this fixed ray the *polar axis*.

(3) For any point P in the plane, except O, measure the distance $r = d(O, P)$.

(4) Measure in radians the angle θ between the polar axis and the ray starting at O and passing through P going from the polar axis in counterclockwise (positive) direction.

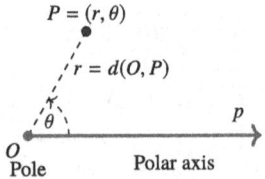

We call the pair (r, θ) the polar coordinates of the point P (with respect to the pole O and the polar axis p).

There is a bijection between the set of all points in the plane, without the pole, and the set $\mathbb{R}^+ \times [0, 2\pi) = \{(r, \theta) : r \in \mathbb{R}^+ \text{ and } \theta \in [0, 2\pi)\}$. This means that each point P in the plane, except O, is uniquely determined by a pair $(r, \theta) \in \mathbb{R}^+ \times [0, 2\pi)$.

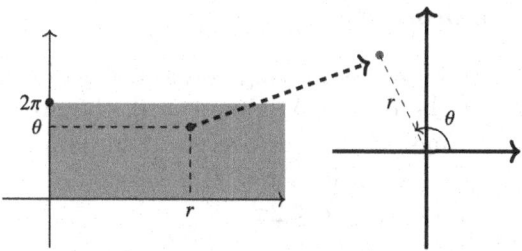

If an xy-rectangular coordinate system in the plane is given, then it is common to take the origin O as the pole and the positive direction of the x-axis (the horizontal axis) as the polar axis. In this case: $x = r\cos\theta$ and $y = r\sin\theta$.

It is common to consider the polar coordinates (r, θ) so that θ is any real number and that r is any non-zero real number. In the case $r > 0$, $(r, \theta) = (r, \theta_0)$, where $\theta_0 \in [0, 2\pi)$ with $\theta = \theta_0 + 2n\pi$ for some $n \in \mathbb{Z}$. If $r < 0$, then $(r, \theta) = (-r, \pi + \theta)$.

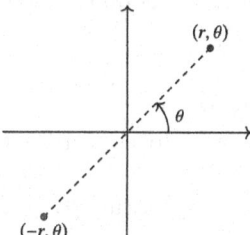

In this setting, the value $r = 0$ corresponds to the pole.

Polar Curves. A *polar curve* is the graph of a function $r = f(\theta)$, $\theta \in I$, i.e., the set of all points P whose polar coordinates are of the form $(f(\theta), \theta)$, $\theta \in I$.

Derivative in Polar Coordinates. Suppose that y is a differentiable function of x and that $r = f(\theta)$ is a differentiable function of θ. Then from $x = r\cos\theta$, $y = r\sin\theta$ it follows that $\dfrac{dy}{dx} = \dfrac{\frac{dy}{d\theta}}{\frac{dx}{d\theta}} = \dfrac{\frac{dr}{d\theta}\sin\theta + r\cos\theta}{\frac{dr}{d\theta}\cos\theta - r\sin\theta}$.

6.2 Parametric Curves

Use the appropriate definitions and your knowledge of limits and derivatives and their properties to solve the following problems.

6.2.1 *Routine problems*

Problem 6.1. An object is moving counterclockwise along a circle with the centre at the origin. At time $t = 0$, the object is at point $A = (0, 5)$ and at time $t = 2\pi$ it is back to point A for the first time. Determine parametric equations $x = f(t)$, $y = g(t)$ that describe the motion for the duration $0 \le t \le 2\pi$.

Problem 6.2. The trajectory of a particle in a plane is a function of time t in seconds and it is determined by the parametric equations $x = 3t^3 + 2t - 3$, $y = 2t^3 + 2$, $t > 0$. Prove that there is exactly one time when the particle crosses the line $y = x$.

Problem 6.3. Determine an equation of the tangent line to the parametric curve defined by $x = t - t^{-1}$, $y = 1 + t^2$, $t \in \mathbb{R}\backslash\{0\}$, at the point $(x(1), y(1))$.

Problem 6.4. Consider the parametric curve C defined by $x(t) = 1 + 3t$, $y(t) = 2 - t^3$, $t \in \mathbb{R}$.

(1) Determine $\frac{dy}{dx}$ in terms of t.
(2) Eliminate t to obtain an expression for C of the form $y = f(x)$.

Problem 6.5. Consider the parametric curve defined by $x = 3(t^2 - 3)$, $y = t^3 - 3t$, $t \in \mathbb{R}$.

(1) Determine the derivative $\frac{dy}{dx}$ in terms of t.
(2) Determine an equation of the tangent line to the curve at the point corresponding to $t = 2$.

Problem 6.6. Let a parametric curve be defined by $x = 2\sin t + 1$, $y = 2t^3 - 3$, $t \in \mathbb{R}$. Determine $\frac{d^2y}{dx^2}$ in terms of t.

Problem 6.7. Consider the parametric curve defined by $x = 3(t^2 - 9)$, $y = t(t^2 - 9)$, $t \in \mathbb{R}$.

(1) Determine all x-intercepts.
(2) Determine the coordinates of the point on the curve corresponding to $t = -1$.

(3) Determine an equation of the tangent line to the curve at the point corresponding to $t = -1$.
(4) Evaluate the second derivative at the point corresponding to $t = -1$.

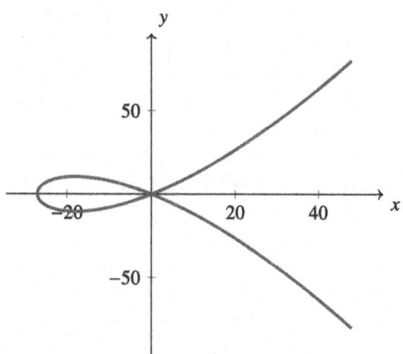

$x = 3(t^2 - 9)$, $y = t(t^2 - 9)$, $t \in \mathbb{R}$

Problem 6.8. Consider the parametric curve defined by $x = t(t^2 - 3)$, $y = 3(t^2 - 3)$, $t \in \mathbb{R}$.

(1) Determine the y-intercepts of the curve.
(2) Determine the points on the curve where the tangent line is horizontal or vertical.
(3) Sketch the curve.

Problem 6.9. Sketch the parametric curve $x = \sin^2 \pi t$, $y = \cos^2 \pi t$, $t \in [0, 2]$. Clearly label the initial and terminal points and describe the motion of the point $(x(t), y(t))$ as t varies in the given interval.

Problem 6.10. A parametric curve is defined by $x = \theta - \sin\theta$, $y = 1 - \cos\theta$, $\theta \in \mathbb{R}$.

(1) Determine $\frac{dx}{dy}$ and $\frac{d^2x}{dy^2}$ in terms of θ.
(2) Determine an equation of the tangent line to the curve at the point of the curve obtained by setting $\theta = \frac{\pi}{3}$.

Problem 6.11. The graphs of functions $x = f(t)$ and $y = g(t)$ with $t \in [0, 2\pi]$ are shown in the following figure.

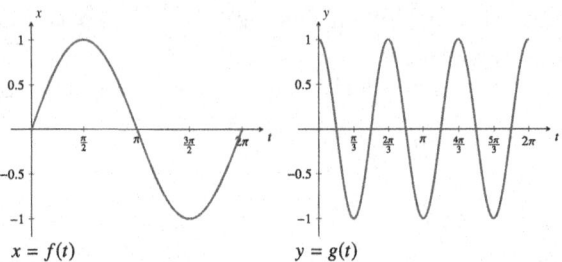

$$x = f(t) \qquad\qquad y = g(t)$$

Identify which is the corresponding parametric curve $\{(f(t), g(t)) : t \in [0, 2\pi]\}$:

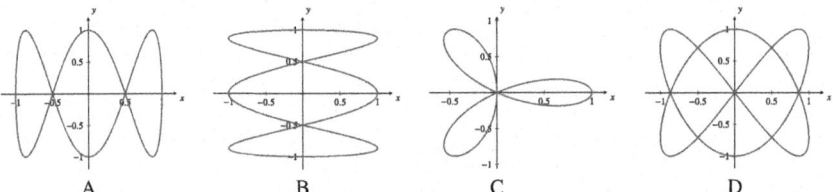

A   B   C   D

Problem 6.12. Consider the parametric curve defined by $x(t) = -2 + 2\cos t$, $y(t) = 1 - 2\sin t$, $t \in [0, 2\pi]$.

(1) State the Cartesian equation $F(x, y) = 0$ of the curve and then sketch the curve. Determine the direction of evolution of the curve for increasing t and indicate it on the graph.

(2) Determine the points on the curve for which the tangent line has a slope of 1.

Problem 6.13. This question concerns the Lissajous curve defined by $x = 2\cos 3t$, $y = 2\sin 2t$, $t \in [0, 2\pi]$.

(1) Determine $\frac{dy}{dx}$ for this curve.

(2) Determine equations of two tangent lines at the origin.

(3) Identify the graph that corresponds to the curve.

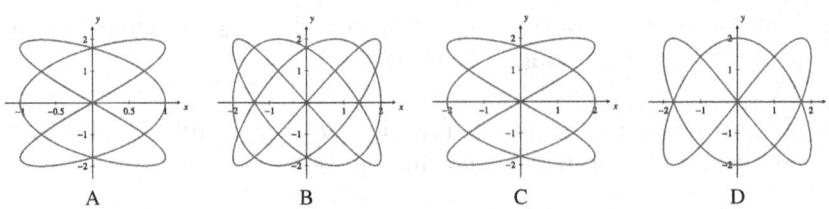

A   B   C   D

Problem 6.14. A parametric curve is defined by $x = t - 1$, $y = t^2 - 2t + 2$, $t \in \mathbb{R}$.

(1) Determine the derivative $\frac{dy}{dx}$ as a function of t.
(2) Eliminate the parametric dependence to determine an expression of the form $y = f(x)$ for the given curve.
(3) Determine an expression for m and b (as functions of x_1) such that the equation of the line $y = mx + b$ is tangent to the curve $y = f(x)$ at $(x_1, f(x_1))$.
(4) Determine an expression for all tangent lines to the curve $y = f(x)$ that pass through the point $(2, 0)$.

Problem 6.15. This question concerns the parametric curve defined by $x = t^3 - 4t$, $y = 2t^2 - 4t$, $t \in \mathbb{R}$.

(1) Which of the two graphs corresponds to the given curve?

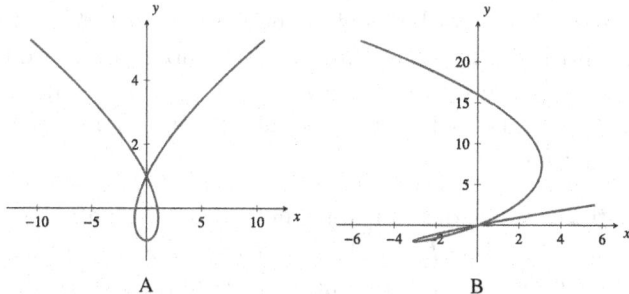

A   B

(2) Evaluate the y-coordinates of all points where the curve crosses the y-axis.
(3) This curve crosses itself at exactly one point. Determine equations of both tangent lines at that point.

Problem 6.16. Consider the parametric curve C defined by $x = e^{-t}$, $y = 2^{2t}$, $t \in \mathbb{R}$.

(1) Determine $\frac{dy}{dx}$ and $\frac{d^2y}{dx^2}$ in terms of t.
(2) Does the graph of C have any points of inflection?
(3) Eliminate t to obtain an expression for C of the form $y = f(x)$.
(4) Sketch the curve C. Plot at least three points of the curve C, labeling the points with the corresponding values of t. Draw an arrow on your curve indicating the direction of motion as t increases toward plus infinity.

Problem 6.17. Consider the parametric curve defined by $x(t) = e^t$, $y(t) = te^t$, $t \in \mathbb{R}$.

(1) Determine $\frac{dy}{dx}$ and $\frac{d^2y}{dx^2}$ in terms of t.
(2) Determine an equation of the tangent line to the curve at the point where $t = 0$.
(3) For which values of t is the curve concave upward?

Problem 6.18. Consider the parametric curve defined by $x(t) = t\sin t$, $y(t) = t\cos t$, $t \in \mathbb{R}$.

(1) Determine the derivative $\frac{dy}{dx}$ in terms of t.
(2) Determine an equation of the tangent line at the point corresponding to $t = \frac{\pi}{2}$.
(3) Determine if the curve is concave upward or downward at $t = \frac{\pi}{2}$.

Problem 6.19. A small ball is fastened to a long rubber band and twirled around in such a way that the ball moves in a circular path determined by the vector function $\vec{r} = \vec{r}(t) = b\langle \cos \omega t, \sin \omega t \rangle$, where b and ω are constants. Determine the velocity, speed, and acceleration of the ball at time t.

Problem 6.20. Consider the parametric curve defined by $x(t) = t - \sin 2t$, $y = 4 - 3\cos t$, $t \in [0, 10]$. Determine the values for the parameter t where the tangent line to the curve is vertical.

Problem 6.21. Sketch the graph of the parametric curve defined by $x = \cos^2 t$, $y = \cos t$, $t \in [0, \pi]$. Write an equation of the tangent line at the point where the slope is $\frac{1}{\sqrt{3}}$.

6.2.2 *Not-so-routine problems*

Problem 6.22. Determine the asymptotes of the parametric curve defined by $x = \frac{1}{t}$, $y = \frac{t}{1+t}$, $t \in \mathbb{R}\backslash\{-1, 0\}$.

Problem 6.23. Determine the asymptotes of the parametric curve defined by $x = \tan t$, $y = 2t + \tan t$, $t \in \left(-\frac{\pi}{2}, \frac{\pi}{2}\right)$.

Problem 6.24. Determine the asymptotes of the parametric curve defined by $x = \frac{2e^t}{t-1}$, $y = \frac{te^t}{t-1}$. $t \in \mathbb{R}\backslash\{1\}$.

Problem 6.25. Determine the asymptotes of the parametric curve defined by $x = \frac{t^3}{1-t^2}$, $y = \frac{1+t^2}{1-t^2}$, $t \in \mathbb{R}\backslash\{-1,1\}$.

Problem 6.26. Consider the Lissajous curve defined by $x = 3\cos 2t$, $y = \sin 4t$, $t \in [0, \pi]$.

(1) Determine points on the curve corresponding to $t = 0$ and $t = \frac{\pi}{4}$.
(2) Draw arrows indicating the direction the curve is sketched as the value of t increases from 0 to π.
(3) Evaluate the slope of the tangent line to the curve at the point, corresponding to $t = \frac{\pi}{12}$.
(4) State the intervals of t for which the curve is concave upward and the intervals of t for which the curve is concave downward.

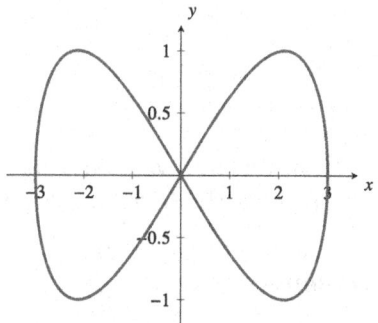

$x = 3\cos(2t)$, $y = \sin(4t)$, $0 \leq t \leq \pi$

Problem 6.27. A cardioid is defined by $x = 2\cos t - \cos 2t$, $y = 2\sin t - \sin 2t$, $t \in [0, 2\pi]$. Sketch its graph by investigating functions $x = x(t)$ and $y = y(t)$ for periodicity, symmetry, and monotonicity.

Problem 6.28. A parametric curve is defined by $x = \sin\frac{t}{2}$, $y = \tan t$, $t \neq \frac{(2k+1)\pi}{2}$, $k \in \mathbb{Z}$. Sketch its graph by investigating functions $x = x(t)$ and $y = y(t)$ for periodicity, symmetry, and monotonicity.

Problem 6.29. Consider the asteroid $x = \cos^3 t$, $y = \sin^3 t$, $t \in [0, 2\pi]$.

(1) Sketch a graph of the asteroid. On the graph draw arrows indicating the direction the curve is sketched as the t-value increases from 0 to 2π.
(2) Determine an equation of the tangent line to the curve at the point corresponding to $t = \frac{3\pi}{4}$.

(3) Determine the concavity of this curve when $t = 1$.

(4) State the intervals of t for which the curve is concave upward and the intervals of t for which the curve is concave downward.

Problem 6.30. The Folium of Descartes is defined by $x = \frac{3t}{1+t^3}$, $y = \frac{3t^2}{1+t^3}$, $t \in \mathbb{R}\backslash\{-1\}$. Sketch its graph by investigating functions $x = x(t)$ and $y = y(t)$ for periodicity, symmetry, and monotonicity.

(1) Sketch a graph of the Folium of Descartes.

(2) By eliminating the parameter t, determine an expression $F(x, y) = 0$ for this curve.

(3) Evaluate the slope of the tangent line to the curve at the point, where $t = 1$.

(4) Evaluate $\frac{d^2y}{dx^2}$ at $t = 1$.

6.3   Polar Curves

Use the appropriate definitions and your knowledge of trigonometry, limits, and derivatives to solve the following problems.

6.3.1   *Routine problems*

Problem 6.31. Determine the polar coordinates for each of the points A, B, C, and D as follows.

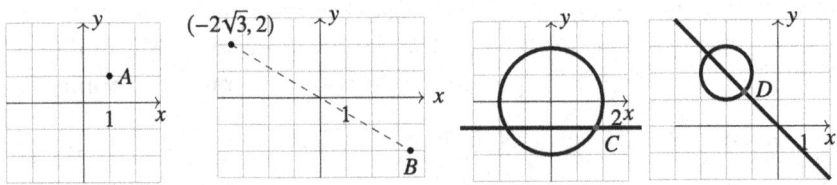

Problem 6.32. For each of the following circles, determine a polar equation, i.e., an equation in terms of r and θ: (1) $x^2 + y^2 = 4$, (2) $(x-1)^2 + y^2 = 1$, (3) $x^2 + (y - 0.5)^2 = 0.25$.

Problem 6.33.

(1) Let C denote the graph of the polar equation $r = 5\sin\theta$. Determine the rectangular coordinates of the point on C corresponding to $\theta = \frac{3\pi}{2}$.

(2) Convert the equation for C into rectangular coordinates. What kind of curve is this?

(3) Express C in parametric form, i.e., express both x and y as functions of θ.

(4) Determine an expression for $\frac{dy}{dx}$ in terms of θ.

(5) Determine an equation of the tangent line to C at the point corresponding to $\theta = \frac{\pi}{6}$.

Problem 6.34. Sketch a polar coordinate plot of each of the following functions: (1) $r = 2\sin\theta$; (2) $r = 2\cos\theta$.

Problem 6.35. Express the polar equation of the rose $r = \cos 2\theta$ in rectangular coordinates.

Problem 6.36. Evaluate the maximum height above the x-axis of the cardioid $r = 1 + \cos\theta$.

Problem 6.37. Evaluate the slope of the tangent line to the polar curve $r = 2$ at the points where it intersects the polar curve $r = 4\cos\theta$.

Problem 6.38. Show that the curve $r = \tan\left(\frac{\theta}{2}\right)$ has an asymptote. Determine an equation of this asymptote in the rectangular coordinate system in which the polar axis is the positive direction of the x-axis and the pole is at the origin.

Problem 6.39. Consider the rose defined by the polar equation $r = 4\cos 3\theta$ for $0 \leq \theta < 2\pi$.

(1) Determine the rectangular coordinates for the point on the curve corresponding to $\theta = \frac{\pi}{3}$.

(2) Evaluate the slope of the tangent line to the curve at the point corresponding to $\theta = \frac{\pi}{3}$.

Problem 6.40. Consider the rose defined by the polar equation $r = 4\sin 3\theta$ for $0 \leq \theta < 2\pi$.

(1) Determine the rectangular coordinates for the point on the curve corresponding to $\theta = \frac{\pi}{6}$.

(2) Evaluate the slope of the tangent line to the curve at the point corresponding to $\theta = \frac{\pi}{6}$.

(3) Indicate which of the graphs below depict roses $r = 4\sin 3\theta$ and $r = 4\cos 3\theta$, respectively.

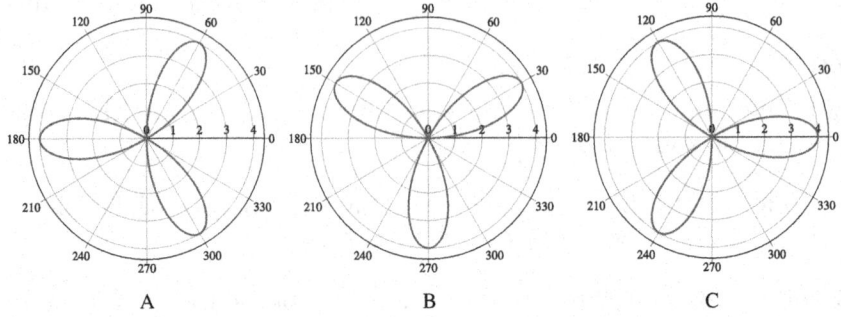

A   B   C

Problem 6.41. Consider the curve defined by the polar equation $r = 1 + 3\cos 2\theta$ for $0 \le \theta < 2\pi$.

(1) Determine the rectangular coordinates for the point on the curve corresponding to $\theta = \dfrac{\pi}{6}$.
(2) Evaluate the slope of the tangent line to the curve at the point corresponding to $\theta = \dfrac{\pi}{6}$.
(3) Indicate which of the graphs below is the graph of the curve $r = 1 + 3\cos 2\theta$.

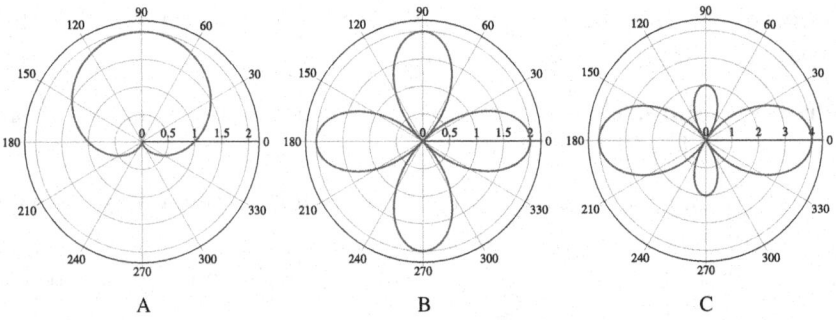

A   B   C

Problem 6.42. Consider the limaçon defined by $r = 1 - 2\sin\theta$ for $0 \le \theta < 2\pi$.

(1) Determine the rectangular coordinates for the point on the curve corresponding to $\theta = \dfrac{3\pi}{2}$.
(2) The curve intersects the x-axis at two points other than the pole. Determine the polar coordinates for these other points.

(3) In the following figure, identify the graphs that correspond to limaçons $r = 1 - 2\sin\theta$ and $r = 1 + 2\sin\theta$, respectively.

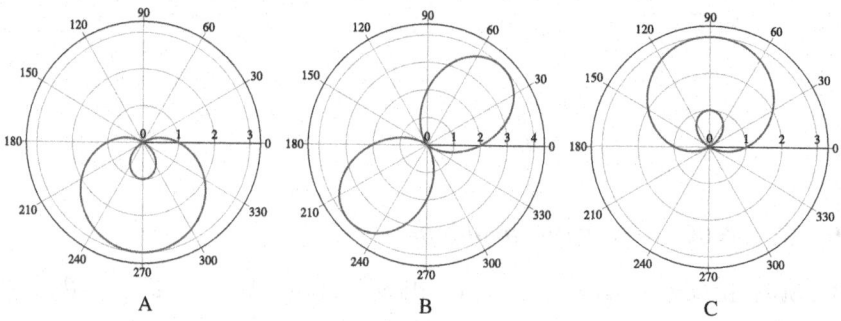

A   B   C

Problem 6.43. Consider the limaçon $r = 1 + 2\cos\theta$ for $0 \le \theta < 2\pi$.

(1) Determine the rectangular coordinates for the point on the curve corresponding to $\theta = \dfrac{\pi}{3}$.

(2) Evaluate the slope of the tangent line at the point corresponding to $\theta = \dfrac{\pi}{3}$.

(3) In the following figure, identify the graph of the curve.

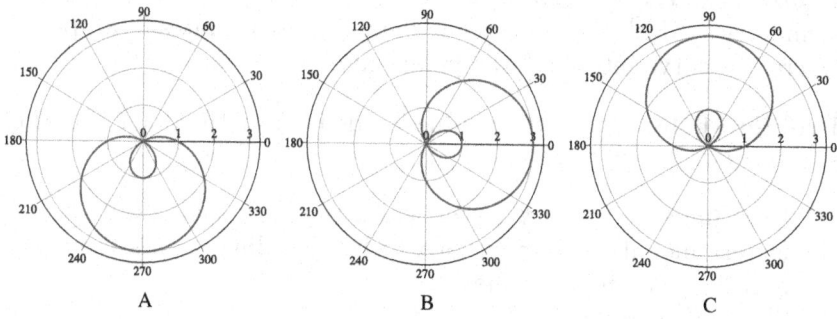

A   B   C

Problem 6.44. The Conchoid of de Sluze is the cubic curve first constructed by René de Sluze in 1662. Its polar equation is $r = a\sec\theta + b\cos\theta$ for $a, b \in \mathbb{R}\backslash\{0\}$. Determine the slope of the tangent line to the Conchoid of de Sluze at the point corresponding to $\theta \ne \dfrac{\pi}{2} + k\pi$, $k \in \mathbb{Z}$.

Problem 6.45. Match curves $r^2 = \cos 2\theta$, $r = 0.2\theta$, and $r = 2(2\cos\theta - \sec\theta)$ with their graphs:

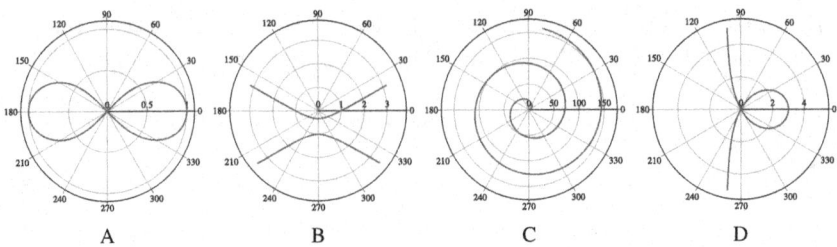

A   B   C   D

6.3.2 *Not-so-routine problems*

Problem 6.46. Consider the cardioid defined by $r = 1 - \cos\theta$, $0 \leq \theta < 2\pi$.

(1) Given a point P on this curve with polar coordinates (r, θ), represent its rectangular coordinates (x, y) in terms of θ.
(2) Evaluate the slope of the tangent line to the curve at the point corresponding to $\theta = \dfrac{\pi}{2}$.
(3) Determine the points on the cardioid where the tangent line is horizontal or vertical.

Problem 6.47. Sketch the cardioid $r = -1 + \cos\theta$, indicating any symmetries. Mark on your sketch the polar coordinates of all points where the curve intersects the polar axis.

Problem 6.48. Consider the limaçon defined by the polar equation $r = 1 + 2\sin\theta$, $-\pi \leq \theta < \pi$.

(1) Determine all $\theta \in [-\pi, \pi)$ for which the radius r is positive.
(2) Determine all $\theta \in [-\pi, \pi)$ for which the radius r takes the maximum and minimum values.
(3) Sketch the curve.

Problem 6.49. Consider the rose defined by the polar equation $r = \cos 2\theta$, $0 \leq \theta < 2\pi$.

(1) Determine $\dfrac{dy}{dx}$ in terms of θ.
(2) Determine the rectangular coordinates for the point on the curve corresponding to $\theta = \dfrac{\pi}{8}$.
(3) Determine an equation of the tangent line to the curve at the point corresponding to $\theta = \dfrac{\pi}{8}$.

(4) Sketch this curve for $0 \le \theta \le \frac{\pi}{4}$ and label the point from part (2) on your curve.

Problem 6.50. Sketch a polar coordinate plot of each of the following functions:

(1) $r = 1 - 2\cos\theta$ (a limaçon)   (2) $r = 0.5 + \sin\theta$ (a limaçon)
(3) $r = 1 + \sin\theta$ (a limaçon)   (4) $r = 4 + 7\cos\theta$ (a limaçon)
(5) $r^2 = -4\sin 2\theta$ (a lemniscate)   (6) $r = \cos 3\theta$ (a rose).

Problem 6.51. Consider the curve C defined by the polar equation $r = 1 + 2\sin 3\theta$, $-\pi \le \theta < \pi$.

(1) Sketch a polar coordinate plot of the curve.
(2) How many points lie in the intersection of the curve C and the unit circle?
(3) *Algebraically*, find all values of θ so that $1 = 1 + 2\sin 3\theta$, $-\pi \le \theta \le \pi$.
(4) Explain in a sentence or two why the answer to part (2) differs from (or is the same as) the number of solutions you found in part (3).

Problem 6.52. Consider the following curve C defined in polar coordinates as $r(\theta) = 1 + \sin\theta + e^{\sin\theta}$, $0 \le \theta \le 2\pi$.

(1) Calculate the value of $r(\theta)$ for $\theta = 0, \frac{\pi}{2}, \frac{3\pi}{2}$.
(2) Sketch a graph of C.
(3) What is the minimum distance from a point on the curve C to the origin?

Problem 6.53. Consider the hyperbolic spiral defined by the polar equation $r\theta = 1$, $\theta \in \mathbb{R}\backslash\{0\}$.

(1) Evaluate the derivative of this hyperbolic spiral when $\theta = \frac{5\pi}{2}$.
(2) Express the polar equation $r \cdot \theta = 1$ in the rectangular coordinates.
(3) Sketch the graph of the curve for $\theta \in [0.5, 3\pi]$.

Problem 6.54. Sketch the graph of the Cissoid of Diocles defined by $r = \sec\theta - \cos\theta$.

Problem 6.55. Sketch the graph of the Conchoid of de Sluze defined by $r = a\sec\theta + b\cos\theta$ if: (1) $a = -1$ and $b = 4$; (2) $a = 1$ and $b = 2$.

Problem 6.56. A bee goes out from its hive in a spiral path determined in polar coordinates by $r = be^{kt}$ and $\theta = ct$, where b, k, and c are positive constants. Show that the angle between the bee's velocity and acceleration remains constant as the bee moves outward.

6.4   Answers, Hints, Solutions

Solution 6.1. $x = -5\sin t$, $y = 5\cos t$, $t \in [0, 2\pi]$. Take t to be the angle between the positive direction of the y-axis and the ray with the initial point at the origin and passing through the point that represents the position of the particle.

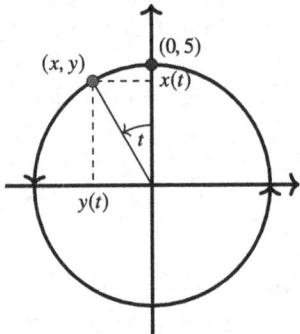

Solution 6.2. We are looking for $t > 0$ such that $3t^3 + 2t - 3 = 2t^3 + 2$, or, equivalently, $t^3 + 2t - 5 = 0$. Since $f(t) = t^3 + 2t - 5$ is a continuous function, by the Intermediate Value Theorem, there is a positive real number t_0 such that $f(t_0) = 0$. Since $f'(t) = 3t^2 + 2 > 0$, the function f is increasing, which implies that t_0 is the unique zero of the function f.

Solution 6.3. $y = x + 2$. Observe that $x(1) = 0$ and $y(1) = 2$. From $\dfrac{dy}{dx} = \dfrac{\frac{dy}{dt}}{\frac{dx}{dt}} = \dfrac{2t}{1+t^{-2}}$, it follows that $\dfrac{dy}{dx}\Big|_{t=1} = 1$.

Solution 6.4. (1) $\dfrac{dy}{dx} = -t^2$. (2) $y = 2 - \dfrac{(x-1)^2}{3}$.

Solution 6.5. (1) $\dfrac{dy}{dx} = \dfrac{t^2-1}{2t}$, $t \neq 0$. (2) $y - 2 = \dfrac{3}{4}(x - 3)$.

Solution 6.6. $\dfrac{d^2y}{dx^2} = \dfrac{\frac{d}{dt}\left(\frac{dy}{dx}\right)}{\frac{dx}{dt}} = \dfrac{3t(2\cos t + t\sin t)}{2\cos^3 t}$, $t \neq \dfrac{\pi}{2} + n\pi$, $n \in \mathbb{Z}$.

Solution 6.7. (1) $(-27, 0)$, $(0, 0)$. Solve $y = 0$ for t. (2) $(-24, 8)$. (3) $y = x + 32$. (4) $\left.\dfrac{d^2 y}{dx^2}\right|_{t=-1} = \left.\dfrac{t^2 + 3}{12 t^3}\right|_{t=-1} = -\dfrac{1}{3}$.

Solution 6.8. (1) $(0, 0)$, $(0, -9)$.
(2) From $\dfrac{dy}{dx} = \dfrac{2t}{t^2 - 1}$, it follows that the tangent line is horizontal at the point $(0, -9)$ and vertical at the points $(-2, -6)$ and $(2, -6)$.
(3) Observe that $(x(-t), y(-t)) = (-x(t), y(t))$. Conclude that the graph is symmetric with respect to the y-axis. Use (1) and (2) to sketch the graph.

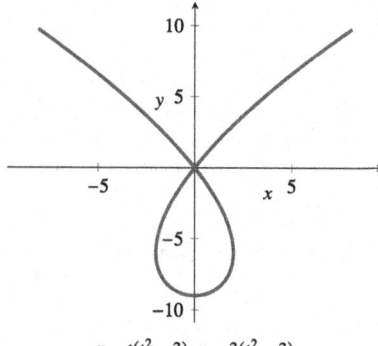

$x = t(t^2 - 3)$, $y = 3(t^2 - 3)$

Solution 6.9. Note that $x + y = 1$ and that $x, y \in [0, 1]$. Observe that $x(0) = x(1) = x(2) = 0$ and $y(0) = y(1) = y(2) = 1$. Conclude that the point starts at $(0, 1)$, goes through $(1, 0)$ two times, namely for $t = 0.5$ and $t = 1.5$, and through every other point four times, before stopping at $(0, 1)$.

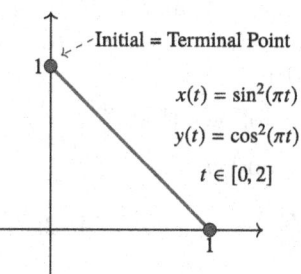

Solution 6.10. (1) $\dfrac{dx}{dy} = \dfrac{1 - \cos\theta}{\sin\theta}$, $\dfrac{d^2 x}{dy^2} = \dfrac{1 - \cos\theta}{\sin^3\theta}$. (2) $y = \sqrt{3}x - \dfrac{\pi\sqrt{3}}{3} + 2$.

Solution 6.11. A. Follow the curve as t increases, starting at the point $(f(0), g(0)) = (0, 1)$ on the curve.

Solution 6.12. (1) Express $\cos t$ and $\sin t$ in terms of x and y to get the circle $(x+2)^2 + (y-1)^2 = 4$. Check which points correspond to $t = 0$ and $t = \frac{\pi}{2}$ to determine the orientation.

(2) Solve $\frac{dy}{dx} = \cot t = 1$ for $t \in (0, 2\pi)$ to obtain two points: $(-2 + \sqrt{2}, 1 - \sqrt{2})$ and $(-2 - \sqrt{2}, 1 + \sqrt{2})$.

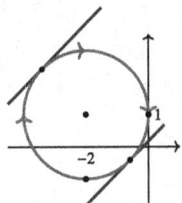

$x = -2 + 2\cos t,\, y = -2 + 2\sin t$

Solution 6.13. (1) $\frac{dy}{dx} = -\frac{2\cos 2t}{3\sin 3t}$.

(2) Note that for both $t = \frac{\pi}{2}$ and $t = \frac{3\pi}{2}$ the curve passes through the origin. Thus, $y = \pm\frac{2x}{3}$.

(3) C. For example, count how many times the curve crosses the x-axis and the y-axis.

Solution 6.14. (1) $\frac{dy}{dx} = 2t - 2$.

(2) $y = x^2 + 1$.

(3) $m = 2x_1$, $b = 1 - x_1^2$.

(4) Note that the point $(2, 0)$ does not belong to the curve. Since all tangent lines to the curve are given by $y = 2x_1 x + 1 - x_1^2$, $x_1 \in \mathbb{R}$, we need to solve the equation $0 = 4x_1 + 1 - x_1^2$ for x_1. Hence, $x_1 = 2 \pm \sqrt{5}$ and the tangent lines are given by $y = 2(2 \pm \sqrt{5})x - 8 \mp 4\sqrt{5}$.

Solution 6.15. (1) B. Note that if $t = 0$ or $t = 2$, then $x = y = 0$.

(2) Solve $x = t^3 - 4t = t(t^2 - 4) = 0$ to obtain two y-intercepts, namely $(0, 0)$ and $(0, 16)$.

(3) From $\frac{dy}{dx} = \frac{4(t-1)}{3t^2 - 4}$, it follows that $\frac{dy}{dx}\big|_{t=0} = 1$ and $\frac{dy}{dx}\big|_{t=2} = \frac{1}{2}$. The tangent lines are $y = x$ and $y = \frac{x}{2}$.

Solution 6.16. (1) $\frac{dy}{dx} = -e^t \cdot 2^{2t+1}\ln 2$; $\frac{d^2y}{dx^2} = e^{2t} \cdot 2^{2t+1} \cdot (1 + 2\ln 2)\ln 2$.

(2) No, the second derivative is never zero.

(3) $y = 2^{-2\ln x} = x^{\frac{-2}{\log_2 e}}$, $x > 0$.

(4) See the following figure.

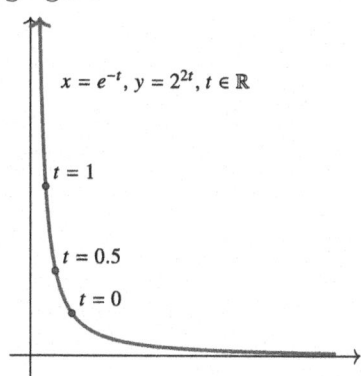

$x = e^{-t}, y = 2^{2t}, t \in \mathbb{R}$

$t = 1$

$t = 0.5$

$t = 0$

Solution 6.17. (1) $\frac{dy}{dx} = 1 + t$; $\frac{d^2y}{dx^2} = e^{-t}$. (2) $y = x - 1$. (3) Observe that $\frac{d^2y}{dx^2} > 0$ for all $t \in \mathbb{R}$ and conclude that the graph of the curve is concave upward everywhere.

Solution 6.18. (1) $\frac{dy}{dx} = \frac{\cos t - t \sin t}{\sin t + t \cos t}$, $\sin t + t \cos t \neq 0$. (2) $y = -\frac{\pi}{2}\left(x - \frac{\pi}{2}\right)$. (3) Since $\frac{d^2y}{dx^2}\Big|_{t=\frac{\pi}{2}} = -2 - \frac{\pi^2}{4} < 0$, the graph of the curve is concave downward at $t = \frac{\pi}{2}$.

Solution 6.19. By definition, the velocity vector is given by $\vec{v} = \frac{d\vec{r}}{dt} = b\omega\langle -\sin\omega t, \cos\omega t\rangle$. It follows that the speed of the ball is constant: $|\vec{v}| = |b\omega|$. Also, $\vec{a} = -b\omega^2\langle\cos\omega t, \sin\omega t\rangle = -\omega^2\vec{r}$.

Solution 6.20. Observe that $\frac{dx}{dt} = 1 - 2\cos 2t$ and $\frac{dx}{dt} = 3\sin t$. Since derivatives $\frac{dx}{dt}$ and $\frac{dy}{dt}$ are never simultaneously zero for $t \in (0, 10)$, it follows that the given curve is smooth. From $\frac{dy}{dx} = \frac{3\sin t}{1 - 2\cos 2t}$, conclude that $\frac{dy}{dx}$ is not defined if $1 - 2\cos 2t = 0$, $0 \leq t \leq 10$. Observe that the first derivative becomes unbounded as $1 - 2\cos 2t \to 0$ and conclude that there is a vertical tangent line at the point $(x(\theta), y(\theta))$ for each $\theta \in \left\{\frac{\pi}{6}, \frac{5\pi}{6}, \frac{7\pi}{6}, \frac{11\pi}{6}, \frac{13\pi}{6}, \frac{17\pi}{6}\right\}$.

Solution 6.21. By eliminating the parameter t, the equation becomes $y^2 = x$, $(x, y) \in [0, 1] \times [-1, 1]$. Observe that $\frac{dy}{dx} = \frac{1}{2\cos t} = \frac{1}{\sqrt{3}}$ implies that $t = \frac{\pi}{6}$. The tangent line is given by $4x - 4\sqrt{3}y + 3 = 0$.

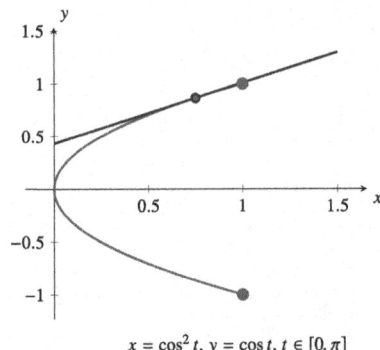

$$x = \cos^2 t, \, y = \cos t, \, t \in [0, \pi]$$

Solution 6.22. Asymptotes are $x = -1$ and $y = 0$. There are two limits of interest: $t \to -1$ and $t \to 0$. Observe that $\lim\limits_{t \to -1} x(t) = -1$ and $\lim\limits_{t \to -1^+} y(t) = \infty$ and conclude that the line $x = -1$ is a vertical asymptote to the graph. What happens when $t \to 0$? Alternatively, express y as a function of x.

Solution 6.23. The slant asymptotes are $y = x + \pi$ and $y = x - \pi$. Observe that $\lim\limits_{t \to \frac{\pi}{2}^-} x(t) = \lim\limits_{t \to \frac{\pi}{2}^-} y(t) = \infty$ and $\lim\limits_{t \to -\frac{\pi}{2}^+} x(t) = \lim\limits_{t \to -\frac{\pi}{2}^+} y(t) = -\infty$. From $\lim\limits_{t \to \frac{\pi}{2}^-} \frac{y(t)}{x(t)} = 1$ and $\lim\limits_{t \to \frac{\pi}{2}^-} (y(t) - x(t)) = \pi$, it follows that the line $y = x + \pi$ is a slant asymptote to the graph of the curve when $t \to \frac{\pi}{2}^-$, i.e., when $x \to \infty$. Similarly, the line $y = x - \pi$ is a slant asymptote to the graph of the curve when $x \to -\infty$.

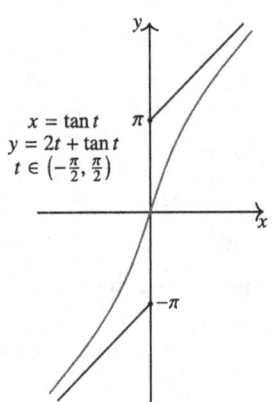

$$x = \tan t$$
$$y = 2t + \tan t$$
$$t \in \left(-\frac{\pi}{2}, \frac{\pi}{2}\right)$$

Solution 6.24. The slant asymptote is the line $y = \frac{x}{2} + e$. There are three limits of interest: $t \to \pm\infty$ and $t \to 1$. Observe that $\lim\limits_{t \to 1^+} x(t) = \lim\limits_{t \to 1^+} y(t) = \infty$ and $\lim\limits_{t \to 1^-} x(t) = \lim\limits_{t \to 1^-} y(t) = -\infty$. From $\lim\limits_{t \to 1} \frac{y(t)}{x(t)} = \frac{1}{2}$ and $\lim\limits_{t \to 1} \left(y(t) - \frac{1}{2} \cdot x(t) \right) = e$, it follows that the line $y = \frac{x}{2} + e$ is a slant asymptote to the graph of the curve when $t \to 1$ (and $x \to \pm\infty$). From $\lim\limits_{t \to \infty} \frac{y(t)}{x(t)} = \infty$ and $\lim\limits_{t \to -\infty} x(t) = \lim\limits_{t \to -\infty} y(t) = 0$, conclude that the graph has a unique asymptote.

Solution 6.25. Two slant asymptotes are $y = \pm 2x + 2$ and a horizontal asymptote is $y = -1$.

Observe that there are four limits of interest, $t \to \pm\infty$ and $t \to \pm 1$. From $\lim\limits_{t \to -1} \frac{y(t)}{x(t)} = -2$ and $\lim\limits_{t \to -1} (y(t) + 2x(t)) = 2$, it follows that the line $y = -2x + 2$ is a slant asymptote to the graph of the curve when $t \to -1$ (and $x \to \pm\infty$.) From $\lim\limits_{t \to 1} \frac{y(t)}{x(t)} = 2$ and $\lim\limits_{t \to 1} (y(t) - 2x(t)) = 2$, it follows that the line $y = 2x + 2$ is another slant asymptote to the graph of the curve when $t \to -1$ (and $x \to \pm\infty$).

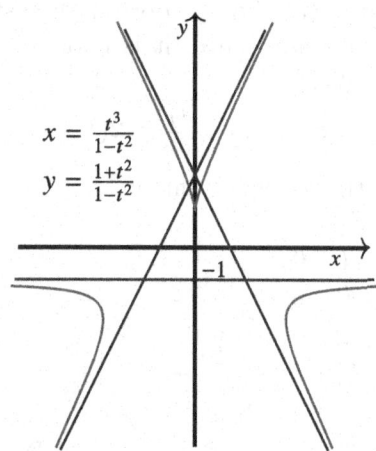

$$x = \frac{t^3}{1 - t^2}$$
$$y = \frac{1 + t^2}{1 - t^2}$$

From $\lim\limits_{t \to \pm\infty} x(t) = \mp\infty$ and $\lim\limits_{t \to \pm\infty} y(t) = -1$, it follows that $y = -1$ is a horizontal asymptote.

Solution 6.26. (1) $(3, 0)$ and $(0, 0)$.
(2) See the following figure.

210

Differential Calculus

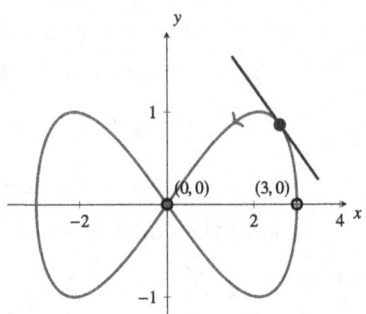

The curve and the line $4x + 6y = 9\sqrt{3}$

(3) The tangent line is given by $4x + 6y = 9\sqrt{3}$.

(4) Conclude from the graph and from what you have observed in (1)–(3) that the graph of the curve is concave upward if $t \in \left(\frac{\pi}{4}, \frac{\pi}{2}\right) \cup \left(\frac{3\pi}{4}, \pi\right)$. Alternatively, you may discuss the sign of the second derivative $\frac{d^2y}{dx^2} = -\frac{2}{9\sin^2\theta} \cdot (2\sin 4\theta + \cos 4\theta \cot 2\theta)$.

Solution 6.27. Since both functions $x = x(t)$ and $y = y(t)$ are periodic with period 2π, we restrict their domains to $t \in [-\pi, \pi]$. Since $(x(-t), y(-t)) = (x(t), -y(t))$ for all $t \in [0, \pi]$, it follows that the graph of the curve (in the xy-coordinate system) is symmetric with respect to the x-axis. Hence, it is enough to investigate each function only for $t \in [0, \pi]$.

From $\frac{dx}{dt} = 2\sin t(2\cos t - 1)$, $\frac{dy}{dt} = 2(1 - \cos t)(2\cos t + 1)$, and $\frac{dy}{dx} = \frac{\frac{dy}{dt}}{\frac{dx}{dt}}$, we obtain the following table:

t	0		$\frac{\pi}{3}$		$\frac{2\pi}{3}$		π
$\frac{dx}{dt}$	0	+	0	−	$-\frac{3\sqrt{3}}{2}$	−	0
$x(t)$	1	↗	$\frac{3}{2}$	↘	$-\frac{1}{2}$	↘	−3
$\frac{dy}{dt}$	2	+	$\frac{3}{2}$	+	0	−	−4
$y(t)$	0	↗	$\frac{\sqrt{3}}{2}$	↗	$\frac{3\sqrt{3}}{2}$	↘	0
$\frac{dy}{dx}$	ND	+	ND	−	0	+	ND

In particular, in the xy-plane, we have the following:

x	−3	↗	$-\frac{1}{2}$	↗	$\frac{3}{2}$	↘	1
y	0	↗	$\frac{3\sqrt{3}}{2}$	↘	$\frac{\sqrt{3}}{2}$	↘	0

Together with the symmetry with respect to the x-axis, this produces the graph.

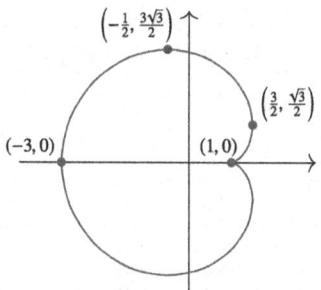

$$x = 2\cos t - \cos 2t, \; y = 2\sin t - \sin 2t$$

Solution 6.28. Since $x = x(t) = \sin\frac{t}{2}$ is periodic with period 4π and since $y = \tan t$ is periodic with period π, we restrict our investigation to $[-2\pi, 2\pi]\backslash\left\{\pm\frac{\pi}{2}, \pm\frac{3\pi}{2}\right\}$. This gives $x \in \left[-1, -\frac{\sqrt{2}}{2}\right) \cup \left(-\frac{\sqrt{2}}{2}, \frac{\sqrt{2}}{2}\right) \cup \left(\frac{\sqrt{2}}{2}, 1\right]$. Since $(x(-t), y(-t)) = (-x(t), -y(t))$ for all $t \in [0, 2\pi]$, it follows that the graph of the curve (in the xy-coordinate system) is symmetric with respect to the origin. This means that it is enough to investigate each function only for $t \in [0, 2\pi]\backslash\left\{\frac{\pi}{2}, \frac{3\pi}{2}\right\}$. Since $x(2\pi - t) = x(t)$ and $y(2\pi - t) = -y(t)$ for $t \in [0, 2\pi]$, we conclude that the graph of the curve is symmetric with respect to the x-axis. Hence, we restrict our investigation to $[0, \pi]\backslash\left\{\frac{\pi}{2}\right\}$. From

$$\lim_{t\to\frac{\pi}{2}} x(t) = \frac{\sqrt{2}}{2}, \quad \lim_{t\to\frac{\pi}{2}^-} y(t) = \infty, \quad \text{and} \quad \lim_{t\to\frac{\pi}{2}^+} y(t) = -\infty,$$

it follows that the line $x = \frac{\sqrt{2}}{2}$ is a vertical asymptote.

From $\frac{dx}{dt} = \frac{1}{2}\cdot\cos\frac{t}{2}$, $\frac{dy}{dt} = \sec^2 t$, and $\frac{dy}{dx} = \frac{\frac{dy}{dt}}{\frac{dx}{dt}}$, we obtain the following table:

t	0		$\frac{\pi}{2}$		π
$\frac{dx}{dt}$	$\frac{1}{2}$	$+$	$\frac{\sqrt{2}}{4}$	$+$	0
$x(t)$	0	$\nearrow$	$\frac{\sqrt{2}}{2}$	$\nearrow$	1
$\frac{dy}{dt}$	1	$+$	ND	$+$	1
$y(t)$	0	$\nearrow^\infty$	ND	$-\infty\,\nearrow$	0
$\frac{dy}{dx}$	2	$+$	ND	$+$	0

In particular, in the xy-plane, we have the following:

x	0	$\nearrow$	$\frac{\sqrt{2}}{2}$	$\nearrow$	1
y	0	$\nearrow^\infty$	ND	$-\infty\,\nearrow$	0

Together with the symmetry with respect to the origin and the symmetry with respect to the x-axis, this produces the graph.

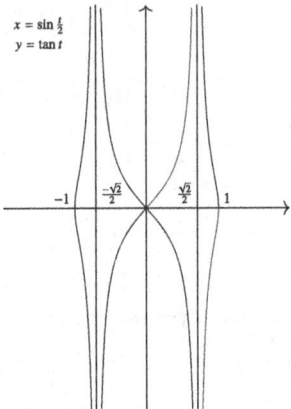

Solution 6.29. (1) Since both functions $x = x(t)$ and $y = y(t)$ are periodic with period 2π, we restrict their domains to $t \in [-\pi, \pi]$. Since $(x(-t), y(-t)) = (x(t), -y(t))$ for all $t \in [0, \pi]$, it follows that the graph of the curve (in the xy-coordinate system) is symmetric with respect to the x-axis. This means that it is enough to investigate each function only for $t \in [0, \pi]$. Since $(x(\pi - t), y(\pi - t)) = (-x(t), -y(t))$ for any $t \in \left[0, \dfrac{\pi}{2}\right)$, it follows that the graph of the curve is symmetric with respect to the y-axis. Hence, it is enough to investigate each function only for $t \in \left[0, \dfrac{\pi}{2}\right]$.

From $\dfrac{dx}{dt} = -3\cos^2 t \sin t$, $\dfrac{dy}{dt} = 3\sin^2 t \cos t$, and $\dfrac{dy}{dx} = \dfrac{\frac{dy}{dt}}{\frac{dx}{dt}} = -3\tan t$, we obtain the following table:

t	0		$\frac{\pi}{2}$
$\frac{dx}{dt}$	0	$-$	0
$x(t)$	1	$\searrow$	0
$\frac{dy}{dt}$	0	$+$	0
$y(t)$	0	$\nearrow$	1
$\frac{dy}{dx}$	ND	$-$	ND

In particular, in the xy-plane, we have the following:

x	1	$\searrow$	0
y	0	$\nearrow$	1

Together with the symmetry with respect to the x-axis and y-axis and the fact that $\frac{d^2y}{dx^2} = \frac{1}{3\cos^4 t \sin t} > 0$ for $t \in \left(0, \frac{\pi}{2}\right)$, this produces the graph.

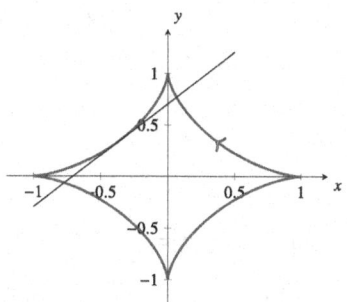

$$x = \cos^3 t, \ y = \sin^3 t$$

(2) $2x - 2y + \sqrt{2} = 0$.

(3) From $\frac{d^2y}{dx^2} = -\frac{\sec^2 t}{-3\cos^2 t \sin t} = -\frac{1}{3\cos^4 t \sin t}$, we obtain $\frac{d^2y}{dx^2}\Big|_{t=1} = \frac{1}{3\cos^4 1 \sin 1} < 0$. Concave downward.

(4) See the graph or discuss the sign of the second derivative. What is the domain of the second derivative?

Solution 6.30. (1) Observe that $\lim\limits_{t \to -1^-} x(t) = \lim\limits_{t \to -1^+} y(t) = \infty$ and $\lim\limits_{t \to -1^+} x(t) = \lim\limits_{t \to -1^-} y(t) = -\infty$. This implies that the range of both functions $x = x(t)$ and $y = y(t)$ is $\mathbb{R}$. From $\lim\limits_{t \to -1} \frac{y(t)}{x(t)} = -1$ and $\lim\limits_{t \to -1} (y(t) + x(t)) = -1$, it follows that the line $y = -x - 1$ is a slant asymptote to the graph of the curve when $x \to \pm\infty$. From $\frac{dx}{dt} = \frac{3(1-2t^3)}{(1+t^3)^2}$, $\frac{dy}{dt} = \frac{2t(2-t^3)}{(1+t^3)^2}$, and $\frac{dy}{dx} = \frac{\frac{dy}{dt}}{\frac{dx}{dt}}$, we obtain the following table:

t		-1		0		$\frac{1}{\sqrt[3]{2}}$		$\sqrt[3]{2}$	
$\frac{dx}{dt}$	$+$	ND	$+$	3	$+$	0	$-$	-1	$-$
$x(t)$	0 $\nearrow^\infty$	ND	$_{-\infty}\nearrow$	0	$\nearrow$	$\sqrt[3]{4}$	$\searrow$	$\sqrt[3]{2}$	$\searrow_0$
$\frac{dy}{dt}$	$-$	ND	$-$	0	$+$	$\frac{2\sqrt[3]{4}}{3}$	$+$	0	$-$
$y(t)$	0 $\searrow_{-\infty}$	ND	$^\infty\searrow$	0	$\nearrow$	$\sqrt[3]{2}$	$\nearrow$	$\sqrt[3]{4}$	$\searrow_0$
$\frac{dy}{dx}$	$-$	ND	$-$	0	$+$	ND	$-$	0	$+$

In the xy-plane, we have the following:

x	$-\infty$	$\nearrow$	0	$\nearrow$	$\sqrt[3]{4}$	$\searrow$	$\sqrt[3]{2}$	$\searrow$	0	$\nearrow^\infty$
y	∞	$\searrow$	0	$\nearrow$	$\sqrt[3]{2}$	$\nearrow$	$\sqrt[3]{4}$	$\searrow$	0	$\searrow_{-\infty}$

Together with the fact that the curve is smooth, this produces the graph.

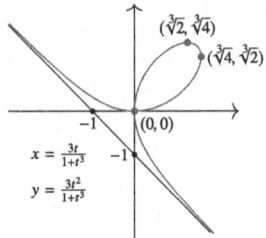

(2) $x^3 + y^3 = 3xy$. Observe that $t = \frac{y}{x}$.

(3) From $\left.\frac{dx}{dt}\right|_{t=1} = -\frac{3}{4}$ and $\left.\frac{dy}{dt}\right|_{t=1} = \frac{3}{4}$, it follows that the slope is $\left.\frac{dy}{dx}\right|_{t=1} = -1$. Alternatively, differentiate the expression in (2) with respect to x.

(4) Use the fact that $\frac{d^2y}{dx^2} = \frac{d}{dx}\left(\frac{dy}{dx}\right) = \frac{1}{\frac{dx}{dt}} \cdot \frac{d}{dt}\left(\frac{dy}{dx}\right)$ to obtain $\left.\frac{d^2y}{dx^2}\right|_{t=1} = -\frac{32}{3}$. Alternatively, differentiate the expression in (2) with respect to x twice.

Solution 6.31. $A = \left(r = \sqrt{2}, \theta = \frac{\pi}{4}\right)$. Observe that the rectangular coordinates of the point A are $(1, 1)$. This implies that the distance between the point and the origin is $r = \sqrt{2}$. The angle between the positive direction of the x-axis and the ray with the initial point at the origin that passes through the point A is $\theta = \frac{\pi}{4}$. $B = \left(4, \frac{11\pi}{6}\right)$, $C = \left(2, \frac{11\pi}{6}\right)$, $D = \left(2\sqrt{2} - 1, \frac{3\pi}{4}\right)$. Observe that the point D belongs to the intersection of the circle $(x + 2)^2 + (y - 2)^2 = 1$ and the line $y = -x$. Calculate the distance between D and the origin as the difference of the distance between the centre of the circle and the origin and the radius of the circle. Hence, $r = 2\sqrt{2} - 1$.

Solution 6.32. (1) $r = 2$. (2) $r = 2\cos\theta$. (3) $r = \sin\theta$. Use the fact that the rectangular and the polar coordinates of a point in the plane are related by $x = r\cos\theta$ and $y = r\sin\theta$.

Solution 6.33. (1) $(0, 5)$. Observe that $r\left(\frac{3\pi}{2}\right) = -1 < 0$, so the point lies on the positive ray of the y-axis.
(2) $x^2 + y^2 = 5y$. This is the circle $x^2 + (y - 2.5)^2 = 6.25$.
(3) $x = 5\sin\theta\cos\theta$, $y = 5\sin^2\theta$, $\theta \in [0, 2\pi]$.

(4) $\frac{dy}{dx} = \frac{2\sin\theta\cos\theta}{\cos^2\theta-\sin^2\theta} = \tan 2\theta.$

(5) $2\sqrt{3}x - 2y = 5.$

Solution 6.34. Observe that both curves are circles.

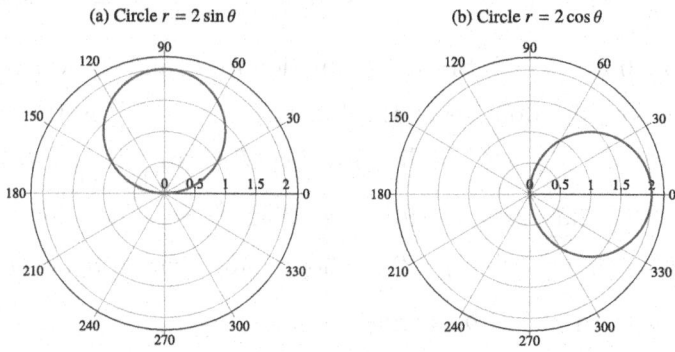

(a) Circle $r = 2\sin\theta$ (b) Circle $r = 2\cos\theta$

Solution 6.35. $(x^2+y^2)^3 = (x^2-y^2)^2$. Recall that $\cos 2\theta = \cos^2\theta - \sin^2\theta$. By multiplying the given expression by r^2, we obtain $r^3 = r^2\cos^2\theta - r^2\sin^2\theta$.

Solution 6.36. $\frac{3\sqrt{3}}{4}$. On the given cardioid, $x = (1+\cos\theta)\cos\theta$ and $y = (1+\cos\theta)\sin\theta$, $\theta \in [-\pi, \pi]$. The question is to evaluate the maximum value of y. Note that the inequality $y > 0$ is equivalent to the inequality $\sin\theta > 0$.

From $\frac{dy}{d\theta} = 2\cos^2\theta + \cos\theta - 1$, it follows that the critical numbers of the function $y = y(\theta)$ are the values of θ for which $\cos\theta = \frac{-1\pm 3}{4}$. Hence, the critical numbers are the values of θ for which $\cos\theta = -1$ or $\cos\theta = \frac{1}{2}$. Since $y_{\max} > 0$, it follows that $\sin\theta = \sqrt{1-\left(\frac{1}{2}\right)^2} = \frac{\sqrt{3}}{2}$.

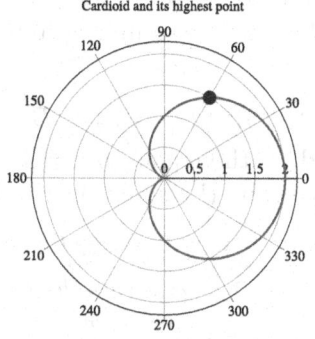

Cardioid and its highest point

The maximum height is $y = \frac{3\sqrt{3}}{4}$.

Solution 6.37. $\pm\frac{\sqrt{3}}{3}$. Solve $2 = 4\cos\theta$ to obtain $\left(2, \pm\frac{\pi}{3}\right)$, the intersection points (in polar coordinates) of the two curves. Note that the circle $r = 2$ is determined by its parametric equations $x = 2\cos\theta$ and $y = 2\sin\theta$. The result follows from $\frac{dy}{dx} = \frac{2\cos\theta}{-2\sin\theta} = -\cot\theta$.

Solution 6.38. $y = 2$. Since the function $\theta \mapsto \tan\frac{\theta}{2}$ is periodic with a period 2π, the domain of the function $r = \tan\frac{\theta}{2}$ is the interval $(-\pi, \pi)$. From $\lim\limits_{\theta\to\pi^-} x(\theta) = \lim\limits_{\theta\to\pi^-} \tan\frac{\theta}{2}\cdot\cos\theta = \infty$, $\lim\limits_{\theta\to-\pi^+} x(\theta) = -\infty$, $\lim\limits_{\theta\to\pi^-} y(\theta) = \lim\limits_{\theta\to\pi^-} \tan\frac{\theta}{2}\cdot\sin\theta = \lim\limits_{\theta\to\pi^-} 2\sin^2\frac{\theta}{2} = 2$, and $\lim\limits_{\theta\to-\pi^+} y(\theta) = 2$, it follows that the line $y = 2$ is a horizontal asymptote of the curve.

Solution 6.39. (1) $(-2, -2\sqrt{3})$. (2) $-\frac{\sqrt{3}}{3}$.

Solution 6.40. (1) $(2\sqrt{3}, 2)$. (2) $-\sqrt{3}$. (3) C. Use the fact that the point $(-2, -2\sqrt{3})$ lies on the curve $r = 4\cos 3\theta$ to conclude that this polar equation represents the rose in graph C. Similarly, the rose in graph B matches the equation $r = 4\sin 3\theta$. *Note:* The 3-leaf rose in graph A is determined by the equation $r = -4\cos 3\theta$.

Solution 6.41. (1) $\left(\frac{5\sqrt{3}}{4}, \frac{5}{4}\right)$. (2) $\frac{\sqrt{3}}{23}$. (3) C. Observe that $r(0) = r(\pi) = 4$ and $r\left(\frac{\pi}{2}\right) = r\left(\frac{3\pi}{2}\right) = -2$. *Note:* The cardioid in graph A is determined by the equation $r = 1 + \sin\theta$. The 4-leaf rose in graph B is determined by the equation $r = 2\cos 2\theta$.

Solution 6.42. (1) $(0, -3)$.
(2) $(\pm 1, 0)$. Solve $y = (1 - 2\sin\theta)\sin\theta = 0$.
(3) Graph A corresponds to $r = 1 - 2\sin\theta$ and graph C corresponds to $r = 1 + \sin 2\theta$. *Note:* The curve in graph B is determined by the equation $r = 2 + 2\sin 2\theta$.

Solution 6.43. (1) $(1, \sqrt{3})$. (2) $\frac{1}{3\sqrt{3}}$. (3) B. *Note:* The limaçon in graph A is determined by the equation $r = 1 - 2\sin\theta$. The limaçon in graph C is determined by the equation $r = 1 + 2\sin\theta$.

Solution 6.44. Let $\theta \in \mathbb{R}$, $\theta \neq \frac{\pi}{2} + k\pi$, $k \in \mathbb{Z}$. From $x = r\cos\theta = a + b\cos^2\theta$ and $y = r\sin\theta = a\tan\theta + b\sin\theta\cos\theta$, it follows that $\frac{dy}{dx} = \frac{\frac{a}{\cos^2\theta} + b(\cos^2\theta - \sin^2\theta)}{-2b\sin\theta\cos\theta} = -\frac{a + a\tan^2\theta + b\cos 2\theta}{b\sin 2\theta}$.

Solution 6.45. A – The lemniscate $r^2 = \cos 2\theta$. C: The Archimedean spiral $r = 0.2\theta$. D: The right strophoid $r = 2(2\cos\theta - \sec\theta)$. *Note*: The hyperbola in graph B is determined by the equation $r(1 - 2\sin\theta) = 1$.

Solution 6.46. (1) $x = (1 - \cos\theta)\cos\theta$, $y = (1 - \cos\theta)\sin\theta$.

(2) From $\frac{dy}{dx} = \frac{\cos\theta - \cos 2\theta}{\sin 2\theta - \sin\theta}$, it follows that the slope is $\frac{dy}{dx}\Big|_{\theta = \frac{\pi}{2}} = -1$.

(3) Recall the trigonometric identities $\cos\alpha - \cos\beta = -2\sin\frac{\alpha+\beta}{2}\sin\frac{\alpha-\beta}{2}$ and $\sin\alpha - \sin\beta = 2\cos\frac{\alpha+\beta}{2}\sin\frac{\alpha-\beta}{2}$. Conclude that $\frac{dy}{dx} = \frac{\sin\frac{3\theta}{2}\sin\frac{\theta}{2}}{\cos\frac{3\theta}{2}\sin\frac{\theta}{2}}$, $\theta \notin \left\{0, \frac{\pi}{3}, \pi, \frac{5\pi}{3}\right\}$. Since $\frac{dy}{dx} = 0$ if $\sin\frac{3\theta}{2} = 0$ and $\theta \neq 0$, the horizontal tangent lines are at the points on the curve that corresponds to $\theta = \frac{2\pi}{3}$ and $\theta = \frac{4\pi}{3}$. The rectangular coordinates of those points are $\left(-\frac{3}{4}, \pm\frac{3\sqrt{3}}{4}\right)$. From $\lim\limits_{\theta \to \frac{\pi}{3}^-}\frac{dy}{dx} = \lim\limits_{\theta \to \frac{\pi}{3}^-}\frac{\sin\frac{3\theta}{2}}{\cos\frac{3\theta}{2}} = -\infty$ and $\lim\limits_{\theta \to \frac{\pi}{3}^+}\frac{dy}{dx} = \infty$, it follows that there is a vertical tangent line at the point on the curve that correspond to $\theta = \frac{\pi}{3}$. The rectangular coordinates of that point are $\left(\frac{1}{4}, \frac{\sqrt{3}}{4}\right)$.

Similarly, there is a vertical tangent line at the point $\left(\frac{1}{4}, -\frac{\sqrt{3}}{4}\right)$ corresponding to $\theta = \frac{5\pi}{3}$, and another vertical tangent line at the point $(-2, 0)$ corresponding to $\theta = \pi$. Observe $\lim\limits_{\theta \to 0}\frac{dy}{dx} = \lim\limits_{\theta \to 0}\frac{\sin\frac{3\theta}{2}}{\cos\frac{3\theta}{2}} = 0$ and $\lim\limits_{\theta \to 0} x(\theta) = \lim\limits_{\theta \to 0} y(\theta) = 0$.

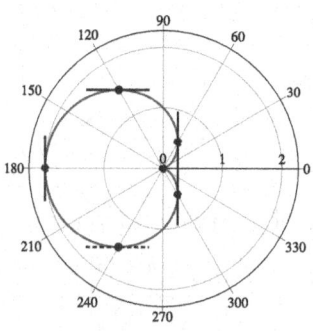

It follows that the point $(0, 0)$ corresponding to $\theta = 0$ is a cusp.

Solution 6.47. Since $\theta \mapsto -1 + \cos\theta$ is a periodic function with period 2π, we take $[-\pi, \pi]$ as the domain of the polar equation $r = r(\theta) = -1 + \cos\theta$. From $r(-\theta) = r(\theta)$, conclude that the graph of the cardioid is symmetric with respect to the x-axis (polar). Start by drawing the graph of the curve when $\theta \in [0, \pi]$. From $\frac{dy}{dx} = \frac{\sin\frac{3\theta}{2}\sin\frac{\theta}{2}}{\cos\frac{3\theta}{2}\sin\frac{\theta}{2}}$, conclude that the cardioid is a smooth curve for $\theta \in (0, \pi)$. Observe that $\lim_{\theta \to \pi^-} \frac{dy}{dx} = \infty$ and $\lim_{\theta \to 0^+} \frac{dy}{dx} = 0$. Together with the symmetry with respect to the x-axis, this means that the curve has a vertical tangent line at the point that corresponds to $\theta = \pi$. Note that the polar coordinates of this point are $(r(\pi), \pi) = (-2, \pi)$ and that its rectangular coordinates are $(r(\pi)\cos(\pi), r(\pi)(\sin(\pi)) = (2, 0)$. The point $(0, 0)$ corresponding to $\theta = 0$ is a cusp.

We obtain the graph of the cardioid in three steps. In Step 1, we draw a graph of the function $\theta \mapsto r = -1 + \cos\theta$, $\theta \in [0, \pi]$, in the *rectangular $r\theta$-coordinate* system. In Step 2, we use the graph from Step 1 to construct a table that describes the behavior of r while θ increases over the domain. In Step 3, we use the table from Step 2, together with the facts about the cardioid that we have already established (the symmetry with respect to the x-axis, smoothness, a vertical tangent line, a cusp), to draw the cardioid in the polar coordinate system.

Step 1: The graph of $r = -1 + \cos\theta$, $\theta \in [0, \pi]$.

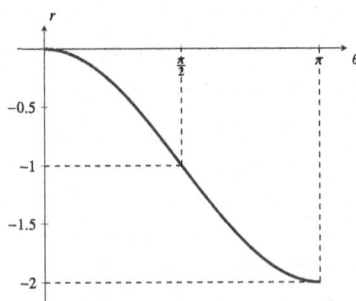

Step 2: The behavior of r while θ increases.

θ	0	↗	$\frac{\pi}{2}$	↗	π
r	0	↘	-1	↘	-2

Step 3: Cardioid.

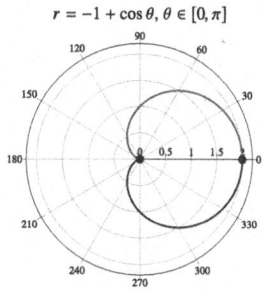

$r = -1 + \cos\theta, \theta \in [0, \pi]$

Solution 6.48. (1) To determine $\theta \in [-\pi, \pi)$ such that $r = r(\theta) > 0$, solve $r = \sin\theta > -\frac{1}{2}$. It follows that $r > 0$ if $\theta \in \left[-\pi, -\frac{5\pi}{6}\right) \cup \left(-\frac{\pi}{6}, \pi\right)$.

(2) Solve $\frac{dr}{d\theta} = 2\cos\theta = 0$ in $[-\pi, \pi)$ to establish that $\theta = \pm\frac{\pi}{2}$ are the only critical numbers. From $r(-\pi) = r(\pi) = 1$, $r\left(-\frac{\pi}{2}\right) = -1$, and $r\left(\frac{\pi}{2}\right) = 3$, by the Extreme Value Theorem, it follows that $r_{\max} = 3$ and $r_{\min} = -1$.

(3) Since $\theta \mapsto r(\pi - \theta) = 1 + 2\sin(\pi - \theta) = 1 + 2\sin\theta = r(\theta)$, the graph of the limaçon is symmetric with respect to the y-axis (i.e., the line $\theta = \pm\frac{\pi}{2}$). This symmetry works in the following way. If $\theta \in \left[0, \frac{\pi}{2}\right]$, then $\pi - \theta \in \left[\frac{\pi}{2}, \pi\right]$ and $r(\theta) = r(\pi - \theta)$. If $\theta \in \left[-\frac{\pi}{2}, 0\right]$, then $-\pi - \theta \in \left[-\pi, -\frac{\pi}{2}\right]$ and $r(\theta) = r(\pi - \theta) = r(\pi - \theta - 2\pi) = r(-\pi - \theta)$. Hence, we take $\theta \in \left[-\frac{\pi}{2}, \frac{\pi}{2}\right]$. From $\frac{dy}{dx} = \frac{\cos\theta \cdot (4\sin\theta + 1)}{-4\sin^2\theta - \sin\theta + 2}$, it follows that the curve has vertical tangent lines at the points that correspond to $\theta = -\arcsin\frac{1+\sqrt{33}}{8} \approx -1$ radian ($\approx -57.29°$) and $\theta = \arcsin\frac{\sqrt{33}-1}{8} \approx 0.634$ radians ($\approx 36.32°$). Hence, the limaçon is a smooth curve. Also, $\frac{dy}{dx} = 0$ for $\theta \in \left\{\pm\frac{\pi}{2}, -\arcsin(0.25) \approx -0.252\right\}$.

Step 1: The graph of $r = 1 + 2\sin\theta$, $\theta \in [-\pi/2, \pi/2]$.

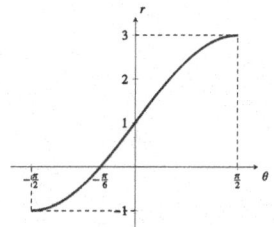

Step 2: The behavior of r while θ increases.

θ	$-\frac{\pi}{2}$	$\nearrow$	$-\frac{\pi}{6}$	$\nearrow$	0	$\nearrow$	$\frac{\pi}{2}$
r	-1	$\nearrow$	0	$\nearrow$	1	$\nearrow$	3

Step 3: Limaçon.

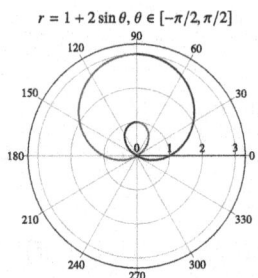

$r = 1 + 2\sin\theta,\ \theta \in [-\pi/2, \pi/2]$

Solution 6.49. (1) $\frac{dy}{dx} = \frac{-2\sin 2\theta \sin\theta + \cos 2\theta \cos\theta}{-2\sin 2\theta \cos\theta - \cos 2\theta \sin\theta} = \frac{2\tan 2\theta \tan\theta - 1}{2\tan 2\theta + \tan\theta}$.

(2) Use the trigonometric identities for half angles to obtain $\cos\frac{\pi}{8} = \frac{\sqrt{2+\sqrt{2}}}{2}$ and $\sin\frac{\pi}{8} = \frac{\sqrt{2-\sqrt{2}}}{2}$. The point corresponding to $\theta = \frac{\pi}{8}$ is $\left(\frac{1}{2}\sqrt{1 + \frac{1}{\sqrt{2}}}, \frac{1}{2}\sqrt{1 - \frac{1}{\sqrt{2}}}\right)$.

(3) Use the fact that the slope of the tangent line is $\frac{dy}{dx}\Big|_{\frac{\pi}{8}} = 7 - 5\sqrt{2}$.

(4) Use (1) to conclude that $r = \cos 2\theta$ is a smooth curve for $\theta \in \left(0, \frac{\pi}{4}\right)$.

Step 1: The graph of $r = \cos 2\theta$, $\theta \in [0, \pi/4]$.

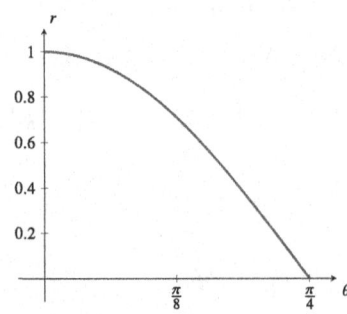

Step 2: The behavior of r while θ increases.

$$\frac{\theta \;\; 0 \;\; \nearrow \;\; \frac{\pi}{4}}{r \;\; 1 \;\; \searrow \;\; 0}$$

Step 3: The graph of the curve.

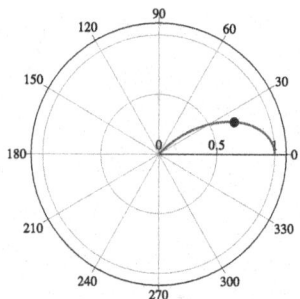

Solution 6.50. See the following figure.

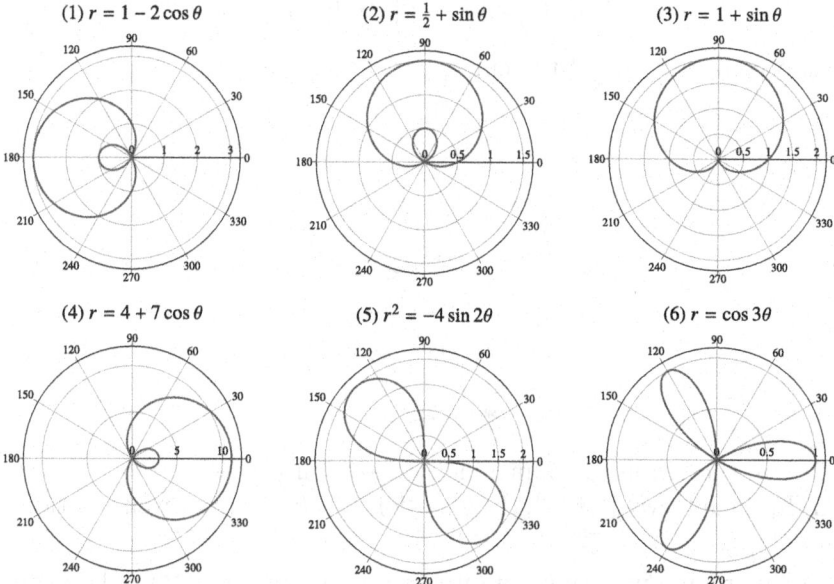

(1) $r = 1 - 2\cos\theta$

(2) $r = \frac{1}{2} + \sin\theta$

(3) $r = 1 + \sin\theta$

(4) $r = 4 + 7\cos\theta$

(5) $r^2 = -4\sin 2\theta$

(6) $r = \cos 3\theta$

Solution 6.51. Observe that the function $\theta \mapsto r = 1 + 2\sin 3\theta$ is periodic with period $\frac{2\pi}{3}$. (1) First obtain the graph of the polar curve $r = 1 + 2\sin 3\theta$, $\theta \in \left[-\frac{\pi}{3}, \frac{\pi}{3}\right]$. Observe that the equation $r = 1 + 2\sin 3\theta = 0$ has two solutions in this interval, namely

$\theta = -\frac{5\pi}{18}$ and $\theta = -\frac{\pi}{18}$. In the next step, we rotate this graph clockwise by $\frac{2\pi}{3}$ radians and $\frac{4\pi}{3}$ radians, respectively.

Step 1: The graph of $r = 1 + 2\sin 3\theta$, $\theta \in [-\pi/3, \pi/3]$.

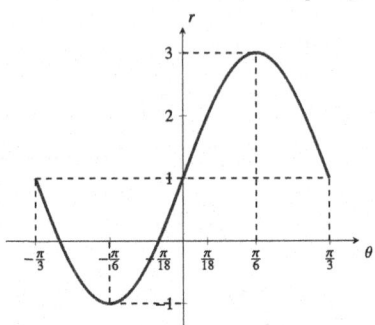

Step 2: The behavior of r while θ increases.

θ	$-\frac{\pi}{3}$	$\nearrow$	$-\frac{5\pi}{18}$	$\nearrow$	$-\frac{\pi}{6}$	$\nearrow$	$-\frac{\pi}{18}$	$\nearrow$	0	$\nearrow$	$\frac{\pi}{6}$	$\nearrow$	$\frac{\pi}{3}$
r	1	$\searrow$	0	$\searrow$	-1	$\nearrow$	0	$\nearrow$	1	$\nearrow$	3	$\searrow$	1

Step 3: The graph of the curve.

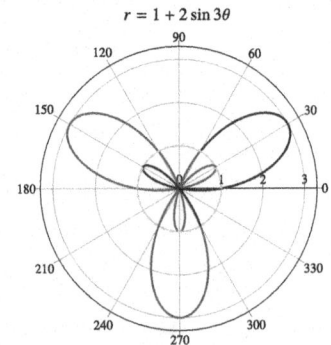

$r = 1 + 2\sin 3\theta$

(2) Use the graph to count nine intersection points.

(3) In the interval $[-\pi, \pi]$, the equation $\sin 3\theta = 0$ has seven solutions: $-\pi, -\frac{2\pi}{3}, -\frac{\pi}{3}, 0, \frac{\pi}{3}, \frac{2\pi}{3}, \pi$.

(4) The remaining points of intersection are obtained by solving the equation $-1 = 1 + 2\sin 3\theta$.

Solution 6.52. (1) $r(0) = 2$, $r\left(\frac{\pi}{2}\right) = 2 + e$, $r\left(\frac{3\pi}{2}\right) = e^{-1}$.

(2) Observe that the function $\theta \mapsto r = 1 + \sin\theta + e^{\sin\theta}$ is periodic with period 2π. From $r(\pi - \theta) = r(\theta)$, conclude that the graph of

the curve is symmetric with respect to the y-axis. Draw the part of the curve for $\theta \in \left[-\frac{\pi}{2}, \frac{\pi}{2}\right]$ first. From $\frac{dr}{d\theta} = (1 + e^{\sin\theta})\cos\theta$, conclude that the function $\theta \mapsto r(\theta)$ is increasing in this interval.

Step 1: The graph of $r = 1 + \sin\theta + e^{\sin\theta}$, $\theta \in [-\pi/2, \pi/2]$.

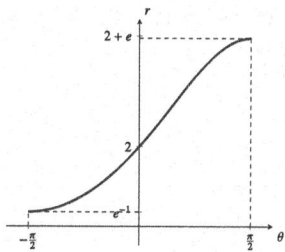

Step 2: The behavior of r while θ increases.

θ	$-\frac{\pi}{2}$	↗	0	↗	$\frac{\pi}{2}$
r	e^{-1}	↗	2	↗	$2+e$

Step 3: The graph of the curve.

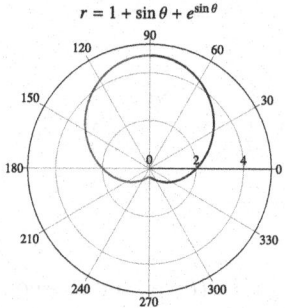

(3) From $\frac{dr}{d\theta} = \cos\theta(1 + e^{\sin\theta}) = 0$, conclude that the critical numbers are $\pm\frac{\pi}{2}$. By the Extreme Value Theorem, the minimum distance equals e^{-1}.

Solution 6.53. (1) The slope is given by $\left.\frac{dy}{dx}\right|_{\theta=\frac{5\pi}{2}}$. From $x = r\cos\theta = \frac{\cos\theta}{\theta}$ and $y = \frac{\sin\theta}{\theta}$, it follows that $\frac{dy}{dx} = \frac{\sin\theta - \theta\cos\theta}{\cos\theta + \theta\sin\theta}$. Thus, $\left.\frac{dy}{dx}\right|_{\theta=\frac{5\pi}{2}} = \frac{2}{5\pi}$.

(2) Recall the relationship between the rectangular and polar coordinates: for a point (x, y), $r = \sqrt{x^2 + y^2}$ and $\tan \theta = \frac{y}{x}$ when $x \neq 0$. From $\theta = \frac{1}{r}$ for $r \neq 0$, it follows that $\frac{y}{x} = \tan \frac{1}{\sqrt{x^2+y^2}}$ for $x \neq 0$.

(3) **Step 1:** The graph of $r\theta = 1$, $\theta \in [0.5, 3\pi)$.

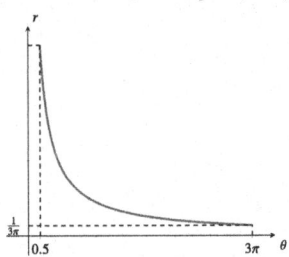

Step 2: The behavior of r while θ increases.

θ	0.5	$\nearrow$	3π
r	2	$\searrow$	$\frac{1}{3\pi}$

Step 3: The graph of the curve.

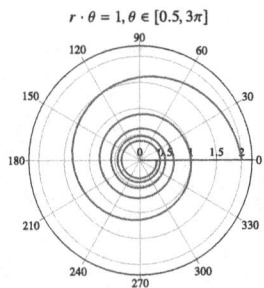

$r \cdot \theta = 1, \theta \in [0.5, 3\pi]$

Solution 6.54. Since $\theta \mapsto r(\theta) = \sec \theta - \cos \theta = \frac{\sin^2 \theta}{\cos \theta}$ is a periodic function with period 2π, take the set $\left[-\pi, -\frac{\pi}{2}\right) \cup \left(-\frac{\pi}{2}, \frac{\pi}{2}\right) \cup \left(\frac{\pi}{2}, \pi\right]$ as the domain of the polar equation. Since $r(\pi + \theta) = -r(\theta)$, the polar coordinates $(r(\theta), \theta)$ and $(r(\pi + \theta), \pi + \theta)$ represent the same point on the graph of the curve. Hence, we reduce the domain to $\left(-\frac{\pi}{2}, \frac{\pi}{2}\right)$. From $r(-\theta) = r(\theta)$, conclude that the graph of the Cissoid of Diocles is symmetric with respect to the x-axis (polar). In the first instance, draw the graph for $\theta \in \left[0, \frac{\pi}{2}\right)$ and then complete the graph by using the symmetry. From $\frac{dy}{dx} = \frac{\tan \theta}{2} \cdot (3 + \tan^2 \theta)$, conclude that the Cissoid of Diocles is a smooth, always increasing, curve for $\theta \in \left(0, \frac{\pi}{2}\right)$. Observe that $\lim\limits_{\theta \to \frac{\pi}{2}^-} x(\theta) = \lim\limits_{\theta \to \frac{\pi}{2}^-} \sin^2 \theta = 1$ and

$\lim\limits_{\theta \to \frac{\pi}{2}^-} y(\theta) = \lim\limits_{\theta \to \frac{\pi}{2}^-} \frac{\sin^3 \theta}{\cos \theta} = \infty$. This implies that the line $x = 1$ is a

vertical asymptote to the curve. Finally, from $\lim\limits_{\theta \to 0^+} \frac{dy}{dx} = \lim\limits_{\theta \to 0^+} x(\theta) = \lim\limits_{\theta \to 0^+} y(\theta) = 0$, it follows that the point $(0,0)$ corresponding to $\theta = 0$ is a cusp.

Step 1: The graph of $r = \sec \theta - \cos \theta$, $\theta \in [0, \pi/2)$.

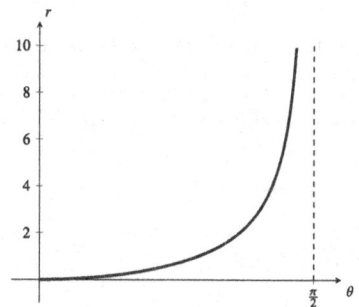

Step 2: The behavior of r while θ increases.

θ	0	↗	$\frac{\pi}{2}$
r	0	↗∞	ND

Step 3: Cissoid of Diocles.

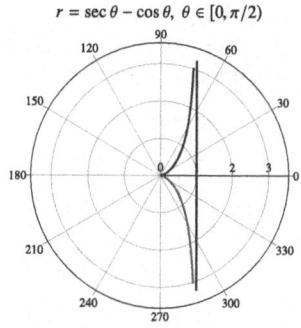

$r = \sec \theta - \cos \theta$, $\theta \in [0, \pi/2)$

Solution 6.55. (1) Start by drawing the graph of the curve for $\theta \in \left[0, \frac{\pi}{2}\right)$ and then complete it by using the symmetry with respect to the x-axis. From $\frac{dy}{dx} = \frac{1+\tan^2\theta-4\cos 2\theta}{4\sin 2\theta}$, conclude that the Conchoid of de Sluze $r = -\sec\theta + 4\cos\theta$ is a smooth curve for $\theta \in \left(0, \frac{\pi}{2}\right)$. Observe that $\lim\limits_{\theta\to\frac{\pi}{2}^-} x(\theta) = \lim\limits_{\theta\to\frac{\pi}{2}^-} (-1 + 4\cos^2\theta) = -1$ and $\lim\limits_{\theta\to\frac{\pi}{2}^-} y(\theta) = \lim\limits_{\theta\to\frac{\pi}{2}^-} \left(-\frac{\sin\theta}{\cos\theta} + 4\sin\theta\cos\theta\right) = -\infty$. This implies that the line $x = -1$ is a vertical asymptote to the curve. In addition, from $\lim\limits_{\theta\to 0^+} \frac{dy}{dx} = -\infty$, it follows that at the point $(3,0)$ corresponding to $\theta = 0$ the curve has a vertical tangent line. Finally, observe that $r\left(\frac{\pi}{3}\right) = 0$ and that the slope of the tangent line at the origin is $\left.\frac{dy}{dx}\right|_{\frac{\pi}{3}} = \sqrt{3}$ for the part of the curve obtained when $\theta \in \left(0, \frac{\pi}{2}\right)$.

Step 1: The graph of $r = -\sec\theta + 4\cos\theta$, $\theta \in [0, \pi/2)$.

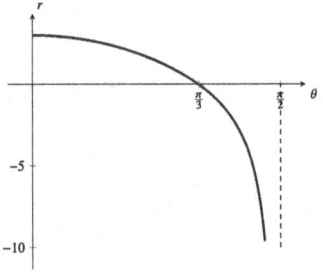

Step 2: The behavior of r while θ increases.

θ	0	↗	$\frac{\pi}{3}$	↗	$\frac{\pi}{2}$
r	3	↘	0	↘$_{-\infty}$	ND

Step 3: Conchoid of de Sluze.

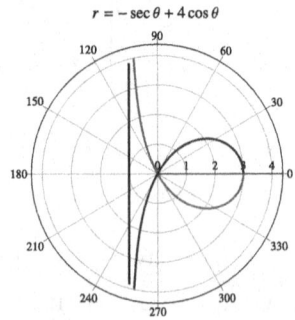

$r = -\sec\theta + 4\cos\theta$

(2) Start by drawing the graph of the curve for $\theta \in \left[0, \frac{\pi}{2}\right)$ and complete it by using the symmetry with respect to the x-axis. From $\frac{dy}{dx} = -\frac{1+\tan^2\theta+2\cos 2\theta}{2\sin 2\theta}$, we conclude that the Conchoid of de Sluze $r = \sec\theta + 2\cos 2\theta$ is a smooth curve for $\theta \in \left(0, \frac{\pi}{2}\right)$. We observe that $\lim\limits_{\theta\to\frac{\pi}{2}^-} x(\theta) = \lim\limits_{\theta\to\frac{\pi}{2}^-} (1 + 2\cos^2\theta) = 1$ and $\lim\limits_{\theta\to\frac{\pi}{2}^-} y(\theta) =$ $\lim\limits_{\theta\to\frac{\pi}{2}^-} \left(\frac{\sin\theta}{\cos\theta} + \sin 2\theta\right) = \infty$. This implies that the line $x = 1$ is a vertical asymptote to the curve. Finally, from $\lim\limits_{\theta\to 0^+} \frac{dy}{dx} = -\infty$, it follows that at the point $(3, 0)$ corresponding to $\theta = 0$ the curve has a vertical tangent line.

Step 1: The graph of $r = \sec\theta + 2\cos\theta$, $\theta \in [0, \pi/2)$.

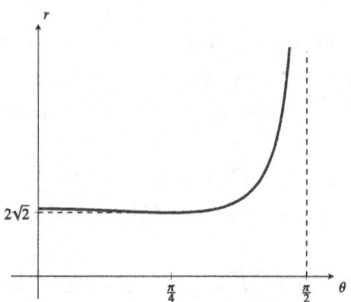

Step 2: The behavior of r while θ increases.

θ	0	↗	$\frac{\pi}{4}$	↗	$\frac{\pi}{2}$
r	3	↘	$2\sqrt{2}$	↗∞	ND

Step 3: Conchoid of de Sluze.

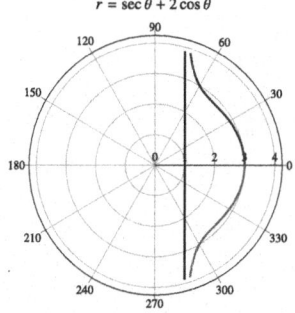

$r = \sec\theta + 2\cos\theta$

Solution 6.56. See the following figure for a graph of the case $b = 1$, $k = 0.01$, and $c = 2$. The position in the $xy-$plane of the bee at time t is determined by a vector function $\vec{s}(t) = \langle be^{kt}\cos ct, be^{kt}\sin ct\rangle$. Recall that the angle α between the velocity and the acceleration is given by $\cos\alpha = \dfrac{\vec{v}\cdot\vec{a}}{|\vec{v}||\vec{a}|}$, where $\vec{v}(t) = \vec{s}'(t)$ and $\vec{a}(t) = \vec{s}''(t)$.

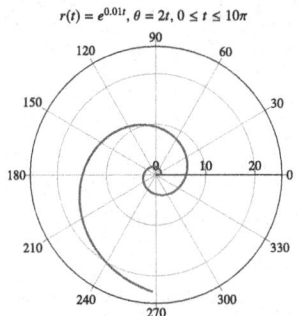

$$r(t) = e^{0.01t}, \theta = 2t, 0 \le t \le 10\pi$$

One way to solve this problem is to consider that the bee moves in the complex plane. In that case, its position is given by $F(t) = be^{kt}\cos ct + i\cdot be^{kt}\sin ct = be^{kt}(\cos ct + i\sin ct) = be^{(k+ic)t}$, where i is the imaginary unit.

Observe that $\vec{v}(t) = \langle\mathrm{Re}(F'(t)), \mathrm{Im}(F'(t))\rangle$ and $\vec{a}(t) = \langle\mathrm{Re}(F''(t)), \mathrm{Im}(F''(t))\rangle$. Next, observe that $F'(t) = (k + ic)F(t)$ and $F''(t) = (k + ic)^2 F(t)$.

From $F''(t) = (k + ic)F'(t)$, it follows that $\mathrm{Re}(F''(t)) = k\cdot\mathrm{Re}(F'(t)) - c\cdot\mathrm{Im}(F'(t))$ and $\mathrm{Im}(F''(t)) = k\cdot\mathrm{Im}(F'(t)) + c\cdot\mathrm{Re}(F'(t))$. Finally, $\vec{v}\cdot\vec{a} = \mathrm{Re}(F'(t))\cdot\mathrm{Re}(F''(t)) + \mathrm{Im}(F'(t))\cdot\mathrm{Im}(F''(t)) = k((\mathrm{Re}(F'(t)))^2 + (\mathrm{Im}(F'(t)))^2) = k|F'(t)|^2$, which immediately implies the required result.

Chapter 7

True–False and Multiple Choice Questions

7.1 Functions

7.1.1 *True or false*

Answer each of the following either TRUE or FALSE.

Problem 7.1. $\sqrt{a^2} = a$ for all $a \in \mathbb{R}$.

Problem 7.2. For a given function f, if $f(s) = f(t)$, then $t = s$.

Problem 7.3. The average rate of change of the function $f(t) = t^2 - 1$ over the interval $[1, 3]$ is 4.

Problem 7.4. The average rate of change of the function $y = f(x)$ from $x = 3$ to $x = 3.5$ is $2(f(3.5) - f(3))$.

Problem 7.5. Every function is either an odd function or an even function.

Problem 7.6. If $y = f(x) = 2^{|x|}$, then the range of the function f is the set of all non-negative real numbers.

Problem 7.7. e^{-x} is negative for some values of x.

Problem 7.8. $e^x \geq e^{-x}$ for all $x \in \mathbb{R}$.

Problem 7.9. $\ln e^{x^2} = x^2$ for all $x \in \mathbb{R}$.

Problem 7.10. $\ln x$ exists for any $x > 1$.

Problem 7.11. $\ln x = \pi$ has a unique solution.

Problem 7.12. If $x > 0$, then $(\ln x)^6 = 6 \ln x$.

Problem 7.13. If $x, y > 0$, then $\ln(x + y) = (\ln x) \cdot (\ln y)$.

Problem 7.14. The function $f(x) = \tan x$ is defined for all $x \in \mathbb{R}$.

Problem 7.15. The function $f(x) = \cosh x$, $x \in \mathbb{R}$ is periodic.

Problem 7.16. The function $f(x) = \arctan x$, $x \in \mathbb{R}$ is periodic.

Problem 7.17. $\sinh^2 x - \cosh^2 x = 1$ for all $x \in \mathbb{R}$.

Problem 7.18. If $f(x) = x^2$ and $g(x) = x+1$, then $f(g(x)) = x^2+1$.

Problem 7.19. If f, g, and h are functions defined on the set of real numbers, then $(g + h) \circ f = g \circ f + h \circ f$.

Problem 7.20. If a function f is one-to-one, then $f^{-1}(x) = \frac{1}{f(x)}$.

Problem 7.21. A function is one-to-one if and only if no horizontal line intersects its graph more than once.

Problem 7.22. Let f^{-1} be the inverse function of the function f. Then $f^{-1}(f(x)) = x$ for *all* real numbers x.

Problem 7.23. If $g(x) = \ln x$, then $g(g^{-1}(0)) = 0$.

Problem 7.24. If $(4, 1)$ is a point on the graph of the function h, then $(4, 0)$ is a point on the graph of the function $f \circ h$, where $f(x) = 3^x + x - 4$.

Problem 7.25. For all x in the domain of the function $y = \sec^{-1} x$, $\sec(\sec^{-1}(x)) = x$.

Problem 7.26. $\sin^{-1}\left(\sin\left(\frac{7\pi}{3}\right)\right) = \frac{7\pi}{3}$.

7.1.2 *Multiple choice questions*

For each of the following questions, choose only one answer.

Problem 7.27. Which of the following is $\arcsin\left(\sin\left(\frac{3\pi}{4}\right)\right)$?

A. 0. **B.** $\frac{\pi}{4}$. **C.** $-\frac{\pi}{4}$. **D.** $\frac{5\pi}{4}$. **E.** $\frac{3\pi}{4}$.

Problem 7.28. Match the function $f(x) = \log_{0.5} x$ with its graph in the following figure.

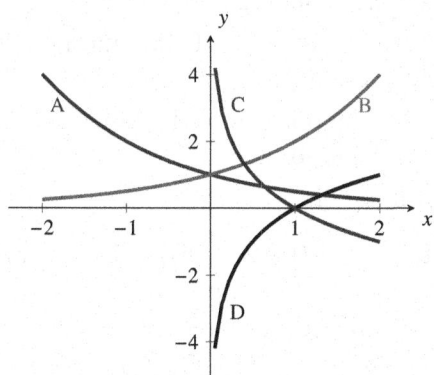

Problem 7.29. What is the value of $\cosh(\ln 3)$?

A. $\frac{1}{3}$. **B.** $\frac{1}{2}$. **C.** $\frac{2}{3}$. **D.** $\frac{4}{3}$. **E.** $\frac{5}{3}$.

Problem 7.30. If $f(4) = 10$ and $g(x) = f(5x - 2)$, which of the following points lie on the graph of the function $y = g(x)$?

A. $(4, 10)$. **B.** $(1.2, 10)$. **C.** $(4, 18)$. **D.** $(10, 4)$. **E.** $(10, 1.2)$.

Problem 7.31. Given $f(5) = -1$, which of the following points is possible to determine on the graph of the function $h(x) = f(x^2 - 3)$?

A. $(-1, 2\sqrt{2})$. **B.** $(-1, 2\sqrt{2})$ and $(-1, -2\sqrt{2})$. **C.** $(-\sqrt{2}, -1)$.
D. $(2\sqrt{2}, -1)$ and $(-2\sqrt{2}, -1)$. **E.** Something else.

Problem 7.32. If $h(2x) = \dfrac{x^2}{4}$, what is $h(x)$?

A. $h(x) = \frac{x}{16}$. **B.** $h(x) = \frac{x^2}{16}$. **C.** $h(x) = \frac{x}{8}$. **D.** $h(x) = \frac{x^2}{8}$.
E. None of these.

Problem 7.33. If $f(x) = 3x - 1$, $g(x) = \sqrt{x}$, and $h(x) = \sin x$, what is the domain of the function $h \circ g \circ f$?

A. $[0, \infty)$. **B.** $(-\infty, \infty)$. **C.** $(-\infty, 1/3]$. **D.** $[1/3, \infty)$. **E.** None of these.

Problem 7.34. What is the domain of the composite function $f \circ g$?

A. A set of all x is in the domain of the function g such that $g(x)$ is in the domain of of the function f. .

B. A set of all x is in the domain of the function g such that x is in the domain of the function f.

C. A set of all x is in the domain of the function f such that $f(x)$ is in the domain of the function g.

D. The set of real numbers.

Problem 7.35. If $F(x) = (\ln{(2x+5)})^2 - 1$, what are three functions f, g, and h such that $F = f \circ g \circ h$?

 A. $f(x) = x^2 - 1$, $g(x) = \ln x$, $h(x) = 2x + 5$. **B.** $f(x) = \ln x$; $g(x) = x^2 - 1$, $h(x) = 2x + 5$. **C.** $f(x) = \ln x$; $g(x) = 2x + 5$, $h(x) = x^2 - 1$. **D.** $f(x) = 2x + 5$; $g(x) = \ln x$, $h(x) = x^2 - 1$. **E.** None of these.

7.2   Limits

7.2.1   *True or false*

Answer each of the following either TRUE or FALSE.

Problem 7.36. If $h(3) = 2$, then $\lim\limits_{x \to 3} h(x) = 2$.

Problem 7.37. If $y = f(x)$ is any function such that $\lim\limits_{x \to 2} f(x) = 6$, then $\lim\limits_{x \to 2^+} f(x) = 6$.

Problem 7.38. An equivalent ε-δ definition of $\lim\limits_{x \to a} f(x) = L$ is: For any $0 < \varepsilon < 0.13$, there is $\delta > 0$ such that if $|x - a| < \delta$, then $|f(x) - L| < \varepsilon$.

Problem 7.39. If $\lim\limits_{x \to 5} f(x) = 0$ and $\lim\limits_{x \to 5} g(x) = 0$, then $\lim\limits_{x \to 5} \frac{f(x)}{g(x)}$ does not exist.

Problem 7.40. If $\lim\limits_{x \to 1} f(x) = 0$ and $\lim\limits_{x \to 1} g(x) = 0$, then $\lim\limits_{x \to 1} \frac{g(x)}{f(x)} = 1$.

Problem 7.41. If $\lim\limits_{x \to 0} f(x) = 0$ and $\lim\limits_{x \to 0} g(x) = 0$, then $\lim\limits_{x \to 0} f(x)^{g(x)} = 1$.

Problem 7.42. If $\lim_{x\to\infty} (f(x) + g(x))$ exists, then $\lim_{x\to\infty} f(x)$ exists and $\lim_{x\to\infty} g(x)$ exists.

Problem 7.43. If $f(x) > 0$ for all x and $\lim_{x\to 0} f(x)$ exists, then $\lim_{x\to 0} f(x) > 0$.

Problem 7.44. $\lim_{x\to 3} \sqrt{x-3} = \sqrt{\lim_{x\to 3}(x-3)}$.

Problem 7.45. If $\lim_{x\to a} f(x)$ exists, then $\lim_{x\to a} \sqrt{f(x)}$ exists.

Problem 7.46. $\lim_{x\to\infty} \frac{(1.00001)^x}{x^{100000}} = 0$.

Problem 7.47. If $-x^3 + 3x^2 + 1 \le g(x) \le (x-2)^2 + 5$ for $x \ge 0$, then $\lim_{x\to 2} g(x) = 5$.

Problem 7.48. The Intermediate Value Theorem may be used to show that $\lim_{x\to 0} x \sin \frac{1}{x} = 0$.

Problem 7.49. If $\lim_{x\to\infty} \frac{k}{f(x)} = 0$ for every number k, then $\lim_{x\to\infty} f(x) = \infty$.

Problem 7.50. $\lim_{x\to 0+} \left(\frac{1}{x} - \frac{1}{x^2} \right) = \lim_{x\to 0+} \frac{1}{x} - \lim_{x\to 0+} \left(-\frac{1}{x^2} \right) = \infty - \infty = 0$.

Problem 7.51. If a function f is defined on the interval $[0, \infty)$ and if its graph has no horizontal asymptotes, then $\lim_{x\to\infty} f(x) = \infty$ or $\lim_{x\to -\infty} f(x) = -\infty$.

Problem 7.52. If the graph of the function $y = f(x)$ has a vertical asymptote at $x = 1$, then $\lim_{x\to 1} f(x) = L \in \mathbb{R}$.

Problem 7.53. If the line $x = 1$ is a vertical asymptote of the graph of the function $y = f(x)$, then f is not defined at 1.

Problem 7.54. The graph of the function $g(x) = \frac{7x^4 - x^3 + 5x^2 + 3}{x^2 + 1}$ has a slant asymptote.

Problem 7.55. The graph of a function can have three different vertical asymptotes.

Problem 7.56. The graph of a function can have two different horizontal asymptotes.

Problem 7.57. The graph of a function can have three different horizontal asymptotes.

Problem 7.58. $\lim\limits_{x \to 0} \dfrac{\sin 2x}{x} = \dfrac{1}{2}$.

Problem 7.59. $\lim\limits_{x \to 0} \dfrac{\sin 2x}{\sin 3x} = \dfrac{2}{3}$.

7.2.2 *Multiple choice questions*

For each of the following questions, choose only one answer.

Problem 7.60. For the function whose graph is shown, what is the value of $\lim\limits_{x \to 1} f(x)$?

A. 1. **B.** 2. **C.** 0. **D.** Does not exist.

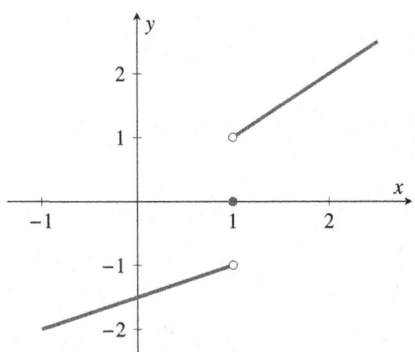

Problem 7.61. For the function defined by $f(x) = \begin{cases} 1 + |x| & \text{if } x < 0 \\ 1 - x & \text{if } 0 < x < 2 \\ (x-2)^2 & \text{if } x > 2 \end{cases}$, what are the values of a for which $\lim\limits_{x \to a} f(x)$ exists?

  A. The set of all real numbers. **B.** All real numbers except 0 and 2. **C.** All real numbers except 0. **D.** All real numbers except 2.
E. None of these.

Problem 7.62. For the function defined by $f(x) = \begin{cases} \sin x + c & \text{if } x < 0 \\ 2 & \text{if } x = 0 \\ \frac{1}{x^2 + c} & \text{if } x > 0 \end{cases}$,

what is the value of c for which $\lim\limits_{x \to 0} f(x)$ exists?

  A. $c = -1$. **B.** $c = 1$. **C.** $c = 1$ and $c = -1$. **D.** There does not exist a value for c.

Problem 7.63. Consider the function $f(x) = \begin{cases} x & \text{if } x \text{ is rational} \\ -x & \text{if } x \text{ is irrational} \end{cases}$.
What can we say about $\lim_{x \to a} f(x)$?

A. There is no a for which $\lim_{x \to a} f(x)$ exists. **B.** $\lim_{x \to a} f(x)$ exists only when $a \neq 0$. **C.** $\lim_{x \to a} f(x)$ exists only when $a = 0$. **D.** Cannot say anything without more information.

Problem 7.64. Two linear functions f and g are shown in the following graph. Observe that $f(1) = g(1) = 0$. Which of the following statements is true?

A. $\lim_{x \to 1} \frac{f(x)}{g(x)} = 1$. **B.** $\lim_{x \to 1} \frac{f(x)}{g(x)} = 0$. **C.** $\lim_{x \to 1} \frac{f(x)}{g(x)} = \infty$.
D. $\lim_{x \to 1} \frac{f(x)}{g(x)}$ exists, but we need more information to determine its value. **E.** $\lim_{x \to 1} \frac{f(x)}{g(x)}$ does not exist.

Problem 7.65. If $\lim_{x \to -1} \frac{x^3+1}{x+1} = \alpha$, $\lim_{x \to 0} \frac{\sqrt{x+9}-3}{x} = \beta$, and $\lim_{x \to 0} \frac{(2+x)^2-4}{x} = \gamma$, what is $3\alpha + 6\beta - 2\gamma$?

A. -2. **B.** 2. **C.** 0. **D.** 1. **E.** It is not defined.

Problem 7.66. If $\lim_{x \to 0} \frac{f(x)}{\sin x} = 1$, what can we say about $\lim_{x \to 0} f(x)$?

A. $\lim_{x \to 0} f(x) = 0$. **B.** $\lim_{x \to 0} f(x) = 1$. **C.** $\lim_{x \to 0} f(x) = \infty$.
D. $\lim_{x \to 0} f(x)$ exists, but we need more information to determine its value. **E.** $\lim_{x \to 0} f(x)$ does not exist.

Problem 7.67. If $f(x) = \frac{1}{x+5}$, what is the value of $\lim_{x \to 0} \frac{f(5+x)-f(5)}{x}$?

A. -0.1. **B.** 0.1. **C.** -0.01. **D.** 0.01. **E.** The limit does not exist.

Problem 7.68. What is the slant asymptote of the graph of the function $f(x) = \frac{x^2+3x-1}{x-1}$?

  A. $y = x + 4$.   **B.** $y = x + 2$.   **C.** $y = x - 2$.   **D.** $y = x - 4$.
E. None of these.

Problem 7.69. How many vertical asymptotes does the graph of the function $f(x) = \ln\left|\frac{x+1}{x^2-7x+12}\right|$ have?

  A. None.   **B.** One.   **C.** Two.   **D.** Three.   **E.** Four.

Problem 7.70. What is the value of $\lim\limits_{x\to 0^+} \frac{\ln x}{x}$?

  A. 0.   **B.** ∞.   **C.** 1.   **D.** -1.   **E.** None of these.

Problem 7.71. What is the value of $\lim\limits_{x\to 0^+} \frac{1-e^x}{\sin x}$?

  A. 1.   **B.** -1.   **C.** 0.   **D.** ∞.   **E.** $\sin e$.

Problem 7.72. What is the value of $\lim\limits_{x\to 0}(1 + \sin x)^{\frac{1}{x}}$?

  A. e.   **B.** 1.   **C.** 0.   **D.** $\ln 2$.   **E.** ∞.

7.3   Continuity

7.3.1   *True or false*

Answer each of the following either TRUE or FALSE.

Problem 7.73. For *all* functions f, if $\lim\limits_{x\to a^-} f(x)$ exists and $\lim\limits_{x\to a^+} f(x)$ exists, then f is continuous at a.

Problem 7.74. It is impossible for a function $y = f(x)$ to be discontinuous at *every* real number x.

Problem 7.75. If f and g are *any* two functions which are continuous on $\mathbb{R}$, then the function $\frac{f}{g}$ is continuous on $\mathbb{R}$.

Problem 7.76. It is possible that functions f and g are *not* continuous at $x_0 \in \mathbb{R}$, but that $f + g$ is continuous at x_0.

Problem 7.77. If p is a polynomial and b is a real number, then $\lim\limits_{x\to b} p(x) = p(b)$.

Problem 7.78. If a function $y = f(x)$ is not a polynomial, then $\lim\limits_{x \to b} f(x) \neq f(b)$.

Problem 7.79. The function $f(x) = \begin{cases} 3 + \frac{\sin(x-2)}{x-2} & \text{if } x \neq 2 \\ 3 & \text{if } x = 2 \end{cases}$ is continuous at all real numbers x.

Problem 7.80. The graph in the following figure exhibits three types of discontinuities.

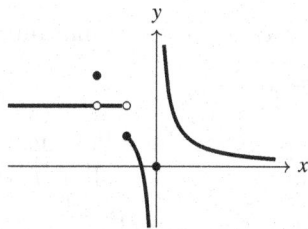

Problem 7.81. Consider the function $f(x) = \begin{cases} x - 4 & \text{if } x < 2 \\ -2 & \text{if } x = 2 \\ x^2 + 7 & \text{if } x > 2 \end{cases}$.
TRUE or FALSE: (1) $f(3) = -1$. (2) $f(2) = 11$. (3) f is continuous at $x = 3$. (4) f is continuous at $x = 2$.

Problem 7.82. For *all* functions f, if $a < b$, $f(a) < 0$, and $f(b) > 0$, then there must be a number c with $a < c < b$ and $f(c) = 0$.

Problem 7.83. The equation $e^{-x} = x$ has a solution in $(0, 1)$.

Problem 7.84. Let $f : [a, b] \to \mathbb{R}$ be a continuous function such that $f(a) < f(b)$. By the Intermediate Value Theorem, $f(a) \leq f(c) \leq f(b)$ for any $c \in (a, b)$.

Problem 7.85. Every function continuous on the interval $(0, 1)$ has a maximum value and a minimum value on $(0, 1)$.

Problem 7.86. A function that is continuous on a closed interval attains an absolute maximum value and an absolute minimum value at numbers in that interval.

Problem 7.87. If the function f does not have an absolute maximum on the interval $[a, b]$, then f is not continuous on $[a, b]$.

Problem 7.88. If the function f has an absolute maximum on the interval $[a, b]$, then f must be continuous on $[a, b]$.

7.3.2 *Multiple choice questions*

For each of the following questions, choose only one answer.

Problem 7.89. Which of the following statements is true?
A. A function f is continuous at a number a if f is defined at a.
B. A function f is continuous at a number a if $\lim\limits_{x\to a} \frac{f(x)-f(a)}{x-a}$ does not exists. **C.** A function f is continuous at a number a if $\lim\limits_{x\to a} f(x)$ exists. **D.** A function f is continuous at a number a if $\lim\limits_{x\to a^+} f(x) = f(a)$ or $\lim\limits_{x\to a^-} f(x) = f(a)$. **E.** A function f is continuous at a number a if $\lim\limits_{x\to a} f(x) = f(a)$.

Problem 7.90. You are given the following information about the functions f and g: $\lim\limits_{x\to 0^-} f(x) = -1$, $\lim\limits_{x\to 0^+} f(x) = 2$, $\lim\limits_{x\to 1} f(x) = 6$, $f(1) = 10$, $g(1) = 8$, and $g(x)$ is continuous at $x = 0$ and at $x = 1$. For which of the following three expressions (I) $\lim\limits_{x\to 0} [2f(x)g(x) - 1]$, (II) $\lim\limits_{x\to 1^+} [f(x) + g(x)]$, and (III) $f(0) - g(1)$, do you have enough information to decide if they exist and to evaluate them if they do exist?

A. Only I. **B.** Only II. **C.** Only III. **D.** Only I and II. **E.** Only II and III.

Problem 7.91. Recall that if a function f is a polynomial, then f is continuous on $\mathbb{R}$. Which of the following is also true?

A. If a function f is not continuous on $\mathbb{R}$, then f is not a polynomial. **B.** If f is continuous on $\mathbb{R}$, then f is a polynomial. **C.** If f is not a polynomial, then f is not continuous on $\mathbb{R}$.

Problem 7.92. Which function has a removable discontinuity at $x = 0$?
A. $f(x) = \frac{\sin x}{x}$ if $x \neq 0$ and $f(0) = 1$. **B.** $f(x) = \frac{\sin x}{x}$ if $x \neq 0$ and $f(0) = 2$. **C.** $f(x) = \cot x$ if $x \neq 0$ and $f(0) = 3$. **D.** $f(x) = \frac{x}{|x|}$ if $x \neq 0$ and $f(0) = 0$. **E.** $f(x) = \frac{x}{|x|}$ if $x \neq 0$ and $f(0) = 1$.

Problem 7.93. For what value of the number a is the function
$$f(x) = \begin{cases} \cosh(x) & \text{if } x < 0 \\ a - \cos(x) & \text{if } x \geq 0 \end{cases}$$ continuous on its domain?
A. 4. **B.** 0. **C.** -1. **D.** 1. **E.** 2.

Problem 7.94. Consider the function $f(x) = \frac{(x-1)(x^2-9)}{x-a}$. For what value of the number a is the function f continuous for all real numbers x?

    A. None.   **B.** 1.   **C.** 3.   **D.** -3.

Problem 7.95. The rational function $g(x) = \frac{x^3+3x^2-4x}{x^2+x-12}$ might have one or more of the following types of discontinuities:

(I) A removable discontinuity, (II) An infinite discontinuity (or vertical asymptote), and (III) A jump discontinuity.

Which of these types of discontinuities does the function g have?

    A. Only I.   **B.** Only II.   **C.** Only III.   **D.** Only I and II.
E. Only II and III.

Problem 7.96. If f is a continuous function on the closed interval $[a, b]$ and N is a number between $f(a)$ and $f(b)$, then there is $c \in [a, b]$ such that $f(c) = N$. Which theorem is this?

    A. Fermat's Theorem.   **B.** The Intermediate Value Theorem.
C. The Mean Value Theorem.   **D.** Rolle's Theorem.   **E.** The Extreme Value Theorem.

Problem 7.97. If f is a continuous function on the closed interval $[a, b]$, then f must attain a maximum and a minimum, each at least once. Which theorem is this?

    A. The Extreme Value Theorem.   **B.** The Intermediate Value Theorem.   **C.** The Mean Value Theorem.   **D.** Rolle's Theorem.
E. None of these.

Problem 7.98. Which of the following must be true for a continuous function on (a, b)?

    A. The function achieves its maximum on (a, b).   **B.** The function is bounded on $[a, b]$.   **C.** The function is differentiable on (a, b).
D. If $f(a) = 2$ and $f(b) = 5$, then $f(c) = 3$, for some $c \in (a, b)$.
E. None of these.

7.4   Derivatives

7.4.1   *True or false*

Answer each of the following either TRUE or FALSE.

Problem 7.99. If $g(x) = x^5$, then $\lim\limits_{x \to 2} \dfrac{g(x)-g(2)}{x-2} = 80$.

Problem 7.100. If the graph of the function f is to the left, then the graph to the right must be that of the function f'.

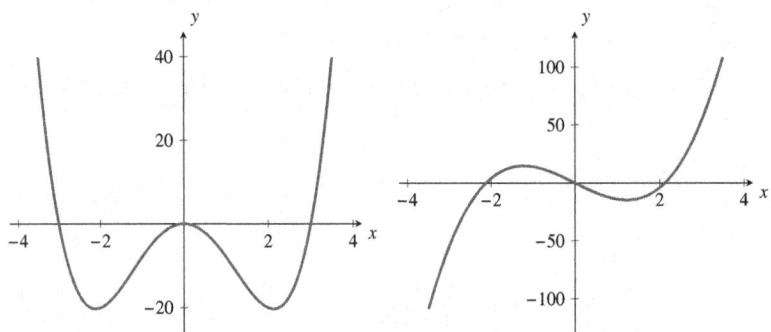

Problem 7.101. If a function f is differentiable on $(-1, 1)$, then it is continuous at $x = 0$.

Problem 7.102. If $f'(c)$ exists, then $\lim\limits_{x \to c} f(x) = f(c)$.

Problem 7.103. If a function f is continuous at the number x, then it is differentiable at x.

Problem 7.104. The function $f(x) = |x|$ is differentiable for all x in its domain.

Problem 7.105. If $f'(x) = g'(x)$ for $0 < x < 1$, then $f(x) = g(x)$ for $0 < x < 1$.

Problem 7.106. For *all* functions f, if $f'(x)$ exists for all x, then $f''(x)$ exists for all x.

Problem 7.107. If the graph of the function f has a vertical tangent line at the point $(c, f(c))$, then f is not differentiable at c.

Problem 7.108. The derivative of an even function is an even function.

Problem 7.109. The derivative of an even function is an odd function.

Problem 7.110. If $y = e^2$, then $y' = 2e$.

Problem 7.111. $\frac{d}{dx}(\ln 10) = \frac{1}{10}$.

Problem 7.112. If $f(x) = 7x + 8$, then $f'(2) = f'(17.38)$.

Problem 7.113. If $f(x) = (1 + x)(1 + x^2)(1 + x^3)(1 + x^4)$, then $f'(0) = 1$.

Problem 7.114. The derivative of the function $f(x) = \tan x$ is a periodic function.

Problem 7.115. The 99th derivative of $y = \sin x$ is $y = \cos x$.

Problem 7.116. If $f'(3) = 4$ and $g'(3) = 5$, then the graph of the function $f(x) + g(x)$ has a slope of 9 at $x = 3$.

Problem 7.117. Let a be a positive number not equal to 1 and let $f(x) = a^x$. Then $f'(x) = a^x$.

Problem 7.118. If n is a real number and $f(x) = x^n$, then $f'(x)$ exists for *all* x in the domain of f and $f'(x) = nx^{n-1}$.

Problem 7.119. The function $f(x) = e^x$ is the *only* function with the property that it is its own derivative.

Problem 7.120. If f and g are differentiable functions for all $x \in \mathbb{R}$, then $\frac{d}{dx}(f(g(x))) = f'(g(x))g'(x)$.

Problem 7.121. If $w = f(x)$, $x = g(y)$, and $y = h(z)$ are differentiable functions, then $\frac{dw}{dz} = \frac{dw}{dx} \cdot \frac{dx}{dy} \cdot \frac{dy}{dz}$.

Problem 7.122. If f is a differentiable function for $x > 0$, then $\frac{d}{dx}f(\sqrt{x}) = \frac{f'(x)}{2\sqrt{x}}$.

Problem 7.123. For all real values of x, we have that $\frac{d}{dx}|x^2 + x| = |2x + 1|$.

Problem 7.124. $\frac{d}{dx}(\tan^2 x) = \frac{d}{dx}\left(\sec^2 x\right)$.

Problem 7.125. $\frac{d}{dx}\left(\frac{\ln 2\sqrt{x}}{\sqrt{x}}\right) = 0$, $x > 0$.

Problem 7.126. $\frac{d}{dx}\left(\frac{\log x^2}{\log x}\right) = 0$ for all $x > 1$.

Problem 7.127. If $y = 10^x$, then $y' = x10^{x-1}$.

Problem 7.128. $\frac{d}{dx}\left(x^x\right) = xx^{x-1}$, $x > 1$.

Problem 7.129. $\frac{d}{du}\left(\frac{1}{\csc u}\right) = \frac{1}{\sec u}$.

Problem 7.130. Suppose that f is a differentiable function that has an inverse function f^{-1} and that $f'(x) \neq 0$ for all x in the domain of the function f. Then f^{-1} is a differentiable function and $(f^{-1}(x))' = \frac{1}{f'(f^{-1}(x))}$ for all x in the domain of the function f^{-1}.

Problem 7.131. If a differentiable function $y = f(x)$ is the inverse function of the function $y = g(x)$, then $g'(x) = \frac{1}{f'(x)}$.

Problem 7.132. $\frac{d}{dx}(\sin^{-1}(\cos x)) = -1$ for $0 < x < \pi$.

Problem 7.133. Given $h(x) = g(f(x))$ and the graphs of the functions f and g in the figure, then a good estimate for $h'(3)$ is $-\frac{1}{4}$.

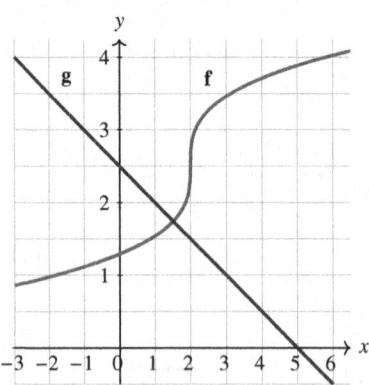

Problem 7.134. If $\lim\limits_{x\to\infty}\frac{f'(x)}{g'(x)}$ does not exist, then $\lim\limits_{x\to\infty}\frac{f(x)}{g(x)}$ does not exist.

7.4.2 *Multiple choice questions*

For each of the following questions, choose only one answer.

Problem 7.135. Which of the following statements is true?

A. A function f is differentiable at a number a if $\lim_{x \to a} f(x) = f(a)$.

B. A function f is differentiable at a number a if $\lim_{x \to a} \frac{f(x)-f(a)}{x-a}$ exists.

C. A function f is differentiable at a number a if f is defined at a.

D. A function f is differentiable at a number a if f is continuous at a. **E.** A function f is differentiable at a number a if we can apply the Intermediate Value Theorem.

Problem 7.136. According to the graph of the function f, which of the following inequalities is true?

A. $f'(a) < f'(b) < f'(c)$. **B.** $f'(c) < f'(b) < f'(a)$. **C.** $f'(b) < f'(c) < f'(a)$. **D.** $f'(b) < f'(a) < f'(c)$.

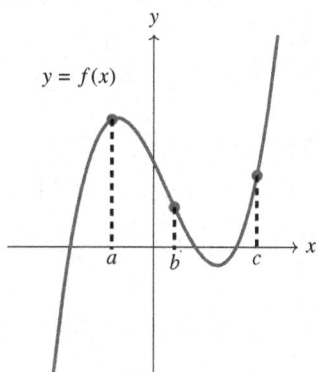

Problem 7.137. Which of the following statements is always true for functions f and g?

(I) If f and g are continuous at $x = a$, then the function $\frac{f}{g}$ is continuous at $x = a$.

(II) If the function $f + g$ is continuous at $x = a$ and $f'(a) = 0$, then g is continuous at $x = a$.

(III) If the function $f + g$ is differentiable at $x = a$, then functions f and g are both differentiable at $x = a$.

A. Only (I). **B.** Only (II). **C.** Only (III). **D.** (I) and (II). **E.** (II) and (III).

Problem 7.138. Which of the following statements is always true for a function f that is differentiable on (a, b)?
(I) $f(x)$ is defined for all x in (a, b).   (II) f is continuous on (a, b).
(III) f is continuous at $x = a$.   (IV) f' is differentiable on (a, b).
A. Only (I) and (II) are true.   **B.** Only (I), (II), and (III) are true.   **C.** Only (I) is true.   **D.** All four statements are true.
E. All four statements are false.

Problem 7.139. The diagonal D of an ice cube changes as the ice melts. What is the instantaneous rate of change of the diagonal with respect to volume V?

A. $\sqrt{3}$.   **B.** $\frac{dV}{dD}$.   **C.** $\frac{dD}{dV}$.   **D.** $\frac{dV}{dD} + \frac{dD}{dV}$.   **E.** $\frac{dV}{dD} \cdot \frac{dD}{dV}$.

Problem 7.140. The area of an equilateral triangle $A = \frac{a^2\sqrt{3}}{4}$ changes as its side a changes. If the side changes with respect to time t, what is the change in area with respect to time?

A. $\frac{dA}{da} = \frac{a\sqrt{3}}{2}$.   **B.** $\frac{dA}{dt} = \frac{a\sqrt{3}}{2} + \frac{da}{dt}$.   **C.** $\frac{dA}{dt} = \frac{a\sqrt{3}}{2} \cdot \frac{da}{dt}$.
D. None of these.

Problem 7.141. If $f(0) = 1$ and $f'(0) = 3$, what is $\frac{d}{dx}\left(\frac{f(x)}{x^2+1}\right)\Big|_{x=0}$?
A. -2.   **B.** -3.   **C.** 0.   **D.** 1.   **E.** 3.

Problem 7.142. If $h(x) = \ln(1 - x^2)$, where $-1 < x < 1$, what is $h'(x)$?

A. $\frac{1}{1-x^2}$.   **B.** $\frac{1}{1+x} - \frac{1}{1-x}$.   **C.** $\frac{2}{1-x^2}$.   **D.** None of these.

Problem 7.143. If $f(x) = x^2 \tan x$, what is $f'(x)$?
A. $2x \sec^2 x$.   **B.** $2x \tan x + x^2 \cot x$.   **C.** $2x \tan x + (x \sec x)^2$.
D. None of these.

Problem 7.144. If $\cosh y = x + x^3 y$, what is the slope of the tangent line at the point $(1, 0)$?
A. 0.   **B.** 3.   **C.** -1.   **D.** Is not defined.

Problem 7.145. If $g(x) = e^{\sqrt{x}}$, what is $g'(x)$?
A. $e^{\sqrt{x}}$.   **B.** $\sqrt{x}e^{\sqrt{x}-1}$.   **C.** $\frac{e^{\sqrt{x}}}{2\sqrt{x}}$.   **D.** None of these.

Problem 7.146. Suppose $y'' + y = 0$. Which of the following is a possible function for $y = f(x)$?

A. $y = \tan x$. **B.** $y = \sin x$. **C.** $y = \sec x$. **D.** $y = 1/x$.
E. $y = e^x$.

Problem 7.147. If $y = \sin(x^2)$, what is $\frac{dy}{dx}$?

A. $2x\cos(x^2)$. **B.** $2x\sin(x^2)$. **C.** $2x\cos(x)$. **D.** $2x\cos(2x)$.
E. $2x\cos(2x)$.

Problem 7.148. If $y = x^2 + 2^x + x^{2x}$, what is $y'(2)$?

A. $2x + x2^{x-1} + 2x^{2x-1}$. **B.** $36(1 + \ln 2)$. **C.** $3(1 + \ln 2)$.
D. $32(1 + \ln 2)$. **E.** None of these.

Problem 7.149. If a rock is thrown upward on the planet Calculus with a velocity of 10 m/sec, its height in meters after t seconds is given by $H = 10t - 1.86t^2$. What is the velocity of the rock when $t = 4\,\mathrm{s}$?

A. $4.48\,\mathrm{m/s}$. **B.** $4.88\,\mathrm{m/s}$. **C.** $-4.48\,\mathrm{m/s}$. **D.** $-4.88\,\mathrm{m/s}$.
E. It is not defined.

7.5 Applications of Derivatives

7.5.1 *True or false*

Answer each of the following either TRUE or FALSE.

Problem 7.150. If the function f is increasing on the interval I and if $f(x) > 0$ for all $x \in I$, then the function $g = \frac{1}{f}$ is decreasing on I.

Problem 7.151. If the function f is increasing on the interval $[0, 1]$, then the function $f^2 = f \cdot f$ is increasing on $[0, 1]$.

Problem 7.152. If $f(c)$ exists but $f'(c)$ does not exist, then $x = c$ is a critical number of the function f.

Problem 7.153. If c is a critical number of the function f, then $f'(c) = 0$.

Problem 7.154. If $f'(c) = 0$, then the function f has a local maximum or a local minimum at $x = c$.

Problem 7.155. If the function f is continuous at $x = 2$ and if $f'(2) = 0$, then f has either a local maximum or minimum at $x = 2$.

Problem 7.156. If a differentiable function f has a local extremum at $x = c$, then $f'(c) = 0$.

Problem 7.157. If $f'(c)$ does not exist and $f'(x)$ changes from negative to positive as x increases through c, then the function f has a local minimum at $x = c$.

Problem 7.158. If $f''(c) = 0$ and $f'''(c) > 0$, then the function f has a local minimum at $x = c$.

Problem 7.159. If the function f has an absolute minimum at $x = c$, then $f'(c) = 0$.

Problem 7.160. If $f''(2) = 0$, then $(2, f(2))$ is an inflection point on the graph of the function $y = f(x)$.

Problem 7.161. If $f'(x) < 0$ and $f''(x) > 0$ for all $x \in \mathbb{R}$, then the graph of the function $y = f(x)$ is concave downward.

Problem 7.162. The slope of the tangent line to the graph of the function $y = f(x)$ at the point $(a, f(a))$ is given by $\frac{f(a+h)-f(a)}{h}$.

Problem 7.163. The graph of the function $y = 2x^3 + 4x + 5$ has no tangent line with slope 3.

Problem 7.164. An equation of the tangent line to the parabola $y = x^2$ at $(-2, 4)$ is $y - 4 = 2x(x + 2)$.

Problem 7.165. If $f'(g) = \frac{1}{(3-g)^2}$, then the equation of the tangent line to the graph of the function $y = f(g)$ at $\left(0, \frac{1}{3}\right)$ is $y = \frac{1}{9}g + \frac{1}{3}$.

Problem 7.166. A curve may have two vertical tangents but no horizontal ones.

Problem 7.167. Let f be a differentiable function and let c be a number in its domain. The graph of the linearization of f at c is the tangent line to the graph of the function $y = f(x)$ at the point $(c, f(c))$.

Problem 7.168. The linearization $y = L(x)$ of a function $y = f(x)$ near the point $x = a$ is given by $L(x) = f'(a) + f(a)(x - a)$.

Problem 7.169. If a function f is not differentiable at c, then its graph has a vertical tangent line at the point $(c, f(c))$.

Problem 7.170. The conclusion of the Mean Value Theorem says that the graph of the function f has at least one tangent line in (a, b), whose slope is equal to the average slope of the function on the interval $[a, b]$.

Problem 7.171. If a function f is differentiable on the set of real numbers and has two roots, then the function f' has at least one root.

Problem 7.172. The Mean Value Theorem is a special case of Rolle's Theorem.

Problem 7.173. Suppose that a function f is differentiable on an interval I. If f is not injective on I, then there exists $c \in I$ such that $f'(c) = 0$.

Problem 7.174. Let f and g be *any* two functions which are continuous on $[0, 1]$ with $f(0) = g(0) = 0$ and $f(1) = g(1) = 10$. Then there must exist $c, d \in [0, 1]$ such that $f'(c) = g'(d)$.

Problem 7.175. If the function f is differentiable on the open interval (a, b), then by the Mean Value Theorem, there is a number $c \in (a, b)$ such that $(b - a)f'(c) = f(b) - f(a)$.

Problem 7.176. There exists a function f such that $f(1) = -2$, $f(3) = 0$, and $f'(x) > 1$ for all $x \in \mathbb{R}$.

Problem 7.177. There is a differentiable function $y = f(x)$ with the property that $f(1) = -2$, $f(5) = 14$, and $f'(x) < 3$ for every real number x.

Problem 7.178. If $y = f(x)$ is a differentiable function and the equation $f'(x) = 0$ has exactly two solutions, then the equation $f(x) = 0$ has no more than three solutions.

Problem 7.179. If a function $y = f(x)$ has a zero at $x = r$, then Newton's Method will find r given an initial guess $x_0 \neq r$ that is close enough to r.

Problem 7.180. A particle whose instantaneous velocity is zero is not accelerating.

7.5.2   *Multiple choice questions*

For each of the following questions, choose only one answer.

Problem 7.181. The edge of the cube is increasing at a rate of $2\,\text{cm/h}$. How fast is the cube's volume changing when its edge is $\sqrt{2}\,\text{cm}$ in length?

   A. $6\,\text{cm}^3/\text{h}$.   **B.** $12\,\text{cm}^3/\text{h}$.   **C.** $3\sqrt{2}\,\text{cm}^3/\text{h}$.   **D.** $6\sqrt{2}\,\text{cm}^3/\text{h}$.
E. None of these.

Problem 7.182. What is the interval on which the function $f(x) = x^{\frac{4}{3}}(7 - x)$ is increasing?

   A. $(-\infty, 4)$.   **B.** $(-\infty, 0)$.   **C.** $\left(0, \frac{7}{4}\right)$.   **D.** $(0, \infty)$.   **E.** $(0, 4)$.

Problem 7.183. Suppose that a function f is continuous on $[a, b]$ and differentiable on (a, b). If $f'(x) > 0$ for all $x \in (a, b)$, which of the following is *necessarily* true?

   A. f is decreasing on $[a, b]$.   **B.** f has no local extrema on (a, b).
C. f is a constant function on (a, b).   **D.** The graph of f is concave upward on (a, b).   **E.** f has no zero on (a, b).

Problem 7.184. Which of the following statements is the definition of a critical number?

   A. A critical number of a function f is a number c in the domain of the function f such that $f'(c) = 0$.   **B.** A critical number of a function f is a number c in the domain of the function f such that $f(c)$ is a local extremum.   **C.** A critical number of a function f is a number c in the interior of the domain of the function f such that either $f'(c) = 0$ or $f'(c)$ is not defined.   **D.** A critical number of a function f is a number c in the domain of the function f such that $(c, f(c))$ is an inflection point.   **E.** A critical number of a function f is a number c in the domain of the function f such that we can apply the Extreme Value Theorem in the neighborhood of the point $(c, f(c))$.

Problem 7.185. Which of the following functions has a critical number that is neither a local maximum nor a local minimum?

   A. $f(x) = |x|$.   **B.** $f(x) = x^3$.   **C.** $f(x) = \cos x$.   **D.** $f(x) = x^4$.
E. $f(x) = \ln x$.

Problem 7.186. Which of the following functions has a local minimum at which the second derivative is 0?

   A. $f(x) = |x|$.   **B.** $f(x) = x^3$.   **C.** $f(x) = \cos x$.   **D.** $f(x) = x^4$.
E. $f(x) = e^x$.

Problem 7.187. Let f be a function such that $f'(x) = 4x^3$ and $f''(x) = 12x^2$. Which of the following is true?

A. f has a local maximum at $x = 0$ by the First Derivative Test.
B. f has a local minimum at $x = 0$ by the First Derivative Test.
C. f has a local maximum at $x = 0$ by the Second Derivative Test.
D. f has a local minimum at $x = 0$ by the Second Derivative Test.
E. f has an inflection point at $x = 0$.

Problem 7.188. A designer wants to introduce a new line of pants: She wants to make at least 100 pairs of pants but not more than 2000 of them. The designer predicts that the cost of producing x pairs of pants is modelled by the function $C = C(x)$. Assume that C is a differentiable function. If a total of x pairs is produced, which of the following must she do to determine the minimum average cost per pair of pants $c(x) = \frac{C(x)}{x}$?

  (I) Determine all points where $c'(x) = 0$ and evaluate $c(x)$ there.
 (II) Compute $c''(x)$ to check which of the critical points in (I) are local maxima.
(III) Check the values of c at the endpoints of its domain.

   A. Only (I).   **B.** Only (I) and (II).   **C.** Only (I) and (III).
D. (I), (II), and (III).

Problem 7.189. A post office can accept a cylindrical package for shipping if the sum of its length and girth (the circumference of its cross-section) is at most 30π cm. What is the maximum volume that can be shipped?

   A. Less than $8500\,\mathrm{cm}^3$.   **B.** Between $8500\,\mathrm{cm}^3$ and $9000\,\mathrm{cm}^3$.
C. Between $9000\,\mathrm{cm}^3$ and $9500\,\mathrm{cm}^3$.   **D.** Between $9500\,\mathrm{cm}^3$ and $10{,}000\,\mathrm{cm}^3$.   **E.** Greater than $10{,}000\,\mathrm{cm}^3$.

Problem 7.190. What is the maximum possible area of a rectangle of perimeter $40\,\mathrm{m}$?

   A. $40\,\mathrm{m}^2$.   **B.** $80\,\mathrm{m}^2$.   **C.** $100\,\mathrm{m}^2$.   **D.** $20\,\mathrm{m}^2$.   **E.** $160\,\mathrm{m}^2$.

Problem 7.191. What is the maximum value attained by the function $f(x) = \frac{(x-2)^2}{x+4}$ on the interval $[0,5]$?

    A. 1.   **B.** $\frac{1}{5}$.   **C.** $\frac{2}{5}$.   **D.** 0.   **E.** 2.

Problem 7.192. Which of the following statements is true?

  (I) If the graph of a function f is concave upward on the interval (a,b), then f has to have a local minimum on (a,b).

 (II) It is possible for a function f to have an inflection point at $(a, f(a))$ even if $f'(a)$ does not exist.

(III) If f is a function defined on an interval that contains a, then it is possible for $(a, f(a))$ to be both a critical point and an inflection point on the graph of f.

    A. (I) and (II).   **B.** Only (III).   **C.** (I), (II), and (III).
    D. (II) and (III).   **E.** Only (I).

Problem 7.193. Let $f(x) = (3x+9)e^{-x}$. What is the largest interval containing $x = 12$ on which the graph of f is concave upward?

A. $(11, \infty)$.

B. $(2, 25)$.

C. $(-2, \infty)$.

D. $(-1, \infty)$.

E. $(-1, 25)$.

Problem 7.194. Let $f(x) = x^3 - 3x^2 - 189x + 2$. What is the x-coordinate of a point of inflection on the graph of the function f?

    A. $x = 9$.   **B.** $x = 1$.   **C.** $x = -7$.   **D.** $x = -1$.   **E.** $x = -6$.

Problem 7.195. The function f has the property that $f(3) = 2$ and $f'(3) = 4$. What is a linear approximation of the number $f(2.9)$?

    A. 1.4.   **B.** 1.6.   **C.** 1.8.   **D.** 1.9.   **E.** 2.4.

Problem 7.196. What is a linearization of the function $f(x) = \sqrt[3]{x}$ at $a = 8$?

    A. $L(x) = 2$.   **B.** $L(x) = \frac{x+16}{12}$.   **C.** $L(x) = \frac{1}{3x^{2/3}}$.   **D.** $L(x) = \frac{x-2}{3}$.   **E.** $L(x) = \sqrt[3]{x} - 2$.

Problem 7.197. Let a function f be continuous on $[a, b]$, differentiable on (a, b), and such that $f(b) = 10$ and $f(a) = 2$. On which of the following intervals $[a, b]$ would the Mean Value Theorem guarantee a $c \in (a, b)$ such that $f'(c) = 4$?

A. $[0, 4]$.   **B.** $[0, 3]$.   **C.** $[2, 4]$.   **D.** $[1, 10]$.   **E.** None of these.

Problem 7.198. If a function f is continuous on the closed interval $[a, b]$ and differentiable on the open interval (a, b), then there exists $c \in (a, b)$ such that $f(b) - f(a) = f'(c)(b - a)$. Which theorem is this?

A. The Extreme Value Theorem.   **B.** The Intermediate Value theorem.   **C.** The Mean Value Theorem.   **D.** Rolle's Theorem. **E.** None of these.

Problem 7.199. Suppose that the Mean Value Theorem, applied to $f(x) = x^2 + 23x + 7$ on the interval $[3, a]$, can be used to establish the existence of a number b satisfying $3 < b < a$ and such that $f'(b) = 38$. What is the value of the number a?

A. $a = 6$.   **B.** $a = 7$.   **C.** $a = 9$.   **D.** $a = 10$.   **E.** $a = 12$.

Problem 7.200. Let f be a function that is differentiable everywhere. Suppose that $f(3) = -5$ and $f'(x) \leq 7$ for all values of x. How large can $f(11)$ possibly be?

A. 21.   **B.** 31.   **C.** 41.   **D.** 51.   **E.** 61.

Problem 7.201. How many real roots does the equation $x^5 + 10x + 3 = 0$ have?

A. No real roots.   **B.** Exactly one real root.   **C.** Exactly two real roots.   **D.** Exactly three real roots.   **E.** Exactly five real roots.

Problem 7.202. Let $A = A(t)$ denote the amount of a certain radioactive material left after time t. Assume that $A(0) = 16$ and $A(1) = 12$. How much material is left after time $t = 3$?

A. $\frac{16}{9}$.   **B.** 8.   **C.** $\frac{9}{4}$.   **D.** $\frac{27}{4}$.   **E.** 4.

Problem 7.203. Let $P = P(t)$ be a function which gives the population as a function of time. Assume that P satisfies the natural growth equation and that at some point in time t_0 we have $P(t_0) = 500$ and $P'(t_0) = 1000$. What is the value of the growth rate constant k?

A. $-\frac{1}{2}$.   **B.** $\ln\left(\frac{1}{2}\right)$.   **C.** $\frac{1}{2}$.   **D.** 2.   **E.** $\ln 2$.

7.6   Antiderivatives

7.6.1   *True or false*

Answer each of the following either TRUE or FALSE.

Problem 7.204. A function F is called an antiderivative of f on an interval I if $F'(x) = f(x)$ for all x in I.

Problem 7.205. If a continuous function $y = f(x)$ is negative for all x in its domain, then its antiderivatives are increasing functions.

Problem 7.206. If $y = F(x)$ is an antiderivative of a function $y = f(x)$ on an interval I, then all antiderivatives of $y = f(x)$ on I are of the form $y = F(x) + C$, where C is an arbitrary constant.

Problem 7.207. If F and G are antiderivatives of a function f on the set of all real numbers, then $F(2) > G(2)$ implies $F(4) > G(4)$.

Problem 7.208. If F is an antiderivative of a function f and if G is an antiderivative of a function g, then $F \cdot G$ is an antiderivative of the function $f \cdot g$.

Problem 7.209. The general antiderivative of the function $f(x) = 3x^2$ is the function $F(x) = x^3$.

Problem 7.210. If f is a function such that $f'(x) = -x^3$ and $f(4) = 3$, then $f(3) = 2$.

Problem 7.211. $\int \frac{dx}{x^2+1} = \ln(x^2 + 1) + C$, $C \in \mathbb{R}$.

Problem 7.212. $\int \frac{dx}{3-2x} = \frac{1}{2} \ln|3 - 2x| + C$, $C \in \mathbb{R}$.

Problem 7.213. If f is a function such that $f'(x) = 2x$, then $f(x) = x^2$.

Problem 7.214. If $y = f(x)$ is a function such that $y' = 2y$, then $f(x) = e^{2x}$.

Problem 7.215. If functions f and g are such that $f(3) = g(3)$ and $f'(x) = g'(x)$ for all x, then $f(x) = g(x)$ for all x.

Problem 7.216. If a particle has a constant acceleration, then its position function is a cubic polynomial.

7.6.2 Multiple choice questions

For each of the following questions, choose only one answer.

Problem 7.217. Suppose F is an antiderivative of the function $f(x) = \sqrt[3]{x}$. If $F(0) = \frac{1}{4}$, what is $F(1)$?

A. -1. **B.** $-\frac{3}{4}$. **C.** 0. **D.** $\frac{3}{4}$. **E.** 1.

Problem 7.218. Which of the following functions is an antiderivative of the function $f(x) = x - \sin x + e^x$?

A. $F(x) = 1 - \cos x + e^x$. **B.** $F(x) = x^2 + \ln x - \cos x$.
C. $F(x) = 0.5x^2 + e^x - \cos x$. **D.** $F(x) = \cos x + e^x + 0.5x^2$.
E. None of these.

Problem 7.219. Which of the following statements is the definition of an antiderivative?

A. An antiderivative of a function f is a function F such that $F'(x) = f(x)$. **B.** An antiderivative of a function f is a function F such that $F(x) = f'(x)$. **C.** An antiderivative of a function f is a function F such that $F'(x) = f'(x)$. **D.** An antiderivative of a function f is a function F such that $F(x) = f(x)$. **E.** An antiderivative of a function f is a function F such that $F''(x) = f(x)$.

Problem 7.220. Let a function F be an antiderivative of the function f. Which of the following statements is true?
(I) $f(x) = F(x)$. (II) $f'(x) = F(x)$. (III) $F'(x) = f(x)$.
(IV) $f''(x) = F'(x)$. (V) $F''(x) = f'(x)$.

A. Only (III) is true. **B.** Only (III) and (V) are true. **C.** Only (II) and (IV) are true. **D.** Only (I) is true. **E.** All five statements are false.

7.7   Parametric Equations and Polar Coordinates

7.7.1   *True or false*

Answer each of the following either TRUE or FALSE.

Problem 7.221. A curve may have more than one parametrization.

Problem 7.222. The curve C defined by parametric equations $x = 2\cos t$, $y = 3\sin t$, $t \in [0, 2\pi]$ is a circle.

Problem 7.223. The points described by the polar coordinates $(2, \pi/4)$ and $(-2, 5\pi/4)$ are the same.

Problem 7.224. Any parametric curve $C = \{(f(t), g(t)) : t \in [a, b]\}$ is the graph of a function $y = h(x)$.

Problem 7.225. If $y = f(x)$, $x \in [a, b]$, then the graph of f can be represented as a parametric curve $C = \{(g(t), h(t)) : t \in I\}$ for some functions g and h.

Problem 7.226. Let a curve C be defined by parametric equations $x = x(t)$, $y = y(t)$, $t \in I$. If $\left.\dfrac{dy}{dt}\right|_{t=c} = 0$, then the tangent line to the curve C at the point $(x(c), y(c))$ is horizontal.

Problem 7.227. Let a smooth curve C be defined by parametric equations $x = x(t)$, $y = y(t)$, $t \in I$. If $\left.\dfrac{dx}{dt}\right|_{t=c} \neq 0$, then the tangent line to the curve C at the point $(x(c), y(c))$ is given by $y - y(c) = \dfrac{\left.\frac{dy}{dt}\right|_{t=c}}{\left.\frac{dx}{dt}\right|_{t=c}}(x - x(c))$.

Problem 7.228. Let a curve C be defined by parametric equations $x = x(t)$, $y = y(t)$, $t \in I$, where $x(t)$ and $y(t)$ are twice differentiable functions. Then $\dfrac{d^2y}{dx^2} = \dfrac{\frac{d^2y}{dt^2}}{\frac{d^2x}{dt^2}}$.

Problem 7.229. Parametric curves $C_1 = \{(2\sin t, 2\cos t) : t \in [0, 2\pi]\}$ and $C_2 = \{(2\sin(2t), 2\cos(2t)) : t \in [0, \pi]\}$ have different graphs.

Problem 7.230. The parametric curve $C = \{(\cos(5t), \sin(11t)) : t \in [0, 2\pi]\}$ is called a Lissajous curve.

Problem 7.231. The cycloid $C = \{(r(t - \sin t), r(1 - \cos t)) : t \in [0, \infty)\}$ has no vertical tangent lines.

Problem 7.232. The connection between the polar coordinates (r, θ) and the rectangular coordinates (x, y) of a point in the plane is given by $x = r \cos \theta$, $y = r \sin \theta$.

Problem 7.233. The connection between polar coordinates (r, θ) and the rectangular coordinates (x, y) of a point in the plane is given by $r = \sqrt{x^2 + y^2}$, $\theta = \arctan \frac{y}{x}$.

Problem 7.234. If $\left(2\sqrt{2}, \frac{3\pi}{4}\right)$ are polar coordinates of a point in the plane, then its rectangular coordinates are $(-2, -2)$.

Problem 7.235. Let $a < 0$. If $(0, a)$ are the rectangular coordinates of a point in the plane, then its polar coordinates are $\left(\sqrt{-a}, \frac{3\pi}{2}\right)$.

Problem 7.236. The curve represented by the polar equation $r = 3$ is a circle.

Problem 7.237. The curve represented by the polar equation $r = 3 \sin \theta$ is a circle.

Problem 7.238. The curve represented by the polar equation $r = \sin(3\theta)$ is a circle.

Problem 7.239. The tangent line to the curve $r = 1 + \cos \theta$ at the point $\left(r\left(\frac{\pi}{6}\right), \frac{\pi}{6}\right)$ is parallel to the line $x + y = 0$.

Problem 7.240. The curve $r = 1 + 2\cos \theta$ intersects itself at the pole O.

7.7.2 Multiple choice questions

For each of the following questions, choose only one answer.

Problem 7.241. If a curve $C = \{(x, y) : F(x, y) = 0\}$ is defined by the polar equation $r = 1$, what is $\frac{dy}{dx}$?

   A. $\frac{dy}{dx} = \cot \theta$.   **B.** $\frac{dy}{dx} = -\tan \theta$.   **C.** $\frac{dy}{dx} = 0$.   **D.** $\frac{dy}{dx} = 1$.
E. $\frac{dy}{dx} = -\cot \theta$.

Differential Calculus

Problem 7.242. The parametric curve defined by $x = e^t - 1$, $y = e^{2t}$, $t \in \mathbb{R}$, lies on a graph of which of the following functions?

    A. $y = \ln(x + 1)$.   **B.** $y = (x + 1)^2$.   **C.** $y = (x - 1)^2$.
D. $y = x^2 + 1$.   **E.** $y = \ln(x - 1)$.

Problem 7.243. Which curve is determined by the set $C = \{(t - \sin t, 1 - \cos t) : t \in [0, \infty)\}$?

    A. An ellipse.   **B.** A circle.   **C.** A parabola.   **D.** A hyperbola.
E. A cycloid.

Problem 7.244. Which curve is determined by the set $C = \{(5 \sin t, 5 \cos t) : t \in [0, 2\pi]\}$?

    A. An ellipse.   **B.** A circle.   **C.** A parabola.   **D.** A hyperbola.
E. A cycloid.

Problem 7.245. Which curve is determined by the set $C = \{(5 \cos t, 4 \sin t) : t \in [0, 2\pi]\}$?

    A. An ellipse.   **B.** A circle.   **C.** A parabola.   **D.** A hyperbola.
E. A cycloid.

Problem 7.246. Which curve is determined by the set $C = \{(t^3, t^6) : t \in \mathbb{R}\}$?

    A. An ellipse.   **B.** A circle.   **C.** A parabola.   **D.** A hyperbola.
E. A cycloid.

Problem 7.247. The parametric curve $C = \{(a \cos^3 t, a \sin^3 t) : t \in [0, 2\pi]\}$ is called an asteroid. What is the slope of the tangent line to the curve C at the point $\left(\frac{a\sqrt{2}}{4}, \frac{a\sqrt{2}}{4} \right)$?

    A. -2.   **B.** -1.   **C.** 0.   **D.** 1.   **E.** Something else.

Problem 7.248. The parametric curve $C = \{(2a \tan t, 2a \cos^2 t) : t \in \left(-\frac{\pi}{2}, \frac{\pi}{2} \right)\}$, $a > 0$ is called the Witch of Agnesi. How many inflection points does the Witch of Agnesi have?

    A. Three inflection points.   **B.** Two inflection points.   **C.** One inflection point.   **D.** No inflection point.

Problem 7.249. The parametric curve $C = \left\{ \left(\frac{3at}{1+t^3}, \frac{3at^2}{1+t^3} \right) : t \in \mathbb{R} \backslash \{-1\} \right\}$, $a > 0$ is called the Folium of Descartes. Which of the following statements is true?

(I) The Folium of Descartes has a slant asymptote.   (II) The Folium of Descartes has a vertical asymptote.   (III) The Folium of Descartes has a vertical tangent line.

A. Only (I) and (II);   **B.** Only (III);   **C.** (I), (II), and (III); **D.** (I) and (III);   **E.** Only (I).

Problem 7.250. Which of the following polar equation *does not* represent a circle?

A. $r = \pi$, $\theta \in [0, 2\pi]$.   **B.** $r = \pi \sin\theta$, $\theta \in [0, 2\pi]$.   **C.** $r = \pi \cos\theta$, $\theta \in [0, 2\pi]$.   **D.** $r = 2\pi \sin(2\theta)$, $\theta \in [0, 2\pi]$.   **E.** $r = 2\pi \sin\theta$, $\theta \in [0, 2\pi]$.

Problem 7.251. A curve C is given by its polar equation $r = r(\theta)$, $\theta \in [-\pi, \pi]$. Assume that the pole is at the origin and that the polar axis is the positive direction of the x-axis. If $r(\theta) = r(-\theta)$ for all $\theta \in [0, \pi]$, what can we tell about the symmetry of the graph of C?

A. It is symmetric with respect to the origin.   **B.** It is symmetric with respect to the x-axis.   **C.** It is symmetric with respect to the y-axis.   **D.** It is symmetric with respect to the line $y = x$. **E.** Cannot tell, not enough information is provided.

Problem 7.252. The curve C is given by its polar equation $r = r(\theta)$, $\theta \in [-\pi, \pi]$. Assume that the pole is at the origin and that the polar axis is the positive direction of the x-axis. If $r(\theta) = -r(-\theta)$ for all $\theta \in [0, \pi]$, what can we tell about the symmetry of the graph of C?

A. It is symmetric with respect to the origin.   **B.** It is symmetric with respect to the x-axis.   **C.** It is symmetric with respect to the y-axis.   **D.** It is symmetric with respect to the line $y = x$. **E.** Cannot tell, not enough information is provided.

Problem 7.253. The curve C is given by its polar equation $r = r(\theta)$, $\theta \in [-\pi, \pi]$. Assume that the pole is at the origin and that the polar axis is the positive direction of the x-axis. If $r(\theta - \pi) = r(\theta)$ for all $\theta \in [0, \pi]$, what can we tell about the symmetry of the graph of C?

A. It is symmetric with respect to the origin.   **B.** It is symmetric with respect to the x-axis.   **C.** It is symmetric with respect to the y-axis.   **D.** It is symmetric with respect to the line $y = x$. **E.** Cannot tell, not enough information is provided.

258   *Differential Calculus*

Problem 7.254. What is a Cartesian equation of the tangent line to the curve $r = \sin(3\theta)$ at the point $\left(r\left(\frac{\pi}{6}\right), \frac{\pi}{6} \right)$?

   A. $\sqrt{3}x - y - 2 = 0$.   **B.** $x - \sqrt{3}y + 2 = 0$.   **C.** $\sqrt{3}x + y - 2 = 0$.
D. $x + \sqrt{3}y + 2 = 0$.   **E.** None of these.

Problem 7.255. The curve C is given by its polar equation $r = 1 + 2\sin\theta$, $\theta \in [-\pi, \pi]$. Which value of θ corresponds to the point on the curve C that is furthest from the x-axis?

   A. $\frac{3\pi}{2}$.   **B.** $\frac{\pi}{2}$.   **C.** $\frac{3\pi}{4}$.   **D.** $\frac{\pi}{4}$.   **E.** π.

7.8   Examples, Definitions, and Theorems

Give an example for the each of the following:

Problem 7.256. A function $F = f + g$ is such that the limits of functions f and g at a do not exist and the limit of the function F at a exists.

Problem 7.257. A function $F = f \cdot g$ is such that the limits of functions f and g at a do not exist and the limit of the function F at a exists.

Problem 7.258. A function $F = \frac{f}{g}$ is such that the limits of functions f and g at a do not exist and the limit of the function F at a exists.

Problem 7.259. A function $F = f \cdot g$ is such that the limits of functions F and f at a exist and the limit of the function g at a does not exist.

Problem 7.260. A function f is continuous for all real numbers, but f' is not defined at $x = -2$.

Problem 7.261. A function f has a removable discontinuity.

Problem 7.262. A function f is such that $f(0) = f'(0) = 0$, but $f''(0)$ does not exist.

Problem 7.263. A function f is always decreasing, but its graph changes from concave downward to concave upward at the point $(0, f(0))$.

Problem 7.264. A function f is defined on the interval $[0, 1]$ and is such that $f(0) = -1$ and $f(1) = 1$, but $f(x) \neq 0$, for all $x \in (0, 1)$.

Problem 7.265. A function f has a critical number, but has no local maximum or minimum.

Problem 7.266. A function f has a local minimum at which the second derivative is 0.

Problem 7.267. A function f has a local maximum at which the second derivative is 0.

Problem 7.268. A function f has no inflection points and is such that $f''(0) = 0$.

Problem 7.269. A function f is defined on the interval $[0, 1]$ and has no absolute maximum.

Problem 7.270. A function f does not satisfy the conditions of the Mean Value Theorem on the interval $[0, 1]$.

Problem 7.271. A function f has a vertical tangent line.

Problem 7.272. A function f has an infinite number of vertical asymptotes.

Problem 7.273. A function f has two horizontal asymptotes.

Problem 7.274. A graph of a function f intersects its horizontal asymptote.

Problem 7.275. Complete each definition/theorem by matching it with the correct ending.

(1) The **Mean Value Theorem** states that ...
(2) The **Chain Rule** states that ...
(3) A **critical number** of a function f is a number that ...
(4) The **Extreme Value Theorem** states that ...
(5) **Fermat's Theorem** states that ...
(6) An **antiderivative** of a function f is ...
(7) The **Napier's constant** e is ...
(8) An **inflection point** is a point ...
(9) The **derivative** of a function f at a number a is ...
(10) **L'Hospital's Rule** states that ...

(11) The **Intermediate Value Theorem** states that ...

(12) The **Squeeze Theorem** states that ...

(13) A function f is **continuous** at a number a ...

...and their endings:

(a) ...if f is a continuous function on the closed interval $[a, b]$ and if N is a number between $f(a)$ and $f(b)$, $f(a) \neq f(b)$, then there exists a number c in (a, b) such that $f(c) = N$.

(b) ...if f is a function that satisfies the following hypotheses:

(i) f is continuous on the closed interval $[a, b]$ and

(ii) f is differentiable on the open interval (a, b),

then there is a number c in (a, b) such that $f'(c) = \frac{f(b)-f(a)}{b-a}$.

(c) ...$f'(a) = \lim_{h \to 0} \frac{f(a+h)-f(a)}{h}$ if this limit exists.

(d) ...if f is a continuous function on a closed interval $[a, b]$, then f attains an absolute maximum value $f(c)$ and an absolute minimum value $f(d)$ at some numbers $c, d \in [a, b]$.

(e) ...is in the domain of the function f and such that either $f'(c) = 0$ or $f'(c)$ does not exist.

(f) ...on a continuous curve where the curve changes from concave upward to concave downward or from concave downward to concave upward.

(g) ...the base of the exponential function whose graph has a tangent line of slope 1 at the point $(0, 1)$.

(h) ...if f and g are both differentiable functions, then $\frac{d}{dx}[f(g(x))] = f'(g(x)) \cdot g'(x)$.

(i) ...if functions f, g, and h and a number a are such that $f(x) \leq g(x) \leq h(x)$ when $x \in (a - \varepsilon, a) \cup (a, a + \varepsilon)$ for some $\varepsilon > 0$ and $\lim_{x \to a} f(x) = \lim_{x \to a} h(x) = L$, then $\lim_{x \to a} g(x) = L$.

(j) ...if f is a differentiable function on an interval (a, b) such that $x_0 \in (a, b)$ is a point where f has a local extremum, then $f'(x_0) = 0$.

(k) ...a function F such that $F' = f$.

(l) ...if functions f and g are such that:

• they are differentiable at a point $c \in I$,

• $\lim_{x \to c} f(x) = \lim_{x \to c} g(x) = 0$, or $\lim_{x \to c} f(x) = \pm\infty$ and $\lim_{x \to c} g(x) = \pm\infty$,

- $g'(x) \neq 0$ for all $x \in I \backslash \{c\}$, and
- $\lim\limits_{x \to c} \dfrac{f'(x)}{g'(x)}$ exists,

then $\lim\limits_{x \to c} \dfrac{f(x)}{g(x)} = \lim\limits_{x \to c} \dfrac{f'(x)}{g'(x)}$.

(m) ... $\lim\limits_{x \to a} f(x) = f(a)$.

7.9   Answers, Hints, and Solutions

Solution 7.1. False. Take $a = -3$.

Solution 7.2. False. Take $f(x) = x^2$, $t = -1$, and $s = 1$.

Solution 7.3. True.

Solution 7.4. True.

Solution 7.5. False. Take $f(x) = x^2 + x$.

Solution 7.6. False. The equation $2^{|x|} = 0$ has no solution.

Solution 7.7. False. Graph $f(x) = e^{-x}$.

Solution 7.8. False. Take $x = -1$.

Solution 7.9. True.

Solution 7.10. True. Recall that the domain of the function $f(x) = \ln x$ is the interval $(0, \infty)$.

Solution 7.11. True. $x = e^{\pi}$.

Solution 7.12. False. Take $x = e$. Recall that if $x > 0$, then $6 \ln x = \ln(x^6)$.

Solution 7.13. False. Take $x = y = 1$. But $\ln(xy) = \ln x + \ln y$.

Solution 7.14. False. Recall $\tan x = \dfrac{\sin x}{\cos x}$ and $\cos \dfrac{(2n+1)\pi}{2} = 0$, $n \in \mathbb{Z}$.

Solution 7.15. False. Observe that the equation $\cosh x = 0$ has only one solution.

Solution 7.16. False. Observe that the equation $\arctan x = 0$ has only one solution.

Solution 7.17. False. $\sinh^2 x - \cosh^2 x = -1$.

Solution 7.18. False. $f(g(x)) = (x+1)^2$.

Solution 7.19. True.

Solution 7.20. False. Take $f(x) = x$.

Solution 7.21. True.

Solution 7.22. False. Take $f(x) = \ln x$.

Solution 7.23. True.

Solution 7.24. True.

Solution 7.25. True.

Solution 7.26. False. Observe that $\frac{7\pi}{3} = 2\pi + \frac{\pi}{3}$ does not belong to the range of $f(x) = \sin^{-1}(x)$. In fact, $\sin^{-1}\left(\sin\left(\frac{7\pi}{3}\right)\right) = \frac{\pi}{3}$.

Solution 7.27. B.

Solution 7.28. C.

Solution 7.29. E.

Solution 7.30. B. Note that $g(1.2) = f(4)$.

Solution 7.31. D. Note that $x^2 - 3 = 5$ implies $x = \pm\sqrt{2}$.

Solution 7.32. B. Note that $2x = t$ is equivalent to $x = \frac{t}{2}$. Therefore, $h(t) = h\left(2\left(\frac{t}{2}\right)\right) = \frac{(t/2)^2}{4} = \frac{t^2}{16}$.

Solution 7.33. D. Note that $(h \circ g \circ f)(x) = \sin\sqrt{3x-1}$.

Solution 7.34. A.

Solution 7.35. A.

Solution 7.36. False. Take $h(x) = 1$ if $x \neq 3$, $h(3) = 2$.

Solution 7.37. True.

Solution 7.38. True.

Solution 7.39. False. Take $f(x) = g(x) = x - 5$.

Solution 7.40. False. Take $f(x) = x - 1$ and $g(x) = x^2 - 1$.

Solution 7.41. False. Take $f(x) = e^{-\frac{1}{x^2}}$ and $g(x) = x^2$.

Solution 7.42. False. Take $f(x) = \sin x$ and $g(x) = -\sin x$.

Solution 7.43. False. Take $f(x) = e^{-\frac{1}{x^2}}$.

Solution 7.44. False. Observe that $\lim_{x \to 3} \sqrt{x - 3}$ does not exist because $\lim_{x \to 3^-} \sqrt{x - 3}$ does not exist. On the other hand, $\sqrt{\lim_{x \to 3}(x - 3)} = 0$.

Solution 7.45. False. Take $f(x) = -x^2$ and $a = 1$.

Solution 7.46. False. An exponential function with a base greater than one will always overtake any polynomial. Also, think what would happen with this limit if you apply L'Hospital's Rule 100,000 times.

Solution 7.47. True. Use the Squeeze Theorem.

Solution 7.48. False. Use the Squeeze Theorem. Also, observe that the Intermediate Value Theorem requires that the function is continuous on the interval of consideration.

Solution 7.49. False. Consider $f(x) = -x$.

Solution 7.50. False. $\lim_{x \to 0^+} \left(\frac{1}{x} - \frac{1}{x^2} \right) = \lim_{x \to 0^+} \frac{x-1}{x^2} = -\infty$. Recall that $\lim_{x \to a}(f(x) + g(x)) = \lim_{x \to a} f(x) + \lim_{x \to a} g(x)$ only if both $\lim_{x \to a} f(x)$ and $\lim_{x \to a} g(x)$ exist.

Solution 7.51. False. Consider $f(x) = \sin x$.

Solution 7.52. False. Consider the function $f(x) = \frac{1}{x-1}$ if $x \neq 1$.

Solution 7.53. False. Consider the function $f(x) = \frac{1}{x-1}$, if $x \neq 1$ and $f(1) = 0$.

Solution 7.54. False. Observe that $\lim_{x \to \infty} \frac{g(x)}{mx+b} = \infty$ for any $m \neq 0$.

Solution 7.55. True. Consider $f(x) = \tan x$.

Solution 7.56. True. Consider $f(x) = \arctan x$.

Solution 7.57. False. This would violate the vertical line test.

Solution 7.58. False. The limit is 2.

Solution 7.59. True.

Solution 7.60. D. Observe that $\lim\limits_{x \to 1^-} f(x) = -1 \neq 1 = \lim\limits_{x \to 1^+} f(x)$.
Also, $f(1) = 0$.

Solution 7.61. D. Observe that $\lim\limits_{x \to 0} f(x) = 1$.

Solution 7.62. C. Observe that $\lim\limits_{x \to 0^-} f(x) = c$ and $\lim\limits_{x \to 0^+} f(x) = 1/c$,
so we need to solve the equation $c^2 - 1 = 0$.

Solution 7.63. C. Draw (or imagine), for example, the following points on the graph of the function f: $(\pi, f(\pi))$, $(3.14, f(3.14))$, $(\pi - 0.0001, f(\pi - 0.0001))$, $(3.141, f(3.141))$, $(\pi - 0.00001, f(\pi - 0.00001))$, and $(3.1415, f(3.1415))$. Does $\lim\limits_{x \to \pi} f(x)$ exist? What if you replace π by any other irrational number? Or, if you consider a rational number $a \neq 0$? Now, pick a small positive number ε. What can you say about the number $|f(x)|$ if $x \in (-\varepsilon, \varepsilon)$?

Solution 7.64. D. Use L'Hospital's Rule. Note that the derivative of a function whose graph is a straight line is the slope of that line; thus, to determine the limit we need to know the slopes of these two lines.

Solution 7.65. B. Note that $\alpha = 3$, $\beta = \frac{1}{6}$, and $\gamma = 4$.

Solution 7.66. A. Since the limit exists, it must be an indeterminate form, which implies that $\lim\limits_{x \to 0} f(x) = 0$.

Solution 7.67. C. Observe that the value of the limit is $f'(5)$.

Solution 7.68. A. $y = x + 4$. Observe that $f(x) = x + 4 + \dfrac{3}{x-1}$.

Solution 7.69. D. Three asymptotes are $x = -1$, $x = 3$, and $x = 4$.

Solution 7.70. E. None of these. $\lim\limits_{x \to 0^+} \left(\frac{1}{x} \cdot \ln x \right) = -\infty$.

Solution 7.71. B. -1.

Solution 7.72. A. e. Take $\lim\limits_{x \to 0} \left((1 + \sin x)^{\frac{1}{\sin x}} \right)^{\frac{\sin x}{x}}$.

Solution 7.73. False. Consider $f(x) = \text{sign}\,(x)$.

Solution 7.74. False. Consider the function $f(x) = 1$ if $x \in \mathbb{Q}$ and $f(x) = 0$ if $x \notin \mathbb{Q}$.

Solution 7.75. False. Take $f(x) = 1$ and $g(x) = 0$.

Solution 7.76. True. Take $f(x) = \text{sign}(x)$ and $g(x) = -f(x)$.

Solution 7.77. True. Any polynomial is a continuous function on $\mathbb{R}$.

Solution 7.78. False. Take $f(x) = e^x$.

Solution 7.79. False. Note $\lim_{x \to 2} f(x) = 4 \neq f(2)$.

Solution 7.80. True.

Solution 7.81. (1) False. $f(3) = 16$. (2) False. $f(2) = -2$. (3) True. (4) False. $\lim_{x \to 2^+} f(x) = 11 \neq f(2)$.

Solution 7.82. False. Take $f(x) = \frac{1}{x}$, $a = -1$, and $b = 1$.

Solution 7.83. True. Consider the function $f(x) = e^{-x} - x$ and apply the Intermediate Value Theorem.

Solution 7.84. False. Take $f(x) = 4 - x^2$, $x \in [-2, 1]$. Read the statement of the Intermediate Value Theorem, if necessary.

Solution 7.85. False. Take $f(x) = \tan\left(\pi x - \frac{\pi}{2}\right)$, $x \in (0, 1)$.

Solution 7.86. True. This is the statement of the Extreme Value Theorem.

Solution 7.87. True. Otherwise, we can apply the Extreme Value Theorem.

Solution 7.88. False. Take $f(x) = x$ if $x \in [0, 1)$ and $f(1) = 2$.

Solution 7.89. E.

Solution 7.90. B. Eliminate (I) by observing that $\lim_{x \to 0} f(x)$ does not exist since the left-hand and the right-hand limits are not equal. This implies that $\lim_{x \to 0} [2f(x)g(x) - 1]$ would exist only if $g(0) = 0$, but we are not given this as a fact. We don't know the value of $f(0)$, so it is not possible to determine (III). The value of (II) is 14.

Solution 7.91. A. To eliminate **B** and **C**, take, for example, $f(x) = \sin x$.

Solution 7.92. B. Observe that the function in **A** is continuous, in **C** has an infinite discontinuity, and in **D** and **E** each has a jump discontinuity.

Solution 7.93. E.

Solution 7.94. A. The function is not defined at $x = a$, and therefore not continuous there regardless of the value of a.

Solution 7.95. D. Note that $f(x) = \frac{x(x+4)(x-1)}{(x+4)(x-3)}$.

Solution 7.96. B.

Solution 7.97. A.

Solution 7.98. E. To eliminate **A** and **B**, consider $a = 0$, $b = 1$, and $f(x) = \frac{1}{x}$. To eliminate **C**, consider $a = -1$, $b = 1$, and $f(x) = |x|$. To eliminate **D**, consider $f(x) = -x^2$, $x \in (a, b)$ with $f(a) = 2, f(b) = 5$.

Solution 7.99. True. Observe that $\lim\limits_{x \to 2} \frac{g(x)-g(2)}{x-2} = g'(2)$.

Solution 7.100. True.

Solution 7.101. True.

Solution 7.102. True.

Solution 7.103. False. Consider $f(x) = \sqrt[3]{x}$ at $x = 0$.

Solution 7.104. False. The function f is not differentiable at $x = 0$.

Solution 7.105. False. Take $f(x) = x$ and $g(x) = x + 1$.

Solution 7.106. False. Take $f(x) = x|x|$.

Solution 7.107. True.

Solution 7.108. False. Take $f(x) = x^2$.

Solution 7.109. True. Use the Chain Rule on $f(-x) = f(x)$ to obtain $-f'(-x) = f'(x)$.

Solution 7.110. False. The derivative of a constant function is the zero function.

Solution 7.111. False. The derivative of a constant function is the zero function.

Solution 7.112. True.

Solution 7.113. True. Observe that $f(x) = 1 + x + x^2 + 2x^3 + \cdots + x^{10}$, or use the Product Rule.

Solution 7.114. True.

Solution 7.115. False. Observe that $y^{(4n)} = \sin x$, $y^{(4n+1)} = \cos x$, $y^{(4n+2)} = -\sin x$, $y^{(4n+3)} = -\cos x$, and that $99 = 4 \cdot 24 + 3$.

Solution 7.116. True.

Solution 7.117. False. $f'(x) = a^x \ln a$.

Solution 7.118. False. Take $f(x) = x^{\frac{1}{3}}$.

Solution 7.119. False. Take $f(x) = 0$ or $f(x) = e^{x+c}$, $c \in \mathbb{R}$.

Solution 7.120. True.

Solution 7.121. True.

Solution 7.122. False. $\frac{d}{dx}\left(f(\sqrt{x})\right) = \frac{f'(\sqrt{x})}{2\sqrt{x}}$.

Solution 7.123. False. What about $x = 0$ and $x = -1$?

Solution 7.124. True.

Solution 7.125. True.

Solution 7.126. True.

Solution 7.127. False. $y' = 10^x \ln 10$.

Solution 7.128. False. $\frac{d}{dx}(x^x) = x^x(\ln x + 1)$.

Solution 7.129. True.

Solution 7.130. True.

Solution 7.131. False. $g'(x) = \frac{1}{f'(g(x))}$.

Solution 7.132. True.

Solution 7.133. True. From $h'(x) = g'(f(x)) \cdot f'(x) = -\frac{f'(x)}{2}$ and the graph, it looks like that $h'(3) \approx -\frac{1}{4}$ is a good approximation.

Solution 7.134. False. Take $f(x) = \sin x$ and $g(x) = x$.

Solution 7.135. B.

Solution 7.136. D.

Solution 7.137. B. Observe that (I) is false if $g(a) = 0$ and (III) is false if $f(x) = |x|$, $g(x) = -|x|$, and $a = 0$.

Solution 7.138. A. Observe that (III) is false if $f(x) = \frac{1}{x}$, $a = 0$ and $b = 1$, and that (IV) is false if $f(x) = x|x|$, $a = -1$, and $b = 1$.

Solution 7.139. C.

Solution 7.140. C.

Solution 7.141. E.

Solution 7.142. B.

Solution 7.143. C.

Solution 7.144. C.

Solution 7.145. C.

Solution 7.146. B.

Solution 7.147. A.

Solution 7.148. B.

Solution 7.149. D.

Solution 7.150. True. Note that we do not know if f is a differentiable function so we cannot use the first derivative.

Solution 7.151. False. Take $f(x) = x - 1$ and observe that $f\left(\frac{1}{3}\right) = -\frac{2}{3} < -\frac{1}{2} = f\left(\frac{1}{2}\right)$ but $\left(f\left(\frac{1}{3}\right)\right)^2 = \frac{4}{9} > \frac{1}{4} = \left(f\left(\frac{1}{2}\right)\right)^2$. In general, if we are given that f is a differentiable function such that $f'(x) > 0$, $x \in (0,1)$, then this is not enough to conclude that $\frac{d}{dx}\left([f(x)]^2\right) = 2f(x)f'(x)$ is positive.

Solution 7.152. False. Take a function $f : \{-1, 1\} \to \mathbb{R}$. Recall that a critical number of a function belongs to the interior of function's domain.

Solution 7.153. False. Take $f(x) = |x|$ and $x = 0$.

Solution 7.154. False. Take $f(x) = x^3$ and $x = 0$.

Solution 7.155. False. Take $f(x) = (x - 2)^3$ and $x = 0$.

Solution 7.156. True. This is the claim of Fermat's Theorem.

Solution 7.157. False. Take $c = 0$, $f(x) = -\frac{1}{x^2}$, $x \neq 0$, $f(0) = 0$.

Solution 7.158. False. Take $f(x) = x^3$ and $c = 0$.

Solution 7.159. False. Take $f(x) = |x|$ and $x = 0$.

Solution 7.160. False. Take $f(x) = (x-2)^4$.

Solution 7.161. False. Take $f(x) = e^{-x}$.

Solution 7.162. False. It should be $\lim\limits_{h \to 0} \frac{f(a+h)-f(a)}{h}$.

Solution 7.163. True. Note that $f'(x) = 6x^2 + 4 \geq 4$.

Solution 7.164. False. Observe that $y = 4+2x(x+2)$ is a quadratic polynomial in x, not a linear function. An equation of the tangent line is $y - 4 = -4(x+2)$.

Solution 7.165. True.

Solution 7.166. True. Graph $x^2 = (1-y^2)^2$.

Solution 7.167. True.

Solution 7.168. False. $L(x) = f(a) + f'(a)(x-a)$.

Solution 7.169. False. Take $f(x) = |x|$ and $c = 0$.

Solution 7.170. True.

Solution 7.171. True. Apply Rolle's Theorem or the Mean Value Theorem.

Solution 7.172. False. It is the other way around.

Solution 7.173. True. Apply the Mean Value Theorem.

Solution 7.174. False. Take $f(x) = 10x$ and $g(x) = \begin{cases} -2x, & \text{if } x \in [0, 0.5] \\ 22x - 12, & \text{if } x \in (0.5, 1] \end{cases}$. Observe that, since g is not differentiable, we cannot apply the Mean Value Theorem.

Solution 7.175. False. Take $f(x) = x$ if $x \in [0,1)$ and $f(1) = 0$. Observe that, since f is not continuous on $[0,1]$, we cannot apply the Mean Value Theorem.

Solution 7.176. False. If a function f is differentiable on $\mathbb{R}$ with $f(1) = -2$ and $f(3) = 0$, by the Mean Value Theorem there must be $c \in (1,3)$ such that $f'(c) = \frac{f(3)-f(1)}{3-1} = 1$.

Solution 7.177. False.

Solution 7.178. True. Apply Rolle's Theorem.

Solution 7.179. False. Take $f(x) = \sqrt[3]{x}$ and $x = 0$.

Solution 7.180. False. Consider the trajectory of a stone thrown upwards, $h(t) = t - \frac{g}{2}t^2$, $t \in \left[0, \frac{2}{g}\right]$. Then the velocity equals to zero at $t = \frac{1}{g}$, but the acceleration is due to gravity and it is never zero.

Solution 7.181. B. Observe that $V = x^3$ implies $\frac{dV}{dt} = 3x^2\frac{dx}{dt}$.

Solution 7.182. E.

Solution 7.183. B.

Solution 7.184. C.

Solution 7.185. B.

Solution 7.186. D.

Solution 7.187. B.

Solution 7.188. C.

Solution 7.189. D. Let ℓ be the length of the cylinder and let r be the radius of the cross section. The constraint is $2r\pi + \ell \leq 30\pi$, which is the same as $2r^3\pi^2 + r^2\ell\pi \leq 30r^2\pi^2$. Since the volume of the cylinder is $V = r^2\ell\pi$, it follows that the given constraint may be rewritten as $V \leq 2\pi^2(15r^2 - r^3) = f(r)$. From $f'(r) = 6\pi^2 r(10 - r)$, it follows that the function f attains its maximum at $r = 10$. The maximum possible volume is $V = 1000\pi^2 \approx 9869$ cubic centimeters.

Solution 7.190. C. Set a and b to be the sides of the rectangle. From $a + b = 20$, conclude that the area of the rectangle is $A = a(20 - a)$.

Solution 7.191. A.

Solution 7.192. D. To see that (I) is not true, take $f(x) = e^x$. For (II), consider the function $f(x) = \sqrt[3]{x}$ and $x = 0$. Observe that (II) implies (III).

Solution 7.193. D.

Solution 7.194. B.

Solution 7.195. B. $L(x) = 2 + 4(x - 3)$.

Solution 7.196. B.

Solution 7.197. C. Recall $(b - a)f'(c) = f(b) - f(a)$.

Solution 7.198. C.

Solution 7.199. E. Solve $f(a) - f(3) = 38(a - 3)$.

Solution 7.200. D.

Solution 7.201. B. Observe that the function $f(x) = x^5 + 10x + 3$ is a polynomial of an odd degree. From $f'(x) = 5x^4 + 10 > 0$ for all $x \in \mathbb{R}$, conclude that f is increasing.

Solution 7.202. D. Observe that $A(t) = 16e^{kt}$ for some $k < 0$.

Solution 7.203. D. Recall that $\frac{dP}{dt} = kP$. Note that, for $t = t_0$, $1000 = P'(t_0) = kP(t_0) = 500k$.

Solution 7.204. True.

Solution 7.205. False. They are decreasing functions.

Solution 7.206. True.

Solution 7.207. True. Recall that $F(x) - G(x) = C$ for some $C \in \mathbb{R}$.

Solution 7.208. False. Take $f(x) = 1$ and $g(x) = x$.

Solution 7.209. False. It should be $F(x) = x^3 + C$, $C \in \mathbb{R}$.

Solution 7.210. False. Observe that $f(x) = -\frac{x^4}{4} + 67$.

Solution 7.211. False. It should be $\int \frac{dx}{x^2+1} = \arctan x + C$.

Solution 7.212. False. It should be $\int \frac{dx}{3-2x} = -\frac{1}{2}\ln|3 - 2x| + C$.

Solution 7.213. False. It should be $f(x) = x^2 + C$, $C \in \mathbb{R}$.

Solution 7.214. False. It should be $y = Ce^{2x}$, $C \in \mathbb{R}$.

Solution 7.215. True.

Solution 7.216. False. It is a quadratic polynomial.

Solution 7.217. E.

Solution 7.218. D.

Solution 7.219. A.

Solution 7.220. A. To eliminate (I), (II), and (IV), check the definition of an antiderivative. To eliminate (V), consider $f(x) = |x|$ and observe: (a) f is not differentiable at $x = 0$, and (b) the function $F(x) = \frac{1}{2}x|x|$ is an antiderivative of f.

Solution 7.221. True. For example, the unit circle may be parametrized as $\{(\cos t, \sin t) : t \in [0, 2\pi]\}$ and as $\{(\sin t, \cos t) : t \in [0, 2\pi]\}$.

Solution 7.222. False. This is a parametrization of the ellipse $9x^2 + 4y^2 = 36$.

Solution 7.223. True.

Solution 7.224. False. Consider the unit circle.

Solution 7.225. True. One parametrization is given by $\{(t, f(t)) : t \in [a, b]\}$.

Solution 7.226. False. Recall that $\frac{dy}{dx} = \frac{\frac{dy}{dt}}{\frac{dx}{dt}}$, so for $\frac{dy}{dx}$ to exist, it is necessary that $\frac{dx}{dt} \neq 0$. This information is not provided. As an example you may consider the curve $C = \{(t^2, t^2) : t \in \mathbb{R}\}$.

Solution 7.227. True.

Solution 7.228. False. Take, for example, $x = t$, $y = t^2$, $t \in \mathbb{R}$, a parametrization of the parabola $y = x^2$. In general, since $\frac{dy}{dx}$ is itself a parametric curve, we have that $\frac{d^2y}{dx^2} = \frac{\frac{d}{dt}\left(\frac{dy}{dx}\right)}{\frac{dx}{dt}}$.

Solution 7.229. False. These are two parametrization of the circle $x^2 + y^2 = 4$. Note that, via the substitution $u = 2t$, the second parametrization becomes the first parametrization.

Solution 7.230. True. In general, a Lissajous curve is given by $x = A\sin(at + c)$, $y = B\sin(bt)$.

Solution 7.231. False. From $\frac{dy}{dx} = \frac{\sin t}{1-\cos t}$, it follows, for example, that $\lim_{t \to 2\pi^-} \frac{dy}{dx} = -\infty$ and $\lim_{t \to 2\pi^+} \frac{dy}{dx} = \infty$. Hence, the line $x = 2\pi$ is a vertical tangent line to the graph of the cycloid.

Solution 7.232. True.

Solution 7.233. False. For example, take the point $(x, y) = (-1, -1)$ and observe that its polar coordinates are $(r, \theta) = \left(\sqrt{2}, \frac{3\pi}{4}\right)$. On the other hand, $\arctan \frac{y}{x} = \arctan 1 = \frac{\pi}{4}$.

Solution 7.234. False. The point is in the 2nd quadrant and its rectangular coordinates are $(-2, 2)$.

Solution 7.235. False. The polar coordinates are $\left(-a, \frac{3\pi}{2}\right)$.

Solution 7.236. True.

Solution 7.237. True.

Solution 7.238. False. This is a 3-leaf rose.

Solution 7.239. True. Note, $\left.\frac{dy}{dx}\right|_{\theta=\frac{\pi}{6}} = -\left.\frac{\cos\theta+\cos(2\theta)}{\sin\theta+\sin(2\theta)}\right|_{\theta=\frac{\pi}{6}} = -1$.

Solution 7.240. True. Note that $r\left(\frac{2\pi}{3}\right) = r\left(\frac{4\pi}{3}\right) = 0$.

Solution 7.241. E. Note that $r = 1$ implies $x = \cos\theta$ and $y = \sin\theta$.

Solution 7.242. B. Note that $x = e^t - 1$ implies $t = \ln(x+1)$.

Solution 7.243. E. A cycloid.

Solution 7.244. B. This is a parametrization of the circle $x^2 + y^2 = 25$.

Solution 7.245. A. This is a parametrization of the ellipse $\frac{x^2}{25} + \frac{y^2}{16} = 1$.

Solution 7.246. C. This is a parametrization of the parabola $y = x^2$.

Solution 7.247. B. Observe that the given point corresponds to $t = \frac{\pi}{4}$ and conclude that the slope of the tangent line is $\left.\frac{dy}{dx}\right|_{t=\frac{\pi}{4}} = -1$.

Solution 7.248. B. Observe that $\frac{dy}{dx} = -2\sin t\cos^3 t$ implies that $\frac{d^2y}{d^2x} = \frac{1}{a}\cos^4 t(3\sin^2 t - \cos^2 t)$. Hence, $t = \pm\frac{\pi}{6}$ correspond to two inflection points, $\left(\pm\frac{2a}{\sqrt{3}}, \frac{3a}{2}\right)$.

Solution 7.249. D. Observe that $\lim\limits_{t\to-1^+} x(t) = -\infty$ and $\lim\limits_{t\to-1^-} x(t) = \infty$ and conclude that $\lim\limits_{x\to-\infty} \frac{y}{x} = \lim\limits_{t\to-1^+} \frac{\frac{3at^2}{1+t^3}}{\frac{3at}{1+t^2}} = -1$. Similarly $\lim\limits_{x\to\infty} \frac{y}{x} = -1$. From $\lim\limits_{x\to-\infty}(y-(-1)x) = \lim\limits_{t\to-1^+}\left(\frac{3at^2}{1+t^3} + \frac{3at}{1+t^3}\right) =$

$-3a$ and $\lim\limits_{x\to\infty}(y+x) = -3a$, conclude that the line $y = -x - 3a$ is a slant asymptote. Hence, (I) holds. To eliminate (II), observe that the curve is defined and continuous for any $t \in \mathbb{R}\backslash\{-1\}$. From $\frac{dy}{dx} = \frac{t(2-t^3)}{1-2t^3}$, we conclude that there is a vertical tangent line at the point $(a\sqrt[3]{4}, a\sqrt[3]{2})$, corresponding to $t = \frac{1}{\sqrt[3]{2}}$.

Solution 7.250. D. This is a 4-leaf rose.

Solution 7.251. B. Observe that $x(-\theta) = r(-\theta)\cos(-\theta) = x(\theta)$ and $y(-\theta) = r(-\theta)\sin(-\theta) = -y(\theta)$ for $\theta \in [0, \pi]$. Points $(x(\theta), y(\theta))$ and $(x(-\theta), y(-\theta)) = (x(\theta), -y(\theta))$ are symmetric with respect to the x-axis.

Alternatively, note that $(r(-\theta), -\theta) = (r(\theta), -\theta)$ in polar coordinates.

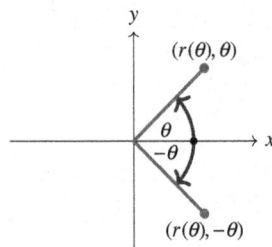

Solution 7.252. C. Observe that $x(-\theta) = r(-\theta)\cos(-\theta) = -x(\theta)$ and $y(-\theta) = r(-\theta)\sin(-\theta) = y(\theta)$ for $\theta \in [0, \pi]$. Points $(x(\theta), y(\theta))$ and $(x(-\theta), y(-\theta)) = (-x(\theta), y(\theta))$ are symmetric with respect to the y-axis.

Alternatively, note that $(r(-\theta), -\theta) = (-r(\theta), -\theta)$ in polar coordinates.

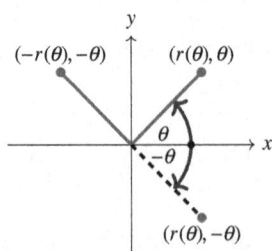

Solution 7.253. A. Observe that $x(\theta - \pi) = r(\theta - \pi)\cos(\theta - \pi) = r(\theta)\cdot(-\cos(\theta)) = -x(\theta)$ and $y(\theta-\pi) = r(\theta-\pi)\sin(\theta-\pi) = -y(\theta)$ for

$\theta \in [0, \pi]$. Points $(x(\theta), y(\theta))$ and $(x(\theta-\pi), y(\theta-\pi)) = (-x(\theta), -y(\theta))$ are symmetric with respect to the origin.

Alternatively, note that $(r(\theta - \pi), \theta - \pi) = (r(\theta), \theta - \pi)$ in polar coordinates.

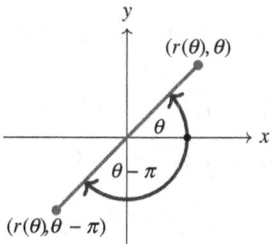

Solution 7.254. C. Observe that $\theta = \frac{\pi}{6}$ determines the point $\left(\frac{\sqrt{3}}{2}, \frac{1}{2}\right)$ in rectangular coordinates. The slope is given by $\left.\frac{dy}{dx}\right|_{\theta=\frac{\pi}{6}} = \left.\frac{3\cos(3\theta)\sin\theta+\sin(3\theta)\cos\theta}{3\cos(3\theta)\cos\theta-\sin(3\theta)\sin\theta}\right|_{\theta=\frac{\pi}{6}} = -\sqrt{3}$.

Solution 7.255. A. The question is to determine the absolute maximum of the function $f(\theta) = |y(\theta)| = |(1 + 2\sin\theta)\sin\theta|$. We can use calculus to evaluate the maximum value of f, but, in this particular case, we can just observe that both factors take their maximum values for $\theta = \frac{\pi}{2}$.

Solution 7.256. Consider $f(x) = \text{sign}(x)$, $g(x) = -\text{sign}(x)$, and $a = 0$. Then $F(x) = f(x) + g(x) = 0$ for all $x \in \mathbb{R}$.

Solution 7.257. Consider the function $f(x) = g(x) = \text{sign}(x)$ and $a = 0$. Then $F(x) = f(x) \cdot g(x) = 1$ for all $x \in \mathbb{R}\backslash\{0\}$.

Solution 7.258. Consider the function $f(x) = g(x) = \text{sign}(x)$ and $a = 0$. Then $F(x) = \frac{f(x)}{g(x)} = 1$ for all $x \in \mathbb{R}\backslash\{0\}$.

Solution 7.259. Consider the function $f(x) = 0$ for all $x \in \mathbb{R}$ and $g(x) = \text{sign}(x)$. Then $F = f(x) \cdot g(x) = 0$ for all $x \in \mathbb{R}$.

Solution 7.260. Consider the function $f(x) = |x + 2|$.

Solution 7.261. Consider the function $f(x) = \frac{\sin x}{x}$ if $x \neq 0$ and $f(0) = 0$. Recall that $\lim\limits_{x \to 0} \frac{\sin x}{x} = 1$.

Solution 7.262. Consider the function $f(x) = x|x|$. Observe that $f'(x) = 2|x|$.

Solution 7.263. Consider the function $f(h) = -\sqrt[3]{x}$.

Solution 7.264. Consider the function f such that $f(x) = -1$ if $x \in [0,1)$ and $f(1) = 1$.

Solution 7.265. Consider $f(x) = x^3$.

Solution 7.266. Consider $f(x) = x^4$.

Solution 7.267. Consider $f(x) = -x^4$.

Solution 7.268. Consider $f(x) = x^4$.

Solution 7.269. Consider the function f such that $f(x) = x$ if $x \in [0,1)$ and $f(1) = 0$.

Solution 7.270. Consider $f(x) = |x - \frac{1}{2}|$.

Solution 7.271. Consider $f(x) = \sqrt[3]{x}$.

Solution 7.272. Consider $f(x) = \tan x$.

Solution 7.273. Consider $f(x) = \arctan x$.

Solution 7.274. Consider the function $f(x) = \frac{\sin x}{x}$, $x \in [1,\infty)$. Observe that $\lim_{x\to\infty} f(x) = 0$ and $f(k\pi) = 0$ for any $k \in \mathbb{N}$.

Solution 7.275. [(a) → (2)], [(b) → (8)], [(c) → (5)], [(d) → (4)], [(e) → (10)], [(f) → (11)], [(g) → (7)], [(h) → (6)], [(i) → (3), [(j) → (12)], [(k) → (1)], [(l) → (9)], [(m) → (13)].

Chapter 8

Recommendations to Thrive in Mathematics

The following is a list of recommendations concerning habits around learning, exam preparations, and well-being. This list was put together for all students who are thinking about their actions, values, and choices that affect their studies in academia. By making a commitment to regularly assess your values and choices in life, you not only discover what you value and how you choose, but also how these choices impact your educational goals and their attainment, which hopefully bring about the right actions. Bear in mind that your values and choices are bound to change as you go through your various life stages. However, by actively engaging in this sort of reflection and assessment exercise, you learn to manage change in your academic career and beyond.

8.1 Tips for Reading these Recommendations

- Do not be overwhelmed with the number of items on this list. You may not want to read the whole chapter at once. Instead, choose some recommendations that appeal to you.
- You may want to make changes in your habits and study approaches after reading the recommendations. Our advice is to take small steps. Small changes are easier to make and chances are those changes will stick with you and become part of your habits.

• Take the time to reflect on the recommendations. Look at the people in your life you respect and admire for their groundedness and accomplishments. Do you believe the recommendations are reflected in their approach to life?

8.2 Habits of a Thriving Student

• **Act responsibly and respectfully**

  ○ Be accountable for your own behavior.
  ○ Respect diversity in people, ideas, and opinions.
  ○ Read documents that are passed on by the instructor and act on them, such as the syllabus, a lecture outline, or an assignment guideline.
  ○ Take an active role in your education, for example, regularly attend class and come prepared, complete assignments to the best of your ability, and seek out an academic advisor for course and program guidance.
  ○ Uphold and be committed to honesty, trust, fairness, respect, responsibility, and courage, which are the fundamental values set out by the International Centre for Academic Integrity, even when situations become tough or stressful. In other words, do not cheat and encourage academic integrity in others.

• **Set goals**

  ○ Set attainable goals based on specific information, such as the syllabus, the academic calendar, or academic advisor.
  ○ Be motivated to reach your educational goals.
  ○ Be committed to thrive.
  ○ Be aware that your physical, mental, and emotional well-being influences how well you can perform academically.

• **Be reflective**

  ○ Understand that deep learning comes out of reflective activities.
  ○ Reflect on your learning by revisiting assignments, midterm exams, and quizzes and comparing them against posted solutions.
  ○ Reflect on why certain concepts and knowledge are more readily or less readily acquired.

- o Reflect on your achievements and your failures to bring about change.
- **Be inquisitive**
  - o Be active in a course and ask questions that aid your learning and build your knowledge base.
  - o Seek out your instructor after class and during office hours to clarify concepts and content and to find out more about the subject area.
  - o Show an interest in your program of studies that drives you to do well.
- **Practice communication**
  - o Articulate questions. This is one of the best ways to probe your own understanding and gain new knowledge.
  - o Speak about the subject matter of your courses, for example, by explaining concepts to your friends.
  - o Take good notes that pay attention to detail but still give a holistic picture.
  - o Observe how mathematics is written and attempt to use a similar style in your written work.
  - o Pay attention to new terminology and use it in your written and oral work.
- **Find ways to enjoy learning**
  - o Be passionate about your program of study.
  - o Be able to cope with a course you don't like because you see the bigger picture.
  - o Be a student because you made a positive choice to be one.
  - o Have a habit of reviewing, such as study notes, lecture material, or the textbook.
  - o Work through assignments individually at first and way before the due date.
  - o Do extra problems as needed.
  - o Read unassigned material related to the course with an eye spend on time.
- **Be resourceful**
  - o Use the resources made available for your course, such as the virtual course container or website, textbook, tutorial, or drop-in help centre.

- Trust that the course is well constructed and contains all necessary material and activities to attain your learning goals.
- Use the library and internet thoughtfully and purposefully to find additional resources for a certain area of study, but do not gorge yourself on outside sources thereby wasting time.
- Research support structures offered through your post-secondary institution – often through the library, teaching and learning centre, or health and counseling services.

- **Be organized**

  - Employ a calendar to organize class time, due dates, exam dates, and other time-sensitive activities.
  - Create a weekly to-do list to stay on top of required tasks.
  - Adopt a particular method for organising lecture notes and extra material that aids your way of thinking and learning. If you have not figured this out yet, then experiment to find out.
  - Organize your learning space as an inviting place for studying.

- **Manage your time effectively**

  - Be in control of your time.
  - Follow a schedule that not only includes time for study and research, but also time for eating, social, and physical activities.

- **Be involved**

  - Be informed about your program of study and your courses and take an active role in them.
  - Visit the instructor or academic advisor to get help.
  - Join a study group and use the support that is being offered in your program.
  - See the bigger picture and find ways to be involved in more than just studies by joining clubs offered at your post-secondary institution.
  - Look for volunteer opportunities, for example, by becoming a peer tutor.

- **Be resilient**

  - Be kind to yourself when you face challenges and setbacks.
  - Withstand stressful experiences by addressing them, by adjusting your mode of operating, and perhaps by also rearranging the environment that you live and study in.

○ Navigate challenges by trying new approaches and by reaching out to others.

○ Bounce back from setbacks by reminding yourself of your accomplishments. We often forget our achievements when we feel disappointed or rejected.

○ Realize that challenges and setbacks are a normal part of life and opportunities to assess your actions, values, and choices. By actively engaging in this assessment, you not only learn to manage ups and downs across the course of your life, but you also become resilient.

8.3 How to Prepare for Exams

• Start preparing for an exam on the FIRST DAY of class!

• Come to all classes and listen for where the instructor stresses material or points to classical mistakes. Make a note about these pointers.

• Treat each chapter with equal importance, but distinguish among items within a chapter.

• Study your lecture notes in conjunction with the textbook because it was chosen for a reason.

• Pay particular attention to technical terms from each lecture. Understand them and use them appropriately yourself. The more you use them, the more fluent you will become.

• Pay particular attention to definitions, theorems, and formulas from each class. Know the major ones by heart.

• Create a "study sheet" that summarizes terminology, definitions, theorems, and formulas. You should think of a study sheet as a condensed form of lecture notes that organizes the material to aid *your* understanding. However, you may not take this study sheet into an exam unless the instructor specifically says so.

• Check your assignments against posted solutions. Be critical and compare how you wrote up a solution versus the instructor and the textbook. Do not be afraid to question a solution. This is a sign of wanting to understand.

• Read through or even work through the assignments and quizzes, if any, a second time. However, be selective the second time around and zero in on material that is of higher importance or where you need another go at it to improve your understanding.

- Study the examples in your lecture notes in detail. Ask yourself why they were offered by the instructor.
- Work through some of the examples in your textbook. Then compare your solution to the detailed solution offered by the textbook, if any, or seek out the instructor or a teaching assistant.
- Does your textbook come with a review section for each chapter or grouping of chapters? Make use of it. This may be a good starting point for a study sheet. There may also be additional practice questions.
- Practice writing exams by doing old midterm and final exams under the same constraints as a real midterm or final exam: strict time limit, no interruptions, no notes, and no other aids unless specifically allowed.
- Study how old exams are set up! How many questions are there on average? What types of questions are being asked? What would be a topic header for each question? Rate the level of difficulty of each question. Now come up with an exam of your own making and have a study partner do the same. Exchange your created exams, write them, and then discuss the solutions.

8.4   Getting and Staying Connected

- Stay in touch with family and friends:
  - A network of family and friends can provide security, stability, support, encouragement, and wisdom.
  - This network may consist of people that live nearby or far away. Technology – in the form of phones and social media – allows us to stay connected no matter where we are. However, it is up to us at times to reach out and stay connected.
  - Do not be afraid to talk about your accomplishments and difficulties with people that are close to you and you feel safe with to get different perspectives.

- Create a study group or join one:

  o Both the person being explained to and the person doing the explaining benefit from this learning exchange.

  o Study partners are great resources! They can provide you with notes and important information if you miss a class. They may have found a great book, website, or other resource for your studies.

  o Most importantly, you will learn that other students also encounter struggles and difficulties, so you are not alone in this journey. Perhaps you will even get some tips how to overcome some of the struggles and difficulties.

- Go to your faculty or department and find out what student groups there are:

  o Is there a math student union or club that promotes student interests within the Department of Mathematics? This is often a great place to find like-minded people and to get connected within mathematics.

  o Student groups or unions may also provide you with connections after you complete your program and are seeking either employment or further areas of study.

- Go to your faculty or department and find out what undergraduate outreach programs there are:

  o Is there an organized group in the Department of Mathematics that prepares students for the William Lowell Putnam Mathematical Competition held annually the first Saturday in December?

  o Are there opportunities to apply for an undergraduate research assistant-ship? These are typically offered through government agencies and supported in the Department of Mathematics (or other science departments and faculties) such as the Canadian NSERC USRA or US NSF REU. These are great opportunities to not only earn some money, but to also embark on some mathematics research being guided by a professor.

  o Is there an undergraduate seminar that presents a variety of topics concerning mathematics? This is a great way to learn about mathematics in action either in research, industries, business, or other math-driven endeavors.

8.5   Staying Healthy

A healthy mind, body, and soul make you thrive. Create a healthy lifestyle by taking an active role in this lifelong process.

- Mentally:
  - Feed your intellectual hunger! Choose a program of study that suits your talents and interests. You can benefit from visiting with an academic advisor.
  - Take breaks from studying! This clears your mind and energizes you.
- Physically:
  - Eat well! Have regular meals and make them nutritious.
  - Exercise! You may want to get involved in a recreational sport.
  - Get out, rain or shine! Your body needs sunshine to produce vitamin D, which is important for healthy bones.
  - Sleep well! Have a bed time routine that will relax you so that you get good sleep and enough of it. Brain functions such as memory maintenance and building depend on sleep.
- Socially:
  - Be respectful, inclusive, and supportive of others, and appreciate diversity.
  - Stay in touch with loved ones.
  - Make friends! Friends are good for listening, help you to study, and make you feel connected.
  - Get involved! Join a university club or student union.
  - Manage challenging relationships.
- Emotionally:
  - Accept yourself for who you are. We all have strengths and weaknesses.
  - Look after yourself and treat yourself to something enjoyable.
  - Ask your instructor, student colleagues, family, friends, or institutional resources for help when you need it.
  - Recognize and manage stressful experiences.

o Be optimistic when facing stressful experiences, this helps you cope with them.

- Spiritually:

o Be in touch with your own values and beliefs.

o Respect the values and beliefs of others.

o Be present, mindful, and aware how you breathe.

o Connect the complex interrelationship between your mind and body for improved mental and physical functionality, for example, through yoga.

8.6 Resources

- How to Ace Calculus: The Streetwise Guide by Colin Adams, Abigail Thompson, and Joel Hass published by Macmillan.
- Paul's Online Notes by Paul Dawkins:
  https://tutorial.math.lamar.edu/.
- Desmos Graphing Calculator:
  https://www.desmos.com/calculator.
- Symbolab Math Solver – Step by Step calculator:
  https://www.symbolab.com/.
- WolframAlpha – Computational Knowledge Engine:
  https://www.wolframalpha.com/examples/mathematics/.
- 16 Habits of Mind (1-page summary):
  https://www.chsvt.org/wdp/Habits_of_Mind.pdf.
- *AMSI Vacation Research Scholarships.* (2022). Australian Mathematical Sciences Institute.
  https://vrs.amsi.org.au/.
- *NSERC Undergraduate Student Research Award.* (2022). Natural Sciences and Engineering Research Council of Canada.
  https://www.nserc-crsng.gc.ca/students-etudiants/ug-pc/usra-brpc_eng.asp.
- *NSF Research Experiences for Undergraduates.* (2022). National Science Foundation.
  https://www.nsf.gov/crssprgm/reu/.

Bibliography

Borowski, E. J. and Borwein, J. M. (1989). *Collins Dictionary of Mathematics* (HarperCollins, London).

Burrill, S., Jungić, V., and Kouzniak, N. (2020). Differential Calculus Course Resources (Unpublished).

Costa, A. and Kallick, B. (2005). *Habits of Mind* (Hawker Brownlow).

Demidovich, B. P. (1969). *Problems in Mathematical Analysis* (Gordon & Breach).

Fundamental Values (2022). International Center for Academic Integrity, https://academicintegrity.org/.

Health & Counselling. (2022). Simon Fraser University, https://www.sfu.ca/students/health/.

Jungić, V., Menz, P., and Pyke, R. (2020). *A Collection of Problems in Differential Calculus*, http://www.sfu.ca/~vjungic/Zbornik2020/Zbornik2020.pdf.

Jungić, V. and Mulholland, J. (2018). *Differential Calculus. Lecture Notes*, http://www.sfu.ca/~vjungic///Calculus1/Calculus1.pdf.

Putnam Competition (2022). Mathematical Association of America, https://www.maa.org/math-competitions/putnam-competition.

Stewart, J., Clegg, D. K., and Watson, S. (2020). *Calculus: Early Transcendentals*, 9th edn. (Brooks Cole).

Thien, S. and Bulleri, A. (1996). Successful students: Guidelines and thoughts for academic success. *The Teaching Professor*, 10(9), pp. 1–2.

Uščumlić, M. and Miličić, P. (1977). *Zbirka Zadataka iz Više Matematike*, 6th edn. (Naučna Knjiga, Belgrade).

Vene, B. (2013). *Zbirka Rešenih Zadataka iz Matematike*, 44th edn. (Zavod za Udžbenike, Belgrade).

Index

Printed in the United States
by Baker & Taylor Publisher Services